Hydration of Ions in Aqueous Solution

Hydration of Ions in Aqueous Solution

Guest Editor

Cory Pye

MDPI

Basel • Beijing • Wuhan • Barcelona • Belgrade • Novi Sad • Cluj • Manchester

Guest Editor
Cory Pye
Department of Chemistry
Saint Mary's University
Halifax, NS
Canada

Editorial Office
MDPI AG
Grosspeteranlage 5
4052 Basel, Switzerland

This is a reprint of the Special Issue, published open access by the journal *Liquids* (ISSN 2673-8015), freely accessible at: https://www.mdpi.com/journal/liquids/special_issues/hydration_ions.

For citation purposes, cite each article independently as indicated on the article page online and as indicated below:

Lastname, A.A.; Lastname, B.B. Article Title. *Journal Name* **Year**, *Volume Number*, Page Range.

ISBN 978-3-7258-7204-6 (Hbk)
ISBN 978-3-7258-7205-3 (PDF)
https://doi.org/10.3390/books978-3-7258-7205-3

Contents

About the Editor

Cory Pye

Cory Pye is an Associate Professor at Saint Mary's University in Halifax, Nova Scotia, Canada. His current interests include solvation and hydration of small cations, anions, molecules, and ion pairs, as well as chemical education related to spectroscopy. He has been involved in projects where these interests have been applied to hydrothermal chemistry, for example, in environments such as nuclear reactors. Some achievements include the initial development of the COSMO and COSMO-RS modules in the Amsterdam Density Functional software package (part of the Amsterdam Modeling Suite).

Preface

This reprint of the virtual Special Issue brings together a collection of reviews and original research articles that further improve our understanding of ions in aqueous solutions. The reprint contains two reviews and eight original research articles. The review by Persson examines methods used to investigate the coordination chemistry of metal ions in solution and compares these structures with those obtained in the solid state. The review by Mondal and co-workers focuses on the Raman spectroscopy of the hydration shell of ions, showing how metal ions perturb the vibrational spectra of water, particularly the water stretching and bending modes. The article by Chialvo et al. critically examines the criteria used for classifying solutes as (solvent) structure-making or structure-breaking. The AIMD studies by Smiechowski et al. suggest that hexaaquazinc(II) undergoes dissociative exchange via pentaaquazinc(II). Hofer uses QM/MM MD simulations to study hydrated fluoride, chloride, and bromide, using a high-level RIMP2 QM method, and examines the hydration structure, first shell water exchange, and the effect of halide–water hydrogen bonding on the first hydration shell. Pathak et al. study the hydration of phosphate ions using QM and MD methods and calculate the self-diffusion constant, conductivity, and coordination number. Yamaguchi et al. study magnesium chloride solutions using X-ray and neutron diffraction up to 4 GPa and find that hexaaquamagnesium(II) may transform into chloropentaaquamagnesium(II) at high pressures.

Finally, the three contributions by the guest editor Pye use ab initio methods to examine the hydration of tin(II), the hydration of antimony(III), and the hydration of tetrahedral oxoanions and molecules, respectively, with a special focus on structure and vibrational spectroscopy. Tin(II) appears to be tricoordinate and trigonal pyramidal. If the aqua ion of antimony(III) exists, it would appear to be a five-coordinate square pyramid. For the tetrahedral species, the B3LYP/6-31+G* level of theory reproduces the experimental results well, and predictions are made for the previously unknown FeO_4, NbO_4^{3-}, and TaO_4^{3-} species.

Cory Pye
Guest Editor

liquids

MDPI

Review

Structures of Hydrated Metal Ions in Solid State and Aqueous Solution

Ingmar Persson

Department of Molecular Sciences, Swedish University of Agricultural Sciences, P.O. Box 7015, SE-750 07 Uppsala, Sweden; ingmar.persson@slu.se

Abstract: This review article summarizes the reported crystallographically determined structures of compounds containing a hydrated metal ion and the reported structures of hydrated metal ions in aqueous solution. A short overview of the methods available to study structures of metal complexes in solution is given.

Keywords: review; hydrated metal ions; structure; solid state; aqueous solution

1. Introduction

Structures of hydrated and solvated metal ions and complexes in solution are of utmost importance in order to understand the chemical behavior in their reaction medium. It is common that structures of metal ion complexes determined in the solid state are extrapolated to have the same or similar structure in solution. This paper gives an overview of reported structures of hydrated metal ions in the solid state and aqueous solution and discusses some cases where the structure in aqueous solution differs from the observations in the solid state or where the results are under scientific discussion. A short presentation of the available methods for accurate structure determination in solution is given. Hydrated metal ions are used as examples as they have simple composition and are of general great importance and interest. Metal ions with low charge density, such as the heavier alkali metal ions and thallium(I), are weakly hydrated and solvated and crystallize in most cases without a complete first hydration or solvation shell. A second hydration shell can only be studied in aqueous solution, as no hydrated metal ion has been reported to crystallize with a second hydration sphere. Symmetry plays an important role in crystallization, and high-symmetry complexes may crystallize from a solution dominated by complexes with low symmetry. Another aspect is the random structural disorder in the solid state. The outcome of a crystallographic investigation is the mean of all irradiated unit cells. If a low-symmetry complex is randomly oriented in a crystal, the mean structure may have a higher symmetry than the individual complexes. This is discussed in detail for some hydrated metal ions with low symmetry in aqueous solution, but with normally higher symmetry in the solid state. In the supporting material section, the structures of all reported crystal structures containing hydrated metal ions with full hydration shell, reported through 2021, are summarized in Supplementary Tables S1–S16.

2. Methods to Investigate the Coordination Chemistry in Solution

The structural information, concerning bond distances and basic configurations of hydrated and solvated metal ions and complexes in solution, is obtained with Extended X-ray Absorption Fine Structure (EXAFS) spectroscopy, Large Angle X-ray Scattering (LAXS), and Large Angle Neutron Scattering (LANS). These methods give structural information in one dimension as a Fourier transform (EXAFS) or a radial distribution function (RDF) (LAXS and LANS) of the studied sample. It is important to stress that the obtained information is a mean value of the irradiated sample. It is therefore a great advantage, if possible, to study solutions with only one chemical species except the solvent and a counter ion with known structure. The data collection of an EXAFS spectrum requires very high X-ray intensity in a wide energy range, at least 700 eV, ideally 1600 eV (wavelength in Å = 12,398.4/energy in eV). Therefore, synchrotron light is required for accurate EXAFS measurements. EXAFS gives only structure information around the absorbing atom through back-scattering of the outgoing photoelectron wave of the excited atom. [1–4]. Furthermore, dilute solutions down to sub-millimolar concentrations can be studied.

LAXS, also known as WAXS (wide angle X-ray scattering), can be performed in two ways: reflection mode, scattering from a planar liquid surface in a Θ-Θ goniometer, and transmission mode, the scattering of the sample in a capillary. The scattering is determined as a function of the scattering angle, 2Θ, in the angle range of 0–140 degrees. LAXS in reflection mode is formed by an X-ray tube as the X-ray source, as both the X-ray source and detector have to be moved synchronically. LAXS in transmission mode is preferably performed by synchrotron light as a higher X-ray intensity can be obtained. LAXS in reflection mode has been found to superior, as a large scattering angle range can be applied without advanced absorption corrections. On the other hand, LAXS in transmission mode required advanced absorption corrections, especially at scattering angles above 90 degrees.

In contrast to EXAFS, LAXS detects all distances in the sample. The scattered intensity from a pair of atoms, independent of if they are bound or not, is proportional to the number of electrons in the atom pair. This means in practice that the required concentration to distinguish the compound or complex under study from the contribution increases with atom numbers in the complex under study. As a rule of thumb, complexes with a first row transition metal ion require a 0.3–0.5 mol·dm^{-3} solution for accurate measurement.

The EXAFS and LAXS methods are complementary as they give partly different information. EXAFS gives very accurate information of well-defined distances up to 4–5 Å from the absorbing atom, while in principle no information is obtained about longer distances. The reason is that the EXAFS signal is highly damped by increasing distance, R_j, and the Debye–Waller coefficient, σ, accounting for the thermal and configurational disorder. The EXAFS equation in its most simple form is expressed in the following way:

$$\chi(k) = \Sigma \frac{N_j S_o{}^2}{kR^2} \cdot \left| f_{eff}(k) \right| \cdot \exp\left(-\frac{2R_j}{\Lambda j(k)} \right) \cdot \exp\left(-2k^2\sigma^2 \right) \cdot \sin\left[2kR_j + \phi_j(k) \right] \quad (1)$$

where $\chi(k)$ is the EXAFS function, k is the photoelectron wavenumber or wave vector, N, the number of distances, $S_o{}^2$ is the amplitude reduction factor, R is the distance between the absorbing atom and the back-scattering in single scattering and total distance divided by two in multiple scattering, f_{eff} is the effective amplitude function for the scattering path, $\exp(-2k^2\sigma^2)$ is the Debye–Waller factor in harmonic approximation including the Debye–Waller parameter accounting for the thermal and configurational disorder, $\Lambda(k)$ is the photoelectron mean free path and $2kR + \Phi(k)$ the total phase. The theory of the EXAFS method and how it can be applied in different areas of science are given in refs. [1–4].

LAXS detects all distances in the sample, but the angular range is smaller, causing the resolution for short distances to be smaller compared to EXAFS. On the other hand, the damping of the LAXS signal for long distances is much smaller compared to EXAFS. This makes it possible to observe very long distances, as the general bulk order, with LAXS giving structural information out to 15–20 Å. As an example, LAXS can accurately detect a second, and possibly a third hydration shell around a metal ion, while they are hardly detectable by EXAFS. The X-ray scattering of liquids and non-crystalline solids is described by the following equation:

$$i(s) = \Sigma\Sigma\left\{ f_i(s) \cdot f_j(s) + \Delta f_i'' \cdot \Delta f_j'' \right\} \cdot \frac{\sin(s \cdot R_{ij})}{s \cdot R_{ij}} \cdot \exp\left(-\frac{l_j^2}{2} \right) + del(s) \left(\frac{\lambda}{\lambda'} \right)^2 I_{m,incoh}(s) \quad (2)$$

where $i(s)$ is the reduced intensity, s is the scattering vector expressed as $s = 4\pi\sin\Theta/\lambda$, where Θ is the scattering angle divided by two, and λ is the wavelength of the X-ray radiation, f is the scattering factors, including a correction for anomalous dispersions (f), where the $\Delta f''$ is the imaginary part, R is the distance between two atoms, and, in the root-mean-square variation assuming Gaussian distribution, l^2, and $del(s)$ describe the fraction of incoherent scattering passing the monochromator, where λ is the wavelength of X-ray radiation and λ' is the wavelength measured in the X-ray spectrum of the X-ray source, giving $del(s)$ for the scattering angle 2Θ, and where $\lambda' = \lambda + 0.02426(1 - \cos 2\Theta)$. For a more detailed description of the LAXS method and the theory of scattering of liquids and amorphous compounds, refs. [5,6] are recommended.

A comparison of Equations (1) and (2) shows that the amplitude of the structure-dependent intensity contributions decreases more rapidly in the EXAFS equation, $\exp(-2{\cdot}R_{ij}/\lambda(k))$, than in the LAXS one, $\sin(s{\cdot}R_{ij})/s{\cdot}R_{ij}$, and this difference increases significantly with increasing interatomic distances. EXAFS and LAXS give different information about the local structure of metal complexes, EXAFS solely from the absorbing atom and LAXS as part of all distances in the sample. The great advantage of EXAFS is that it can be applied on very dilute solutions, down to sub-millimolar concentrations, and it gives accurate information about short distances around the absorbing atom. Furthermore, the contribution from multiple scattering in EXAFS may give information bond angles around the absorbing atom, e.g., M–O–C bond angles, and high-symmetric coordination geometry. The advantage of LAXS is that long distances, such as, e.g., second hydration spheres and bulk order, can be determined with high accuracy.

Large angle neutron scattering (LANS) of liquids is complementary to LAXS. The main difference is that the scattering ability (scattering factors) of the elements are different. X-rays are scattered by the electrons of an atom, while neutrons are scattered by the nucleus of the atom. This means that light atoms, similarly to hydrogen, can hardly be detected by X-rays, while the scattering ability of, e.g., deuterium is large and can be detected in a LANS experiment. As different isotopes of the same element have different neutron scattering abilities, isotopically pure solutions with only one isotope of each element are required. The main limitation for the extensive application of LANS is the radiation source, as a nuclear reactor or a neutron spallation source is required. A comprehensive description of the LANS method is given in ref. [7].

The scope of this review is the structure of hydrated metal ions in aqueous solution and a solid state. Information such as water exchange kinetics and the number of ligands bound to metal ions can be obtained by NMR spectroscopy [8], and information about the strength of the hydrogen bonds in hydrated metal ions is obtained by double-difference infrared spectroscopy [9] and will not be further discussed in this survey.

3. Structures of Hydrated Metal Ions in Solid State and Aqueous Solution

3.1. *Alkali Metal Ions*

3.1.1. Lithium

The tetrahedral tetraaqualithium ion, $[Li(H_2O)_4]^+$, is by far the most common form of hydrated lithium ions in the solid state with a mean Li–O bond distance of 1.941 Å (a mean of 53 reported crystal structures) (Supplementary Table S1a). Tetrahedral coordination around lithium is also present in the dimeric $[Li_2(H_2O)_6]^{2+}$ and $[Li_2(H_2O)_7]^{2+}$ ions, as well as in the polymeric $[Li_3(H_2O)_6]_n{}^{3n+}$ and $[Li_3(H_2O)_8]_n{}^{3n+}$ ions (Supplementary Table S1a). The dimeric lithium hydrate complexes have either a single ($[Li_2(H_2O)_7]^{2+}$) or double ($[Li_2(H_2O)_6]^{2+}$) aqua bridge with mean Li–O bond distances of 1.962 and 1.988 Å, respectively. In the polymeric $[Li_3(H_2O)_6]_n{}^{3n+}$ complex, all water molecules are bridged with a mean Li–O bond distance of 2.001 Å. The bridging Li–O bond distances are significantly longer than the terminal ones, explaining the difference in the mean Li–O bond distances of mono-, di- and polynuclear hydrated lithium ions (Supplementary Table S1a). Mähler and Persson have reported an analysis of Li–O bond distances of reported crystal structures of monomeric lithium complexes with oxygen donor ligands in four- and six-coordination [10]. The mean Li–O bond distances were found to be 1.942 and 2.132 Å, respectively, where the former is in excellent agreement with the mean value of all structures containing a $[Li(H_2O)_4]^+$ ion reported up to date, and the latter is in reasonable agreement with the only reported solid-state structure containing a $[Li(H_2O)_6]^+$ ion, 2.144 Å (Supplementary Table S1a). By extrapolation, the mean Li–O bond distance in a $[Li(H_2O)_5]^+$ ion should be close to 2.04 Å [10]. The structure of the square pyramidal $[Li(H_2O)_5]^+$ ion has been reported in a couple of studies, but only one of these is close to the expected Li–O bond distance, 2.044 Å, while the other reported Li–O bond distances are much longer than expected for this configuration (Supplementary Table S1a).

The hydrated lithium ion has been studied extensively in aqueous solution by X-ray methods, in spite of the severe difficulties, vide ultra. Smirnov and Trostin reviewed available experimental data for the lithium ion until 2006 [11], and later studies are summarized in later general studies and reviews [10,12–15]. As lithium is a very weak scatterer of X-rays, it is not possible to unambiguously determine the coordination number of the hydrated lithium ion in aqueous solution from present X-ray-based experimental data. The results of the reported simulation studies on the hydrated lithium ion in aqueous solution scatter, and no conclusive picture is obtained. However, four-coordination in the tetrahedral configuration is mainly proposed. A neutron scattering study on the hydrated lithium ion in heavy water has reported a mean Li–O bond distance of 1.96 Å [16] with mean coordination numbers of 4.3–5.0, depending on the concentration. However, the reported mean Li–O bond distance correlates well with four-coordination in tetrahedral fashion, vide ultra. Simulation studies of the hydrated lithium ion in tetrahedral configuration have reported Li–O bond distances of 1.95 [17], 1.97 [18], 1.971 [19], and 2.03 Å [20]. None of the dimeric or polymeric complexes in the solid state have been observed in aqueous solution independent of concentration. As lithium scatters X-rays very weakly, studies in aqueous solution using X-ray techniques will give results with limited accuracy. The most reliable data seem to be crystallographic ones, and four-coordination in a tetrahedral fashion is by far the most common configuration of the hydrated lithium ion. It seems, therefore, most likely that the hydrated lithium ion is four-coordinated in a tetrahedral fashion in aqueous solution with a mean Li–O bond distance of 1.94 Å.

3.1.2. Sodium

The octahedral hexaaquasodium ion, $[Na(H_2O)_6]^+$, is the most common sodium hydrate complex in the solid state with a mean Na–O bond distance of 2.417 Å (28 structures) (Supplementary Table S1b). However, other configurations, including monomeric $[Na(H_2O)_2]^+$, $[Na(H_2O)_4]^+$, $[Na(H_2O)_5]^+$, and $[Na(H_2O)_8]^+$ ions, and a large number of di-, tetra-, and octameric hydrated sodium ions, $[Na_2(H_2O)_6]^{2+}$, $[Na_2(H_2O)_7]^{2+}$, $[Na_2(H_2O)_8]^{2+}$, $[Na_2(H_2O)_9]^{2+}$, $[Na_2(H_2O)_{10}]^{2+}$, $[Na_4(H_2O)_{14}]^{4+}$, $[Na_4(H_2O)_{16}]^{4+}$, and $[Na_4(H_2O)_{18}]^{4+}$ ions, as well as the polymeric $[Na_4(H_2O)_{14}]_n{}^{4n+}$, $[Na_4(H_2O)_{18}]_n{}^{4n+}$, and $[Na_8(H_2O)_{24}]_n{}^{8n+}$ chains have also been reported (Supplementary Table S1b). An analysis of the Na–O bond distances of reported crystal structures of sodium complexes with oxygen donor ligands in five- and six-coordination has been reported [10]. The mean Na–O bond distances are 2.358 and 2.412 Å, respectively, which are in excellent agreement with the mean values of the hydrated sodium ion in five- and six-coordination, 2.366 and 2.417 Å, respectively (Supplementary Table S1b).

The reported experimental studies on the hydrated sodium ion in aqueous solution do not give a conclusive picture with Na–O bond distances in the range of 2.34–2.50 Å and coordination numbers in the range of 4.6–8 [21,22]. The reported theoretical simulation studies of the hydrated sodium ion in aqueous solution propose a mean Na–O bond distance of 2.35–2.40 Å in an octahedral fashion [23–25]. By summarizing the available information on the hydrated sodium ion including LAXS studies, solid-state structures containing hydrated sodium ions, vibration spectroscopic studies, and reported theoretical simulations, the hydrated sodium ion seems to bind six water molecules in an octahedral fashion with a mean bond distance of 2.43 Å in aqueous solution [10]. It has not been possible with the available experimental techniques to determine whether the hydrated sodium has a second hydration sphere or not [10].

None of the di- or polymeric complexes reported in solid structures have been detected in aqueous solution independent of concentration. Sodium scatters X-rays weakly, and studies in aqueous solution using X-ray techniques will give results with limited accuracy. The most reliable data seem to be the crystallographic data, and six-coordination in an octahedral fashion is the most common configuration of the hydrated sodium ion. It seems therefore most likely that the hydrated sodium ion is a six-coordinated octahedral fashion in aqueous solution with a mean Na–O bond distance of ca. 2.42 Å.

3.1.3. Potassium

Only a limited number of crystal structures with a full hydration shell around the potassium ion have been reported, including $[K(H_2O)_6]^+$, $[K(H_2O)_7]^+$, and $[K(H_2O)_8]^+$ (Supplementary Table S1c). However, the reported K–O bond distances are very scattered and not close to the K–O bond distances obtained by X-ray techniques in aqueous solution, nor the sum of the ionic radii of potassium in coordination numbers 6–8 proposed by Shannon [26] and the atomic radius of oxygen in the coordinated water molecule, 1.34 Å [27]. It is not possible to draw any conclusion about the structure of the hydrated potassium ion from the reported crystal structures.

The reported mean K–O bond distance in the hydrated potassium ion in aqueous solution from experimental studies is uncertain, 2.65–2.97 Å [10,28–30]. Theoretical simulation studies propose a K–O distance of ca. 2.80 Å and a coordination number close to seven in aqueous solution [23,24,31]. An EXAFS study on an aqueous potassium chloride solution reports a mean K–O bond distance of 2.73 Å and a coordination number of 6–7 [32], and a LAXS study on an aqueous solution of potassium iodide reports a mean K–O bond distance of 2.81(2) Å in seven-coordination [10]. It was not possible to detect any second hydration shell around the potassium ion from LAXS, and an infrared study showed that the water molecules bound to the potassium ion are less affected than the water molecules in the aqueous bulk [10]. This strongly indicates that there are no water molecules with significant order outside the first hydration shell.

3.1.4. Rubidium

Only two crystal structures with a full hydration shell around the rubidium ion, $[Rb(H_2O)_9]^+$, have been reported (Supplementary Table S1d). However, the reported Rb–O bond distances are much longer than the ones obtained in aqueous solution, vide infra, the sum of the proposed ionic radii in coordination numbers 6–8 [26], and the atomic radius of oxygen in coordinated water molecules [27]. The reported Rb–O bond distances in the hydrated rubidium ion in aqueous solution are in a wide range, 2.83–3.05 Å, mean ca. 2.95 Å, with coordination numbers in the range 5.6–8.0 (Table 1 in ref. [33]). A combined LAXS and EXAFS study of aqueous rubidium iodide solutions reported a mean Rb–O bond distance in the hydrated rubidium ion of 2.98 Å [34]. Only one structure of a solid rubidium compound with water molecules in the coordination sphere, $[Y(C_2O_4)_2Rb(H_2O)_4]_n$, has been reported. The mean Rb–O bond distance in this compound is 2.95 Å [35]. According to the reported results on the hydrated rubidium ion in aqueous solution, it is most likely eight-coordinated in a square antiprismatic fashion with a mean Rb–O bond distance of ca. 2.95 Å. As expected from LAXS and infrared spectroscopic studies, no well-ordered water molecules are present outside the first hydration shell of the hydrated rubidium ion in aqueous solution.

3.1.5. Cesium

Only two crystal structures with a full hydration shell around the cesium ion, $[Cs(H_2O)_{10}]^+$, have been reported (Supplementary Table S1e). However, the reported Cs–O bond distances are much longer than those obtained in aqueous solution, vide infra, and the sum of the proposed ionic radii of cesium in coordination number ten proposed by Shannon [26] and the atomic radius of coordinated water molecules [27]. A survey of experimental and simulation studies of the hydrated cesium ion in aqueous solution reported before 2006 a broad range of mean Cs–O bond distances, 3.0–3.2 Å, mean ca. 3.1 Å, as well as proposed coordination numbers, 5.1–9.6 (Table 1 in ref. [33]). Recent LAXS studies on aqueous solutions of cesium hydroxide, iodide, fluoride, and chloride report mean Cs–O bond distances of 3.08, 3.08, 3.10, and 3.13 Å, respectively [10,36,37]. Summarizing, the reported results on the hydrated cesium ion in aqueous solution mean that the most likely structure is eight-coordination in a square antiprismatic fashion with a mean Cs–O bond distance of ca. 3.10 Å. No ordered water molecules outside the first hydration shell of the hydrated cesium ion have been detected [10]. This is expected as cesium is the most weakly hydrated alkali metal ion.

3.2. *Alkaline Earth Metal Ions*

3.2.1. Beryllium

The tetraaquaberyllium(II) ion, $[Be(H_2O)_4]^{2+}$, is the only reported hydrated beryllium ion in the solid state. Its structure has been determined in 20 crystal structures, giving a mean Be–O bond distance of 1.614 Å (Supplementary Table S2a). Due to the weak X-ray scattering ability of beryllium, the results of experimental studies using X-ray techniques in aqueous solution become uncertain. Only one study is reported in the literature [38], with a reported Be–O bond distance of 1.67 Å. This is significantly longer than the bond distances obtained in the solid state. As the only reported study in aqueous solution is uncertain, the most likely structure of the hydrated beryllium(II) ion in aqueous solution is a tetrahedral tetraaquaberyllium(II) ion with the mean Be–O bond distance found in the solid state to be 1.62 Å. Due to the high charge density of the beryllium(II) ion, it is strongly hydrated and certainly has a well-defined second hydration sphere.

3.2.2. Magnesium

A very large number of crystal structures (511) contain a hexaaquamagnesium ion, $[Mg(H_2O)_6]^{2+}$ (Supplementary Table S2b). The configuration of the hexaaquamagnesium ion is octahedral, and the mean Mg–O bond distance in these structures is 2.066 Å. No crystal structures containing a hydrated magnesium ion with another coordination number than six have been reported.

The reported studies of hydrated magnesium ions in aqueous solution are scattered concerning both Mg–O bond distance and coordination number [28,29]. The mean Mg–O bond of these studies is slightly longer than the distance found in the $[Mg(H_2O)_6]^{2+}$ ions in the solid state. As in the case of the hydrated beryllium ion, the results from X-ray methods on magnesium in an aqueous solution are less accurate than crystallographic determinations. A number of theoretical simulation studies have shown that the hexaaquamagnesium ion is by far the most stable structure of the hydrated magnesium ion in aqueous solution [39–42]. The hydrated magnesium ion binds certainly six water molecules in an octahedral fashion in aqueous solution with a mean Mg–O bond distance of 2.07 Å and a well-defined second hydration sphere.

3.2.3. Calcium

The hydrated calcium ion in solid compounds display different coordination numbers, $[Ca(H_2O)_4]^{2+}$, $[Ca(H_2O)_6]^{2+}$, $[Ca(H_2O)_7]^{2+}$, and $[Ca(H_2O)_8]^{2+}$ with mean Ca–O bond distances of 2.253, 2.324, 2.401, and 2.476 Å, respectively (Supplementary Table S2c). The $[Ca(H_2O)_4]^{2+}$, $[Ca(H_2O)_6]^{2+}$, and $[Ca(H_2O)_8]^{2+}$ complexes have square-planar, octahedral, and square antiprismatic configuration, respectively, while the configuration of the $[Ca(H_2O)_7]^{2+}$ complex is best described as monocapped octahedral. Furthermore, some solid state structures contain dimeric hydrated calcium ions where two or three water molecules bridge between two calcium ions, $[(H_2O)_4Ca(H_2O)_2Ca(H_2O)_4]^{4+}$, $[(H_2O)_4Ca(H_2O)_3Ca(H_2O)_4]^{4+}$, $[(H_2O)_6Ca(H_2O)_2Ca(H_2O)_6]^{4+}$, and $[(H_2O)_5Ca(H_2O)_3Ca(H_2O)_5]^{4+}$, where the calcium ion is six-, seven-, eight-, and eight-coordinated, respectively (Supplementary Table S2c).

The structure of the hydrated calcium ion in aqueous solution has been reported in numerous experimental [28,29,43–52] and theoretical simulation [52–57] studies. The results are scattered with coordination numbers of 6–8 and Ca–O bond distances in the range of 2.35–2.48 Å. No dimeric hydrated calcium ions have been reported in aqueous solution. It seems most likely that the hydrated calcium ion in dilute aqueous solution is eight-coordinated in a square antiprismatic fashion with a mean Ca–O bond distance of ca. 2.48 Å, with a second hydration sphere, Ca(–O–H)$\cdots$O, at ca. 4.58 Å [44]. Seven-coordination cannot completely be ruled out, but the reported Ca–O bond distances in experimental studies in aqueous solution are in most cases significantly longer than the mean Ca–O bond distance for seven-coordination in the solid state. Bond distances are much more accurately determined than the number of distances in LAXS and EXAFS experiments, and the strong correlation between the bond distance and the coordination number strongly indicates that the hydrated calcium ion is eight-coordinated in aqueous solution.

3.2.4. Strontium

The hydrated strontium ion is six- and eight-coordinated in octahedral and square antiprismatic fashions in the solid state with mean Sr–O bond distances of 2.496 and 2.613 Å, respectively (Supplementary Table S2d). A single solid-state structure has been reported to contain dimeric hydrated strontium ions, where four water molecules bridge two strontium ions, $[(H_2O)_5Sr(H_2O)_4Sr(H_2O)_5]^{4+}$, where the strontium ion is nine-coordinated (Supplementary Table S2d).

The structure of the hydrated strontium ion in aqueous solution has been reported in several experimental [58–67] and theoretical simulation [68–74] studies. The results are consistent with a coordination number close to eight, and Sr–O bond distances in the range of 2.57–2.67 Å. The hydrated strontium ion is most likely eight-coordinated in a square antiprismatic fashion in aqueous solution with a mean Sr–O bond distance of ca. 2.62 Å. The hydrated strontium ion has a second hydration sphere, Sr(–O–H)$\cdots$O, at ca. 4.78 Å [58].

3.2.5. Barium

The hydrated barium ion is eight- or nine-coordinated in the solid state with mean Ba–O bond distances of 2.777 Å, in square antiprismatic configuration, and 2.832 Å, respectively (Supplementary Table S2e). Solid state structures containing dimeric hydrated barium ions have been reported with the compositions $[(H_2O)_7Ba(H_2O)_2Ba(H_2O)_7]^{2+}$, $[(H_2O)_5Ba(H_2O)_4Ba(H_2O)_5]^{2+}$, and $[(H_2O)_6Ba(H_2O)_4Ba(H_2O)_6]^{2+}$, where barium is nine-, nine- and ten-coordinated, respectively (Supplementary Table S2e).

A limited number of studies have been reported on the hydrated barium ion in aqueous solution. No study in aqueous solution has detected any hydrated dimeric barium ions. These studies report a mean Ba–O bond distance close to 2.82 Å and a coordination number of eight or nine [58,69,74–76]. The hydrated barium ion has a second hydration sphere, Ba(–O–H)$\cdots$O, at ca. 4.90 Å [58].

3.2.6. Radium

A theoretical simulation of the hydrated radium(II) ion in aqueous solution proposes that it is ten-coordinated with a mean Ra–O bond distance of 2.93 Å [74].

3.3. Transition Metal Ions

3.3.1. Group 3 Elements

Scandium

The hydrated scandium ion displays different coordination numbers in solid compounds, $[Sc(H_2O)_6]^{3+}$, $[Sc(H_2O)_7]^{3+}$, and $[Sc(H_2O)_8]^{3+}$, with mean Sc–O bond distances of 2.09, 2.16, and 2.24 Å, respectively (Supplementary Table S3a). The $[Sc(H_2O)_6]^{2+}$ complex has a regular octahedral configuration, while the $[Sc(H_2O)_7]^{2+}$ complex has lower symmetry with the coordination best described as a monocapped octahedron [77] or a pentagonal bipyramid [78]. The $[Sc(H_2O)_8]^{3+}$ ion has an unusual structure with a dicapped trigonal prismatic configuration, with one of the capping positions empty [79–81]. The compound $[Sc(H_2O)_{8.0}](CF_3SO_3)_3$ displays a phase transition from high to low symmetry between 150 and 100 K, where in the low-temperature phase, the two capping water molecules have different Sc–O bond distances [79], while in the high-temperature phase, the capping waters have the same Sc–O bond distance with an occupancy factor of 2/3. The hydrated scandium(III) ion is out of the center of gravity of the trigonal prism in the low-temperature phase and shifted towards the most strongly bound capping water molecule [81].

EXAFS studies have shown that the structure of the hydrated $[Sc(H_2O)_8]^{3+}$ ion has the same structure in aqueous solution as the low-temperature phase of in the solid state, with Sc–O_{prism}, Sc–O_{cap1}, and Sc–O_{cap2} bond distances of 2.17, 2.32, and ca. 2.5 Å, respectively [81]. It is an interesting fact that the hydrated scandium(III) ion has a water deficit in one of the capping positions also in aqueous solution. The hydrated scandium(III) ion is easily hydrolyzed to the $[(H_2O)_5Sc(OH)_2Sc(H_2O)_5]^{4+}$ complex in aqueous solution [80]. The hydrated scandium(III) ion has a well-defined second hydration sphere, Sc(–O–H)$\cdots$O, around the water molecules in the prism at 4.28 Å [81].

Yttrium

In the solid state, the hydrated yttrium(III) ion in both eight- and nine-coordination are reported, $[Y(H_2O)_8]^{3+}$ and $[Y(H_2O)_9]^{3+}$, in square antiprismatic and tricapped trigonal prismatic configuration, respectively (Supplementary Table S3b). A majority of the reported structures of the hydrated yttrium(III) ions are eight-coordinated with a mean Y–O bond distance of 2.353 Å (Supplementary Table S3b), and structures with tricapped trigonal prismatic coordination have mean Y–O bond distances of 2.364 and 2.481 Å to the oxygen in the prism and the capping positions, respectively (Supplementary Table S3b).

The hydrated yttrium(III) ion is reported to be eight-coordinated in a square antiprismatic fashion with a mean Y–O bond distance of 2.367(5) Å [82]. A second hydration sphere with ca. 16 water molecules at a Y(–O–H)$\cdots$O distance was determined to be 4.40(4) Å [82]. It is an interesting observation that scandium(III) and yttrium(III) ions have the same coordination number in aqueous solution, eight, but they have different coordination geometries as described above. The reason for this difference has not been clarified.

3.3.2. Lanthanoids

The chemistry of the lanthanoids is dominated by the oxidation state +III. However, samarium, europium, thulium, and ytterbium have well-defined chemistry in the oxidation state +II and cerium in +IV. The lanthanoid(II) ions samarium(II), thulium(II), and ytterbium(II) are all very strong reducing agents, reducing water to hydrogen gas. The only lanthanoid(II) ion stable in aqueous solution is europium(II) [83]. An EXAFS study has revealed that the hydrated europium(II) ion is seven-coordinated in aqueous solution, with a mean Eu–O bond distance of 2.584 Å [66]. The lanthanoid(II) ions and complexes are strong reducing agents of large importance in organic chemistry, and their chemistry is mainly described in non-aqueous solvents [84]. Cerium(IV) is a very strong oxidation agent, frequently used in organic chemistry [85]. The aqueous chemistry of cerium(IV) is claimed to be similar to that of actinoid(IV) ions [86]. A combined EXAFS and simulation study in perchloric acidic aqueous solution shows that hydrated cerium(IV) is hydrolyzed to a hydrated dimer with an oxygen bridge, $[(H_2O)_7Ce–O–Ce(OH_2)_7]^{6+}$, with mean Ce–$O_O$ and Ce–O_{aq} bond distances of 2.08 and 2.43 Å, respectively [87].

The solids containing hydrated lanthanoid(III) ions are eight- or nine-coordinated (Supplementary Table S4). The eight-coordinated ions, $Ln(H_2O)_8]^{3+}$, have square antiprismatic configuration, and the nine-coordinated ions, $Ln(H_2O)_9]^{3+}$, have either regular tricapped trigonal prismatic configuration or distorted nine-coordination without any well-defined coordination figure (Supplementary Table S4a–n). The ionic radii of the lanthanoid(III) ions decrease along the lanthanoid series due to the lanthanoid contraction. This is caused by the poor shielding of the nuclear charge by the 4f electrons, and thereby, the electrons in the electron shell outside the fourth shell are drawn towards the nucleus, resulting in a decreasing atomic/ionic radius along the lanthanoid series. There is a clear trend in the number of reported structures with eight- and nine-coordination in the solid state along the lanthanoid series. Nine-coordination dominates for the lighter lanthanoids and eight-coordination for the heavier ones (Supplementary Table S4). The heaviest lanthanoid(III) ions, Ho–Lu, with tricapped trigonal prismatic configuration have an increasing water deficit with an increasing atomic number in the capping positions, $[Ho(H_2O)_{8.91}](CF_3SO_3)_3$, $[Er(H_2O)_{8.96}](CF_3SO_3)_3$, $[Tm(H_2O)_{8.8}](CF_3SO_3)_3$, $[Yb(H_2O)_{8.7}](CF_3SO_3)_3$, and $[Lu(H_2O)_{8.2}](CF_3SO_3)_3$ [79,88]. The ionic radii of the lanthanoid(III) ions in six-, seven-, eight- and nine-coordination have been estimated from crystal structures of homoleptic lanthanoid(III) complexes with neutral oxygen donors and studies in solution using oxygen donor solvents with different space demands at coordination [89]. The derived ionic radii of the lanthanoid(III) ions form a linear relationship with the atomic number for each coordination number. The structure and mean bond distances can be estimated with high accuracy, including the radioactive element promethium [89]. However, the mean Ln–O bond distance for nine-coordinated hydrates deviates from the straight-line relationship due to the increasing water deficit starting at holmium, and thereby a coordination number less than nine.

The structure of the hydrated lanthanoid(III) ions in aqueous solution was heavily debated in the mid-1960s. The origin of this debate was the proposed presence of a "gadolinium break". The structure of the lanthanoid(III) ions in aqueous solution was assumed to change from nine- to eight-coordination at or close to gadolinium. This was based on observed discontinuities in a number of physico-chemical properties along the lanthanoid series, including partial molar volume [90], heat capacity [91], molar entropy [91], and viscosity [92], assumed to be related to the half-filled 4f shell of gadolinium(III) [93,94]. The partial molar volumes were expected to decrease continuously if the hydrated lanthanoid (III) ions had the same coordination geometry throughout the series, considering the lanthanoid contraction. The experimentally determined physico-chemical parameters showed in fact a discontinuity with an increase in the partial molar volumes from samarium to gadolinium. This was rationalized with an expected decrease in the hydration number of the heavier lanthanoid(III) ions [90]. The water exchange rate of the hydrated lanthanoid(III) ions decreases with the increasing atomic number and decreasing size of the lanthanoid (III) ions, as expected if the coordination geometry is maintained along the lanthanoid series. This is the expected behavior due to the lanthanoid contraction. However, the activation volumes at water exchange events are almost constant with a maximum at samarium/europium [95–98]. The complex formation rates for the lanthanoid(III) ions with sulfate and acetate ions show a maximum at samarium [99,100]. Liquid–liquid extraction experiments indicated more complex relationships between the structure and the physico-chemical parameters of the hydrated lanthanoid(III) ions in aqueous solution, proposing that the lanthanoid series can be divided into two octads with gadolinium common to both, or into four tetrads intersecting at neodymium/promethium, gadolinium, and holmium/erbium [101]. The tetrad effect is discerned, e.g., in the stability constants of the EDTA complexes in aqueous solution [102–104].

EXAFS studies have shown that structures of the hydrated lanthanoid(III) ions in the solid trifluoromethanesulfonate salts and in aqueous solution are identical to a tricapped trigonal prismatic configuration of water molecules [88]. The capping positions are fully occupied with water molecules in the first three tetrads (lanthanum–holmium). The mean M–O bond distance in the capping position decreases continuously from 2.53 Å for lanthanum to 2.38 Å for dysprosium, while the capping positions decrease less, from 2.60 Å (lanthanum) to 2.50 Å (dysprosium). This means that the difference in the M–O bond distance between the prismatic and capping positions increases along the first three tetrads [85]. In the last tetrad (holmium–lutetium), the occupancy factor in the capping positions decreases from ca. 2.9 for holmium to ca. 2.2 for lutetium in both the solid state and aqueous solution [88]. Crystallographic studies showed a phase transition for $[Lu(H_2O)_{8.2}](CF_3SO_3)_3$ at 130 K from the hexagonal space group $P6_3/m$ (no. 176) at high temperatures to the rhombohedral (trigonal) space group *R*-3 (No. 148) at temperatures below 130 K [88]. The low-temperature phase showed that one of the capping water molecules has a significantly shorter Lu–O bond distance, 2.40 Å, than the remaining ones, 2.56 Å, both with a mean occupancy factor of 0.6 [88], and that the lutetium(III) ion has left the center of gravity towards to the most strongly bound capping water molecule. EXAFS results showed that the structure of the hydrated lutetium(III) ion aqueous solution is in principle identical to the low-temperature phase of $[Lu(H_2O)_{8.2}](CF_3SO_3)_3$ [85] and that the other lanthanoid(III) ions in the fourth tetrad show a similar bonding pattern to the capping water molecules. It can be assumed that the position of the lanthanoid(III) ions in the prism and the lanthanoid(III) ions in the fourth tetrad switch position with the same rate as the water exchange rate, $k \approx 10^8$ s^{-1} (Table 5 in ref. [8]).

3.3.3. Actinoids

Actinoid(III) and Actinoid(IV) Ions

The ionic radii of the actinoid(III) ions are slightly larger, ca. 0.09 Å, than the corresponding lanthanoid(III) ions [105]. Their ionic radii decrease along the actinoid series for the same reason as for the lanthanoid series but for the 5f electrons. Structure determinations of the solid $[An(H_2O)_9)](CF_3SO_3)_3$ salts, An=U, Np, Pu, Am, Cm, and Cf [106–109], showed that they are isostructural with the $[Ln(H_2O)_9)](CF_3SO_3)_3$ salts (Supplementary Tables S4a–n and S5a–e).

A limited number of studies on aqueous solutions on hydrated uranium(III), neptunium(III), plutonium(III), curium(III), berkelium(III), and californium(III) report mean An–O bond distances similar to those in the solid trifluoromethanesulfonate salts [108–115]. It can be assumed that all hydrated actinoid(III) ions have full nine-coordination in the tricapped trigonal prismatic configuration as the lawrencium(III) ion has approximately the same ionic radius as the dysprosium(III) ion [105]. This strongly indicates that the actinoid(III) ions of uranium–lawrencium are nine-coordinated in the tricapped trigonal prismatic configuration with the capping positions fully occupied. The hydrated actinium(III) ion has been reported to be 11-coordinated in aqueous solution, with a mean Ac–O bond distance of 2.63 Å [116]. The chemistry of actinium has been reviewed recently [117].

Only one crystal structure reports a hydrated thorium(IV) ion, $[Th(H_2O)_{10}]Br_4$ [118], where thorium(IV) is ten-coordinated with a mean Th–O bond distance of 2.498 Å. Other thorium(IV) compounds with many water molecules bound to thorium(IV) include the nine-coordinated compounds $H_5O_2[Th(H_2O)_6(OSO_2CF_3)_3][Th(H_2O)_3(OSO_2CF_3)_6]$ [119] and $[(H_2O)_7Th(OH)_2Th(H_2O)_7](C_2N_8)_2C_2HN_8 \cdot 10H_2O$ [120], and the ten-coordinated $[Th(H_2O)_8(ClO_4)](ClO_4)_3 \cdot H_2O$ [119] and $[Th(H_2O)_8(NO_3)](ReO_4)_3 \cdot C_{36}H_{36}N_{24}O_{12} \cdot 3H_2O$ [121], with mean Th–O bond distances of 2.451, 2.478, 2.499, and 2.501 Å, respectively. The mean Th–O bond distance of the hydrated thorium(IV) ion in acidic aqueous solution has been determined to be 2.45 Å by EXAFS, indicating that it is nine-coordinated, most likely in a tricapped trigonal prismatic fashion [119,122]. A second hydration sphere, Th–(O–H)$\cdots$O, around the water molecules in the prism was observed at 4.55 Å, and traces of a third hydration sphere at 6.85 Å [119]. The ionic radii of the actinoid(IV) ions are ca. 0.09 Å smaller than for corresponding lanthanoid(III) ions (Table 1 in ref. [105]).

For the EXAFS studies of the hydrated actinoid(IV) ions in aqueous solution, only mean M–O bond distances are reported. This is due to the low concentrations possible to obtain for these highly radioactive elements. It can be assumed that the hydrated actinoid(IV) ions in aqueous solution are nine-coordinated in a tricapped trigonal prismatic fashion as in the solid state [110–112,123–125].

Oxoactinoid(V) and Dioxoactinoid(VI)/Actinyl(V) and Actinyl(VI)

No solid state structures containing hydrated oxoactinoid(V) ions have been reported. The structure of the dioxouranium(VI) ion has been reported in several studies (Supplementary Table S5b) with the composition $[UO_2(H_2O)_5]^{2+}$ and mean U=O and U–O_{aq} bond distances of 1.760 and 2.422 Å, respectively, where the oxo ligands form a linear unit with the five water molecules forming a pentagon perpendicular to the oxo ligands. The hydrated dioxoneptunium(VI) and dioxoplutonium(VI) ions have the same structure as the hydrated dioxouranium(VI) ion with Np=O and Np–O_{aq} bond distances of 1.744 and 2.416 Å, and Pu=O and Pu–O_{aq} bond distances of 1.732 and 2.409 Å, respectively (Supplementary Table S5c,d).

The structure of the oxouranium(VI) ion in the solid state is maintained in aqueous solution with U=O and U–O_{aq} bond distances of 1.76 and 2.41 Å, respectively [111,113–115,122], in pentagonal bipyramidal fashion. The structures of the hydrated oxoneptunium(V), dioxoneptunium(VI), oxoplutonium(V), and dioxoplutonium(VI) complexes in aqueous solution have been determined by EXAFS. The Np=O and Np–O_{aq} bond distances in $[NpO_2(H_2O)_5]^+$ and $[NpO_2(H_2O)_5]^{2+}$ have been determined to 1.82 and 2.46, and 1.74 and 2.37 Å, respectively, forming pentagonal bipyramids [111,126]. The Pu=O and Pu–O_{aq} bond distances in $[PuO_2(H_2O)_{<5}]^+$ and $[PuO_2(H_2O)_5]^{2+}$ have both been determined to 1.74 and 2.4 Å, forming (tetra)pentagonal bipyramids [115].

3.3.4. Group 4 Elements

Titanium(III)

The hydrated titanium(III) ion (d^1 electron configuration) is six-coordinated in octahedral fashion with a mean Ti–O bond distance of 2.02 Å in three reported structures (Supplementary Table S6a). Two other six-coordinated titanium(III) complexes support this mean Ti–O bond distance (Supplementary Table S6a).

The mean Ti–O bond distance in the only reported structure of the hydrated titanium(III) ion in aqueous solution is significantly longer, 2.12 Å [127], compared to that observed in the $Ti(H_2O)_6]^{3+}$ complex in the solid state, 2.02 Å. An ESR and electron-spin echo study has indicated that the hydrated titanium(III) ion has a slightly compressed octahedral structure in aqueous solution [128]. The structure of the hydrated titanium(III) ion in aqueous solution can be regarded to have an octahedral configuration, possibly slightly compressed, with a mean Ti–O bond distance close to 2.02 Å.

Titanium(IV)/Titanyl(IV)

The oxotitanium(IV) ion (d^0), or titanyl(IV) ion, $[Ti{=}O]^{2+}$, cannot be prepared in sufficiently high concentrations in aqueous solution for structural characterization due to the formation of the insoluble titanium oxide, TiO_2. However, the oxotitanium(IV) ion can be stabilized, and thereby it becomes more soluble by complexation with, e.g., sulfate ligands [129]. Furthermore, the oxotitanium(IV) ion is stable and more soluble in strongly solvating solvents such as dimethylsulfoxide. The mean Ti=O, Ti–O_{eq}, and Ti–O_{ax} bond distances in the dimethylsulfoxide-solvated oxotitanium(IV) ion, $[TiO(OS(CH_3)_2)_5]^{2+}$, have been determined in the solid state to 1.645, 2.035, and 2.22 Å, respectively [129–131], and in dimethylsulfoxide solution, almost identical Ti–O bond distances have been obtained [129]. The structure of the bis(hydrogensulfonato)oxotitanium(IV) complex in aqueous solution has Ti=O, Ti–O_{eq}, and Ti–O_{ax} bond distances of 1.646, 2.034, and 2.22 Å, respectively [129]. Based on these results it can be assumed that the structure of the hydrated titanyl(IV) ion in aqueous solution has a structure and bond distances close to those in these complexes.

Zirconium(IV) and Hafnium(IV)

The hydrated zirconium(IV) and hafnium(IV) ions (d^0) do hydrolyze very easily to hydrated octahydroxidotetrazirconium(IV) and -hafnium(IV) ions in aqueous solution. It is, therefore, necessary to perform the structure determination of the hydrated zirconium(IV) and hafnium(IV) ions in strongly acidic solution [132]. The mean Zr–O and Hf–O bond distances in the hydrated zirconium(IV) and hafnium(IV) ions have been determined to be 2.19 and 2.16 Å in perchloric acidic aqueous solution [132].

Based on the structures of the hydrated octahydroxidotetrazirconium(IV) and -hafnium(IV) ions in the solid state (Supplementary Table S6b) and in aqueous solution, with mean Zr–O_{OH} and ZrO_{aq} bond distances of 2.15 and 2.27 Å, and mean Hf–O_{OH} and Hf–O_{aq} bond distances of 2.11 and 2.23 Å, respectively [132], the coordination geometry around the hydrated zirconium(IV) and hafnium(IV) ions is assumed to be the same in the hydrolyzed complexes, square antiprismatic [132].

3.3.5. Group 5 Elements

Vanadium(II) and Vanadium(III)

The hydrated vanadium(II) (d^3) and vanadium(III) (d^2) ions have a regular octahedral configuration with mean V–O bond distances of 2.13 and 1.99 Å, respectively, in the solid state (Supplementary Table S7a). An EXAFS study of the hydrated vanadium(III) ion in aqueous solution confirmed this structure also in aqueous solution with the same mean V–O bond distance, 1.99 Å [133]. An EXAFS study of the hydrated vanadium(II) and vanadium(III) ions reported mean V–O bond distances ca. 0.05 Å longer than those determined in the solid state [127]. The most likely V–O bond distances in the hydrated vanadium(II) and vanadium(III) ions in aqueous solution are most likely identical to those obtained in the solid state, 2.13 and 1.99 Å, respectively, with an octahedral configuration.

Oxovanadium(IV) and Dioxovanadium(V)/Vanadyl(IV) and Vanadyl(V)

The structure of the hydrated oxovanadium(IV) or vanadyl(IV) ion, $[VO(H_2O)_5]^{2+}$, (d^1) in the solid state is six-coordinated with a strongly bound oxo group and five water molecules, where the water is opposite to the oxo group, O_{ax}, which is significantly weaker bound than those in the equatorial positions, O_{eq} (Supplementary Table S7a). The mean V=O, V–O_{eq}, and V–O_{ax} bond distances are 1.58, 2.03, and 2.19 Å, respectively. This structure is retained in aqueous solution with mean V=O, V–O_{eq}, and V–O_{ax} bond distances of 1.60, 2.02, and 2.20 Å [134], and 1.70 and 2.01 Å [127], respectively. The vanadyl(IV) ion has a structural configuration similar to the titanyl(IV) ion. The V=O double bond is slightly shorter than the Ti=O one, while the V–O and Ti–O bond distances to the water molecules are almost identical, vide ultra. A well-defined second hydration sphere around the equatorially bound water molecules, V–(O–H)$\cdots$O, is observed at 4.06 Å [134]. The structure of the hydrated *cis*-dioxovanadium(V) ion, $[VO_2(H_2O)_4]^+$ (d^0), is so far only reported in acidic aqueous solution. The V=O, V–O_{cis}, and V–O_{trans} bond distances are determined to be 1.62, 1.99, and 2.22 Å, respectively [134]. A second hydration sphere, V–(O–H)$\cdots$O, is observed at 4.16 and 4.27 Å to the water molecules in the *cis* and *trans* positions versus V=O, respectively [134].

3.3.6. Group 6 Elements

Chromium(II)

The structure of the hydrated chromium(II) ion (d^4) has been reported in three solid-state structures (Supplementary Table S8). The chromium(II) ion is expected to display a Jahn–Teller distorted octahedral configuration due to its high spin d^4 electron configuration with its doubly degenerate ground states. Two of the reported structures display the expected Jahn–Teller distortion, while the third structure does not (Supplementary Table S8). The reason is most likely that the direction of the elongated axial Cr–O bond is randomly distributed, causing a mean structure with higher symmetry and a mean bond distance of all Cr–O, as discussed for the nine-coordinated lanthanoid(III) ion, vide ultra, and the hydrated copper(II) ion, vide infra. The mean Cr–O_{eq} and Cr–O_{ax} bond distances of the reported solid-state structures with Jahn–Teller distortion are 2.08 and 2.33 Å for the water molecules in the equatorial and axial positions, respectively (Supplementary Table S8).

The mean equatorial bond distance, Cr–O_{eq}, has been determined to be 2.08 Å in aqueous solution by EXAFS [133], which is in full agreement with the observation in the solid state. However, the mean axial bond distance was not detected. Whether the Jahn–Teller-distorted octahedral configuration is centrosymmetric or not will be discussed in the section of copper(II), vide infra. It can be assumed that the hydrated chromium(II) ion in aqueous solution has an octahedral Jahn–Teller configuration with Cr–O_{eq} and Cr–O_{ax} bond distances of 2.08 and ca. 2.3 Å, respectively.

Chromium(III)

The hydrated chromium(III) ion (d^3) has a well-characterized regular octahedral configuration with a mean Cr–O bond distance of 1.965 Å in the solid state (Supplementary Table S8). Numerous old studies of the hydrated chromium(III) ion in aqueous solution have been reported with a mean Cr–O bond distance of 1.99 Å [28,29]. A more recent study reports a mean Cr–O bond distance of 1.965 Å and a well-defined second hydration sphere, Cr–(O–H)$\cdots$O, of ca. 12 water molecules at 4.08 Å [135]. Chromium(III) hydrolyzes easily to cationic polymeric complexes in aqueous solution in the pH range of 2–6 with $[Cr_2(OH)_2(H_2O)_8]^{4+}$ and $[Cr_4(OH)_6(H_2O)_{12}]^{6+}$ as dominating species, depending on the total chromium(III) concentration [136]. The hydrated chromium(III) ion, $[Cr(H_2O)_6(H_2O)_{12}]^{3+}$, has a regular octahedral configuration with a mean Cr–O bond distance of 1.97 Å, and a well-defined second hydration sphere at 4.08 Å.

Molybenum(II) and Molybdenum(III)

No structure of the hydrated molybdenum(II) ion (d^4) has been reported in the solid state. However, the molybdenum(II) ion is present as a dimer in aqueous solution with a strong quadruple Mo–Mo bond. Each molybdenum binds four molecules at 2.14 Å in the hydrated $[Mo\text{-}Mo]^{4+}$ ion, and the Mo–Mo bond distance is 2.12 Å [137].

The structure of the hydrated molybdenum(III) ion (d^3) has been reported in one crystal structure, $[Mo(H_2O)_6]^{3+}$, with a Mo–O bond distance of 2.089 Å in a regular octahedral configuration [138]. The structure of the molybdenum(III) ion in aqueous solution is reported to be a hydrolyzed dimeric complex with double hydroxo bridges [137], but this interpretation seems questionable.

3.3.7. Group 7 Elements

Manganese(II)

The structure of the hydrated manganese(II) ion (d^5) has been reported in more than 160 crystal structure determinations to date (Supplementary Table S9). The manganese(II) ion binds six water molecules in a regular octahedron with a mean Mn–O bond distance of 2.174 Å. The same structure is present in aqueous solution with a mean Mn–O bond distance of 2.17 Å. [28,29,139,140] and a well-defined second hydration sphere, Mn–(O–H)$\cdots$O, of ca. 12 water molecules at 4.25 Å [141].

Manganese(III)

The manganese(III) ion with a d^4 electron configuration is expected to be Jahn–Teller distorted similarly to the chromium(II) ion discussed above. The two reported crystal structures containing a hydrated manganese(III) ion report a regular octahedral configuration with mean Mn–O bond distances of 1.91 and 1.99 Å, respectively [27,142]. By comparing with other manganese(III) complexes with oxygen donor ligands in the solid state [143,144], the most reasonable structure of the hydrated manganese(III) ion is an octahedral Jahn–Teller-distorted complex with Mn–O_{eq} and Mn–O_{ax} bond distances of ca. 1.91 and 2.2 Å, respectively. However, this postulation needs to be proven by experimental data and/or theoretical simulations.

3.3.8. Group 8 Elements

Iron(II)

The hydrated iron(II) ion (d^6) is well characterized in the solid state, binding six water molecules in a regular octahedral fashion with a mean Fe–O bond distance of 2.120 Å

(90 reported structures) (Supplementary Table S10a). The same structure is observed in aqueous solution [28,29,145,146]. The hydrated iron(II) ion in aqueous solution has a regular octahedral configuration with a mean Fe–O bond distance of 2.12 Å. The hydrated iron(II) ion is expected to have a well-defined second hydration sphere, Fe–(O–H)···O, of ca. 12 water molecules, but no accurate data have been reported.

Iron(III)

The hydrated iron(III) ion (d^5) is well characterized in a number of crystal structures, having a regular octahedral structure with a mean Fe–O bond distance of 1.995 Å (20 reported structures) (Supplementary Table S10a). The same structure is observed in aqueous solution [28,29,145]. The hydrated iron(III) ion in aqueous solution has a regular octahedral configuration with a mean Fe–O bond distance of 1.995 Å. The hydrated iron(III) ion is expected to have a well-defined second hydration sphere, Fe–(O–H)···O, of ca. 12 water molecules, but no accurate data have been reported.

Ruthenium(II) and Ruthenium(III)

The mean Ru–O bond distances in the octahedral hydrated ruthenium(II) and ruthenium(III) ions (d^6 and d^5, respectively) have been reported at 2.11 and 2.02 Å, respectively (Supplementary Table S10b). No experimental studies have been reported on the hydrated ruthenium(II) and ruthenium(III) ions in aqueous solution. It can be expected that the hydrated ruthenium(II) and ruthenium(III) ions retain the same structures in aqueous solution as in the solid state.

3.3.9. Group 9 Elements

Cobalt(II)

The hydrated cobalt(II) ion (d^7) binds six water molecules in a regular octahedral configuration with a mean Co–O bond distance of 2.087 Å in more than 430 reported solid-state structures (Supplementary Table S11a). A single study also reports a compound with a dimeric cobalt(II) hydrate, $[Co_2(H_2O)_{10}]^{4+}$, where two water molecules bridge the cobalt ions, which are six-coordinated in octahedral fashion. The mean Co–O bond distance to the terminal water molecules is significantly shorter, ca. 2.03 Å, than the mean Co–O bond distance to the bridging water molecules, 2.17 Å (Supplementary Table S12a). No dimeric hydrated cobalt(II) ions have been reported in aqueous solution. The structure of the $[Co(H_2O)_6]^{2+}$ ion observed in the solid state is retained in aqueous solution [28,29,147–150]. The hydrated cobalt(II) ion in aqueous solution has a regular octahedral configuration, with a mean Co–O bond distance of 2.09 Å in a regular octahedral fashion and a well-defined second hydration sphere, Co–(O–H)···O, of ca. 12 water molecules at ca. 4.20 Å [151].

Cobalt(III)

The hydrated cobalt(III) ion (d^6) is a strong oxidant and is reduced in water to cobalt(II) under the release of oxygen. The only crystal structures containing a hexaaquacobalt(III) ion report a Co–O bond distance of 1.87 Å [27] (Supplementary Table S11a). No structure determination of the hydrated cobalt(III) ion in aqueous solution has been reported.

Rhodium(III)

The hydrated rhodium(III) ion is hexahydrated in the solid state forming a regular octahedron with a mean Rh–O bond distance of 2.01 Å (Supplementary Table S11b). This structure is retained in aqueous solution with a mean Rh–O bond distance of 2.03 Å and a well-defined second hydration sphere at 4.02 Å [152–154].

Iridium(III)

A single structure determination in the solid state of a compound containing an octahedral hydrated iridium(III) ion, $[Ir(H_2O)_6]^{3+}$, has been reported. The mean Ir–O bond distance is 2.04 Å (Supplementary Table S11b). It can be assumed that this structure is retained in aqueous solution even though no experimental studies are reported.

3.3.10. Group 10 Elements

Nickel

The hydrated nickel(II) ion (d^8) binds six water molecules in a regular octahedral configuration in the solid state as reported in 376 crystal structures with a mean Ni–O bond distance of 2.055 Å (Supplementary Table S12a). Two studies report compounds with a dimeric nickel hydrate, $[Ni_2(H_2O)_{10}]^{4+}$, where two water molecules bridge the nickel ions, which are six-coordinated in an octahedral fashion. The Ni–O bond distance to the terminal water molecules is significantly longer, ca. 2.06 Å, than the Ni–O bond distance to the bridging water molecules, 1.93 Å (Supplementary Table S12a). The structure of the hydrated nickel(II) ion in aqueous solution has been reported in numerous studies [28,29,155], with a mean Ni–O bond distance of 2.055 Å in a regular octahedral fashion and a well-defined second hydration sphere, Ni–(O–H)$\cdots$O, of ca. 12 water molecules at ca. 4.10 Å [151].

Palladium(II) and Platinum(II)

The hydrated palladium(II) and platinum(II) ions (d^8) in the solid state are four-coordinated in square-planar fashion with mean Pd–O and Pt–O bond distances of 2.03 and 2.01 Å, respectively (Supplementary Table S12b).

Several research groups reported independently of each other that, in addition to the four strongly bound water molecules, one or two water molecules are very weakly bound in the axial positions at ca. 2.75 Å, completing tetragonally elongated square-pyramidal or octahedral configurations in aqueous solution [156–160] from EXAFS and theoretical simulation studies, or a combination of them. The hydrated palladium(II) and platinum(II) ions are five- or six-coordinated in aqueous solution with mean Pd–O and Pt–O bond distances of 2.02 and 2.02 Å, respectively, and with one or two water molecules very loosely bound at 2.7–2.8 Å.

3.3.11. Group 11 Elements

Copper(I)

The hydrated copper(I) ion (d^{10}) disproportionate metallic copper and copper(II) ions in aqueous solution of 2 $Cu^+ \rightleftharpoons Cu(s) + Cu^{2+}$, $K_D = [Cu^{2+}][Cu^+]^{-2}$, in water, $K_D \approx 10^6$ M^{-1} [161]. This means that copper(I) ion can only be the dominating copper species in sub-micromolar aqueous solution, and accurate structure determinations can hardly be performed. In order to estimate the structure of the hydrated copper(I) ion in aqueous solution, a free energy correlation was applied. The K_D values of copper(I) in some solvents, and the difference in the mean Cu–O bond distances in the solvated copper(I) and copper(II) ions in the respective solvent, assuming the copper(I) and copper(II) ions have the same configuration independent of the solvent, were used in the free energy relationship correlation [161]. The hydrated copper(I) ion was estimated from this correlation to bind four water molecules in a tetrahedral fashion with a Cu–O bond distance of ca. 2.14 Å [161].

Copper(II)

Hydrated copper(II) ions with the compositions $[Cu(H_2O)_4]^{2+}$, $[Cu(H_2O)_5]^{2+}$, and $[Cu(H_2O)_6]^{2+}$ have been reported in the solid state (Supplementary Table S13a). The $[Cu(H_2O)_4]^{2+}$ complex is square-planar with a mean Cu–O bond distance of 1.93 Å. The $[Cu(H_2O)_5]^{2+}$ complex has either square-pyramidal (the most common configuration) mean Cu–O bond distances of 1.956 and 2.24 Å in the equatorial and axial positions, respectively, or a trigonal bipyramidal configuration with a mean Cu–O bond distance of 1.96 Å to the strongly bound water molecules in the triangle and ca. 2.24 Å to the weakly bound water molecules in the axial positions (Supplementary Table S13a). The structure of the $[Cu(H_2O)_6]^{2+}$ complex in the solid state displays a variety of configurations: the expected Jahn–Teller-distorted octahedral configuration in both centrosymmetric complexes, in which the axial Cu–O_{ax} bond distances are identical, non-centrosymmetric ones, in which the axial Cu–O bond distances are different, regular octahedrons, and irregular octahedrons with pair-wise significantly different Cu–O bond distances (Supplementary Table S13a). The Jahn–Teller-distorted $[Cu(H_2O)_6]^{2+}$ complexes have mean Cu–O_{eq} and Cu–O_{ax} bond distances of 1.98 and 2.33 Å, independent of centrosymmetry or not (Supplementary Table S13a). The regular octahedral $[Cu(H_2O)_6]^{2+}$ complexes have a mean Cu–O bond distance of 2.08 Å (Supplementary Table S13a). The presence of regular $[Cu(H_2O)_6]^{2+}$ complexes is unexpected. An EXAFS study of a series of regular octahedral copper(II) complexes was carried out, where the lattice-independent EXAFS method proved that these complexes in fact are Jahn–Teller-distorted, with the expected splitting in short equatorial and long axial Cu-ligand bonds [162]. The reason for the observed regular octahedral structure of the Jahn–Teller-distorted $[Cu(H_2O)_6]^{2+}$ complex is that the direction of the Cu–O_{ax} bonds are randomly distributed in the structure, causing the mean structure to have higher symmetry than the individual complexes, and thereby an incorrect description the complex. In a recent study, it was shown that solid compounds crystallizing in a non-centrosymmetric space group or with the copper ion in the $[Cu(H_2O)_6]^{2+}$ complex not in the center of symmetry all have significantly different Cu–O_{ax} bond distances. This is also the case in all compounds with $[Cu(H_2O)_5$(O-ligand)] complexes with an octahedral configuration, which cannot be centrosymmetric [163]. This strongly indicates that the $[Cu(H_2O)_6]^{2+}$ ion is non-centrosymmetric and that the axial Cu–O_{ax} bonds are different.

The structure of the hydrated copper(II) ion in aqueous solution has been the theme of scientific discussion for more than 20 years. This discussion concerns whether the hydrated copper(II) ion is five- or six-coordinated, and if it is six-coordinated, whether the Jahn–Teller-distorted complex is centrosymmetric or not, vide ultra. Linear electric field EPR studies gave the first indications that the hydrated copper(II) ion in aqueous solution is non-centrosymmetric [164,165]. The paper initiating the scientific debate was a combined neutron and molecular dynamics study, which proposed that the hydrated copper(II) ion is five-coordinated in aqueous solution [166]. This study was followed by a large number of publications, mainly using EXAFS and XANES spectroscopy and DFT and QMCF/MD simulations to study the coordination chemistry of the hydrated copper(II) ion in aqueous solution [163,167–180]. This scientific discussion continues as no full consensus on the structure of the hydrated copper(II) ion in aqueous solution has been reached. However, it seems most likely that it has a non-centrosymmetric Jahn–Teller-distorted octahedral structure with mean Cu–O_{eq}, Cu–O_{ax1}, and Cu–O_{ax2} bond distances of ca. 1.96, 2.15, and 2.32 Å, respectively, according to the most recent studies [163,180].

Silver

The structure of the hydrated silver(I) ion (d^{10}) in the solid state is two-coordinated in a linear fashion with a mean Ag–O bond distance of 2.13 Å, and one structure where silver is three-coordinated in a triangular fashion, but the reported Ag–O bond distance is unexpectedly long (Supplementary Table S13b).

The structure of the hydrated silver(I) ion in aqueous solution has been debated for many years as different coordination numbers and Ag–O bond distances have been proposed, with a Ag–O bond distance of ca. 2.4 Å in a tetrahedral configuration as the first one [28,29]. A combined LAXS and EXAFS proposed a linearly distorted four-coordination with two Ag–O bonds at 2.32 Å and two very weakly bound at ca. 2.5 Å [181]. A recent combined EXAFS, LAXS, and simulation study proposed a similar structure but without the two weakly bound water molecules. A mean Ag–O bond distance of 2.34 Å and a diffuse hydration shell at a longer distance, and a second hydration sphere, Ag–(O–H)$\cdots$O, at 4.65 Å was proposed [182]. It is interesting to note that this diffuse hydration shell caused a lengthening of the linearly bound water molecules with ca. 0.2 Å, independent of the distance to the weakly bound water molecules.

Gold

Only one structure in the solid state has reported a hydrated gold(III) ion [183]. Gold(III) binds four water molecules in a square-planar fashion with a mean Au–O bond distance of 2.155 Å (Supplementary Table S13c). It is expected that this structure is retained in aqueous solution.

3.3.12. Group 12 Elements

Zinc

Hydrated zinc(II) ions (d^{10}) with the compositions $[Zn(H_2O)_4]^{2+}$, $[Zn(H_2O)_5]^{2+}$, and $[Zn(H_2O)_6]^{2+}$ have been reported in the solid state (Supplementary Table S14a). The $[Zn(H_2O)_4]^{2+}$ complex is tetrahedral, with a mean Zn–O bond distance of 2.03 Å. The $[Zn(H_2O)_5]^{2+}$ complex has a trigonal bipyramidal configuration with a mean Zn–O bond distance of 2.02 Å, and the $[Zn(H_2O)_6]^{2+}$ complex has a regular octahedral configuration with a mean Zn–O bond distance of 2.088 Å (Supplementary Table S14a). The Zn–O bond distances in the two reported $[Zn(H_2O)_4]^{2+}$ complexes are longer than the $[Zn(H_2O)_5]^{2+}$ complexes and significantly longer than tetrahedral zinc(II) complexes with oxygen donor ligands, ca. 1.95 Å [143,144]. The hydrated zinc(II) ion in aqueous solution binds six water molecules at a mean Zn–O bond distance of 2.09 Å in a regular octahedral fashion [28,29]. The hydrated zinc(II) ion has a second hydration sphere, Zn–(O–H)$\cdots$O, of ca. 12 water molecules at ca. 4.2 Å [28,29].

Cadmium

The hydrated cadmium(II) ion (d^{10}) binds six water molecules with a mean Cd–O bond distance of 2.266 Å in a regular octahedral fashion in the solid state (Supplementary Table S14b). A single study reports a hydrated cadmium(II) ion with the composition $[Cd(H_2O)_4]^{2+}$, but the reported Cd–O bond distance is far too long to be considered correct (Supplementary Table S14b). A single study also reports a compound with a dimeric cadmium(II) hydrate, $[Cd_2(H_2O)_{10}]^{4+}$, where two water molecules bridge the cadmium ions which are six-coordinated in an octahedral fashion. The terminal Cd–O bonds are significantly shorter than the bridging ones (Supplementary Table S14b).

The structure of the hydrated cadmium(II) ion in aqueous solution has been determined in several early studies with a mean Cd–O bond distance of ca. 2.29 Å [28,29]. A combined LAXS and ab initio MD simulation study of the hydrated cadmium ion in aqueous solution showed unusually broad bond distance distribution. The mean Cd–O bond distance in the hydrated cadmium ion in aqueous solution was reported to be 2.27 Å, with a possibility of an equilibrium between six- and seven-coordination [184]. The hydrated cadmium(II) ion has a second hydration sphere, Cd–(O–H)$\cdots$O, of ca. 12 water molecules at ca. 4.28 Å [184].

Mercury(I)

The mercury(I) ion (d^9) is dimeric and binds two water molecules in a linear fashion, $[H_2O–Hg–Hg–OH_2]^{2+}$, with mean Hg–O and Hg–Hg bond distances of 2.14 and 2.50 Å in the solid state (Supplementary Table S14c). A larger coordination number around each mercury, with three water molecules binding to each mercury, is observed in aqueous solution with mean Hg–O and Hg–Hg bond distances of 2.26 and 2.525 Å, respectively [185]. This completes a distorted tetrahedral configuration around each mercury in aqueous solution. The hydrated mercury(I) ion has a second hydration sphere, Hg–(O–H)$\cdots$O, of ca. six water molecules around each $Hg(H_2O)_3$ unit at ca. 4.52 Å [185]. A QMCF MD simulation study supports the experimentally obtained results [186].

Mercury(II)

The only crystal structure containing a fully hydrated mercury(II) ion is $[Hg(H_2O)_6](ClO_4)_2$ (Supplementary Table S14c). The $[Hg(H_2O)_6]^{2+}$ ion is reported to have a regular octahedral configuration with a Hg–O bond distance of 2.342 Å. However, an EXAFS study of this compound showed that it has the same wide and asymmetric bond distribution as the hydrated mercury(II) ion in aqueous solution [187], vide infra.

EXAFS studies on the hydrated mercury(II) ion (d^{10}) in aqueous solution have reported two different coordination numbers, six [187] and seven [188], both in wide and highly asymmetric bond distance distributions, with mean Hg–O bond distances of 2.33 and 2.32 Å, respectively [187,188]. The mean Hg–O bond distance reported for the proposed seven-coordinated mercury(II) hydrate, 2.32 Å [188], should be significantly longer than the mean Hg–O bond distance observed in octahedral complexes [26]. A crystal structure of a compound containing a dimethylsulfoxide (dmso)-solvated mercury(II) ion, $[Hg(OS(CH_3)_2)_6](ClO_4)_2$, crystallizes in a low-symmetry group (*P*-1). It shows that four of the dmso molecules has significantly shorter Hg–O bond distance than the remaining two, 2.319 and 2.376 Å, respectively [189]. The EXAFS data of six-coordinated homoleptic mercury(II) complexes in both solid state and solution all show a significant asymmetric bond distance distribution, strongly indicating a splitting of the Hg–O bond distances in the same way as in the solid $[Hg(OS(CH_3)_2)_6](ClO_4)_2$ [187]. This indicates that all six-coordinated mercury(II) complexes have this bond distance distribution, but it is only observed in solid compounds crystallizing in low-symmetry space groups as discussed for hydrated copper(II), lanthanoid(III), and scandium ions, vide ultra. This bond distance distribution can be described as a so-called pseudo-Jahn–Teller effect [190]. The pseudo-Jahn–Teller effect is a destabilization of six-coordinated metal complexes, including mercury(II), with small monodentate ligands [190–192]. This asymmetric coordination figure is caused by a contribution of the mercury(II) $5d_{z2}$ atomic orbital to the bonding molecular orbitals by vibronic coupling. The significant asymmetry of the Hg–O bond distance distribution in the hydrated mercury(II) ion in both aqueous solution and solid $[Hg(H_2O)_6](ClO_4)_2$ strongly supports that the hydrated mercury(II) has a pseudo-Jahn–Teller-distorted octahedral configuration [187]. The LAXS study showed a second hydration sphere, Hg–(O–H)$\cdots$O, of ca. 12 water molecules at ca. 4.20 Å [187].

3.4. *Group 13 Elements*

Aluminum, Gallium Indium, and Thallium(III)

The hydrated trivalent aluminum, gallium, indium, and thallium(III) ions, $[M(H_2O)_6]^{3+}$, all have a well-defined regular octahedral configuration in the solid state, with mean M–O bond distances of 1.88, 1.95, 2.12, and 2.17 Å, respectively (Supplementary Table S15); the Tl–O bond distance is shorter than expected in comparison to the observations in aqueous solution, vide infra. The structure of the hydrated aluminum, gallium, indium, and thallium(III) ions retain their structures observed in the solid state in aqueous solution with mean M–O bond distances of 1.89, 1.96, 2.14, and 2.23 Å, respectively [28,29,135,193,194]. Second hydration spheres, M–(–H)$\cdots$O, with ca. 12 molecules have been detected at 4.05, 4.13, and 4.2 Å for gallium, indium, and thallium(III), respectively [135,193].

3.5. *$d^{10}s^2$ Metal Ions in Groups 13, 14, and 15*

The coordination chemistry of metal ions with a $d^{10}s^2$ electron configuration, thallium(I), tin(II), lead(II), and bismuth(III), is dependent on the hybridization of these orbitals.

The $d^{10}s^2$ metal ions have an inert electron pair, which occupies a hybrid orbital formed by mixing the *ns* and *np* orbitals on the metal ion, according to the valence bond theory. It may be stereo-chemically active when mixing the *ns* and *np* orbitals, and a gap is formed in the coordination sphere, or it could be a pure s^2 electron pair and thereby stereo-chemically inactive, where all positions in the coordination geometry are occupied [195]. A lone electron pair acts as an additional ligand, which normally occupies more space than an ordinary ligand [196,197]. However, by applying the molecular orbital theory to the classical concept of *ns*/*np* orbital hybridization in $d^{10}s^2$ metal ions, it is regarded as incorrect. The stereo-chemical activity of the outer s^2 electron pair is instead a result of anti-bonding metal ion-donor atom interactions [198]. This results in two general geometries, a *holo-directed* one that is stereo-chemically inactive *without* a gap for the anti-bonding orbital, and a *hemi-directed* one being stereo-chemically active *with* a gap for the anti-bonding orbital [195].

The bonding character in the metal ion complex has a large impact on the electron density of the antibonding orbital and thereby on the size of the created void [199]. For ligands forming mainly electrostatic interactions, such as oxygen-donor ligands, the antibonding M(ns)–ligand(*n*p) orbital couples with the M(ns) orbital [199–201].

3.5.1. Thallium(I)

No solid compounds have been reported to contain a fully hydrated thallium(I) ion. Due to the weak hydration, only an indication of two bond distances, 2.73 and 3.18 Å, with a broad bond distance distribution, could be detected in a combined LAXS and EXAFS study [202]. It was not possible from this study to reveal any coordination geometry around thallium(I). As the thallium(I) ion has a $d^{10}s^2$ electron configuration, it can be assumed that it has a large gap in the coordination sphere with a hemi-directed configuration.

3.5.2. Tin(II)

The structure of the hydrated tin(II) ion has only been reported in two studies of solid $[Sn(H_2O)_3](ClO_4)_2$ (Supplementary Table S16a). The $[Sn(H_2O)_3]^{2+}$ ion has a mean Sn–O bond distance of 2.20 Å, and a huge gap in the coordination sphere with all O–Sn–O bond angles ca. 77°. This means that less than one-quarter of the coordination sphere is occupied by binding water molecules and thereby has a huge impact on the electrons in the anti-bonding orbitals.

Early studies reported the hydrated tin(II) ion in aqueous solution to have coordination numbers close to six and two different Sn–O bond distances [28,29], but this interpretation has not been supported by other studies. A recent combined EXAFS and XANES study reported an identical structure of the hydrated tin(II) ion in aqueous solution as in the solid state [203].

3.5.3. Lead(II)

The structure of the hydrated lead(II) ion is reported in four studies, where three contain a three-coordinate complex, $[Pb(H_2O)_3]^{2+}$, and the remaining one contains a four-coordinate $[Pb(H_2O)_4]^{2+}$ complex (Supplementary Table S16b). These structures are strongly affected by anti-bonding orbitals displaying all O–Pb–O bond angles close to 77° in the $[Pb(H_2O)_3]^{2+}$ complex, and in the $[Pb(H_2O)_4]^{2+}$ complex, the nearby O–Pb–O angles are close to 77°, and the diagonal angles are ca. 105° (Supplementary Table S16b). The mean Pb–O bond distances in the $[Pb(H_2O)_3]^{2+}$ and $[Pb(H_2O)_4]^{2+}$ complexes are 2.44 and 2.40 Å, respectively.

An early NMR study reported the hydration number of lead(II) to be six in aqueous solution [204]. A number of computer simulation studies on the structure of the hydrated lead(II) ion have been reported, but they all have difficulties describing the coordination chemistry and structural effects of anti-bonding orbitals. The mean Pb–O bond distance has been reported to be 2.50 Å in an EXAFS study [205]. A combined LAXS and EXAFS study of the hydrated lead(II) reported a mean Pb–O bond distance of 2.54 Å in a very broad bond distance distribution. A second hydration sphere, Pb–(O–H)···O, of ca. 12 water molecules at 4.42 Å was detected [206]. The EXAFS and LAXS studies of the hydrated lead(II) ion have shown a very broad bond distance distribution of around 2.5 Å, and that the number of water molecules cannot be determined accurately but should be in the range of 5–7.

3.5.4. Bismuth(III)

The only reported structure containing a hydrated bismuth(III) ion in the solid state, $[Bi(H_2O)_9](CF_3SO_3)_3$, is nine-coordinated in a tricapped trigonal prismatic configuration with mean Bi–O bond distances of 2.44 (prism) and 2.58 Å (capping positions) (Supplementary Table S16c). The structure of the hydrated bismuth(III) ion in aqueous solution was proven to have a significantly different structure, being eight-coordinated in a square antiprismatic configuration. The mean Bi–O bond distance was reported to be 2.41 Å, and a second hydration sphere, Bi–(O–H)···O, of ca. 16 water molecules at 4.48 Å, was observed [207].

4. Summary

Table 1 summarizes the observed hydrate complexes in the solid state and aqueous solution reported so far and predicted coordination numbers of metal ions from structural relationships of the nine-coordinated lanthanoid(III) and actinoid(III) ions based on other physico-chemical or structural relationships. Table 2 summarizes the mean M–O bond distances in hydrated metal ions in the solid state and aqueous solution. It has not for any metal ion been proven that the coordination number or structure in aqueous solution is concentration-dependent as long as any complex formation with, e.g., counter ions or hydrolysis does not take place. Many metal ions have several coordination numbers, configurations, and formations of dimeric and polymeric hydrate complexes with bridging water molecules in the solid state, but in aqueous solution, one coordination number with a specific configuration strongly dominates (Table 1).

The monovalent metal ions, with the exception of the lithium and silver(I) ions, have one well-defined first hydration shell, but the water molecules outside it are not spectroscopically distinguishable from the aqueous bulk [9,10]. Di- and trivalent metal ions have a well-defined second hydration there on average; one water molecule in the first hydration sphere hydrogen binds two water molecules in the second hydration sphere, as can be shown by double difference infrared spectroscopy [9] and the LAXS and LANS methods [28,29]. Four-valent and small trivalent metal ions may have at least traces of a third hydration as, e.g., shown in the hydrated thorium(IV) ion [118].

Table 1. Summary of metal ion aqua complexes reported in the solid state and aqueous solution. Complexes in italics highlight the complexes present in aqueous solution not observed in the solid state, and underlined complexes are either extrapolated from the structures of metal ion solvates of non-aqueous solvents, from physico-chemical relationships.

Metal Ion	Solid State	Aqueous Solution
Lithium(I)	$[Li(H_2O)_4]^+$, $Li(H_2O)_5]^+$, $[Li(H_2O)_6]^+$, $[Li_2(H_2O)_6]^{2+}$, $[Li_2(H_2O)_7]^{2+}$, $[Li_3(H_2O)_6]^{3+}$,$[Li_3(H_2O)_8]^{3+}$	$[Li(H_2O)_4]^+$
Sodium(I)	$[Na(H_2O)_2]^+$, $[Na(H_2O)_4]^+$, $[Na(H_2O)_5]^+$, $[Na(H_2O)_6]^+$, $[Na(H_2O)_8]^+$, $[Na_2(H_2O)_6]_n{}^{2+}$, $[Na_2(H_2O)_7]_n{}^{2+}$, $[Na_2(H_2O)_8]_n{}^{2+}$, $[Na_2(H_2O)_8]^{2+}$, $[Na_2(H_2O)_9]^{2+}$, $[Na_2(H_2O)_{10}]^{2+}$, $[Na_2(H_2O)_{13}]^{2+}$, $[Na_4(H_2O)_{14}]_n{}^{4+}$, $[Na_4(H_2O)_{14}]_n{}^{4+}$, $[Na_4(H_2O)_{18}]_n{}^{4+}$, $[Na_5(H_2O)_{14}]^{5+}$, $[Na_8(H_2O)_{24}]_n{}^{8+}$	$[Na(H_2O)_6]^+$
Potassium(I)	$[K(H_2O)_6]^+$, $[K(H_2O)_7]^+$, $[K(H_2O)_8]^+$, $[K_2(H_2O)_9]_n{}^{2+}$	$[K(H_2O)_7]^+$
Rubidium(I)	$[Rb(H_2O)_9]^+$	$[Rb(H_2O)_8]^+$
Cesium(I)	$[Cs(H_2O)_{10}]^+$	$[Cs(H_2O)_8]^+$
Beryllium(II)	$[Be(H_2O)_4]^{2+}$	$[Be(H_2O)_4]^{2+}$
Magnesium(II)	$[Mg(H_2O)_6]^{2+}$	$[Mg(H_2O)_6]^{2+}$
Calcium(II)	$[Ca(H_2O)_6]^{2+}$, $[Ca(H_2O)_7]^{2+}$, $[Ca(H_2O)_8]^{2+}$, $[Ca_2(H_2O)_{10}]^{4+}$, $[Ca_2(H_2O)_{11}]^{4+}$, $[Ca_2(H_2O)_{14}]^{4+}$	$[Ca(H_2O)_8]^{2+}$
Strontium(II)	$[Sr(H_2O)_6]^{2+}$, $[Sr(H_2O)_8]^{2+}$, $[Sr(H_2O)_9]^{2+}$, $[Sr_2(H_2O)_{14}]^{4+}$	$[Sr(H_2O)_8]^{2+}$
Barium(II)	$[Ba(H_2O)_8]^{2+}$, $[Ba(H_2O)_9]^{2+}$, $[Ba_2(H_2O)_{14}]^{2+}$, $[Ba_2(H_2O)_{18}]^{4+}$	$[Ba(H_2O)_{8-9}]^{2+}$
Radium(II)		*$[Ra(H_2O)_{\approx 10}]^{2+}$*
Scandium(III)	$[Sc(H_2O)_6]^{3+}$, $[Sc(H_2O)_7]^{3+}$, $[Sc(H_2O)_{8.0}]^{3+}$	$[Sc(H_2O)_{8.0}]^{3+}$

Table 1. *Cont.*

Metal Ion	Solid State	Aqueous Solution
Yttrium(III)	$[Y(H_2O)_8]^{3+}$, $[Y(H_2O)_9]^{3+}$	$[Y(H_2O)_8]^{3+}$
Lanthanum(III)	$[La(H_2O)_6]^{3+}$, $[La(H_2O)_8]^{3+}$, $[La(H_2O)_9]^{3+}$	$[La(H_2O)_9]^{3+}$
Cerium(III)	$[Ce(H_2O)_8]^{3+}$, $[Ce(H_2O)_9]^{3+}$	$[Ce(H_2O)_9]^{3+}$
Cerium(IV)		*$[Ce(H_2O)_8]^{4+}$*
Praseodymium(III)	$[Pr(H_2O)_8]^{3+}$, $[Pr(H_2O)_9]^{3+}$	$[Pr(H_2O)_9]^{3+}$
Neodymium(III)	$[Nd(H_2O)_8]^{3+}$, $[Nd(H_2O)_9]^{3+}$	$[Nd(H_2O)_9]^{3+}$
Prometium(III)		*$[Pm(H_2O)_9]^{3+}$*
Samarium(III)	$[Sm(H_2O)_8]^{3+}$, $[Sm(H_2O)_9]^{3+}$	$[Sm(H_2O)_9]^{3+}$
Europium(II)		*$[Eu(H_2O)_7]^{2+}$*
Europium(III)	$[Eu(H_2O)_8]^{3+}$, $[Su(H_2O)_9]^{3+}$	$[Eu(H_2O)_9]^{3+}$
Gadolinium(III)	$[Gd(H_2O)_8]^{3+}$, $[Gd(H_2O)_9]^{3+}$	$[Gd(H_2O)_9]^{3+}$
Terbium(III)	$[Tb(H_2O)_8]^{3+}$, $[Tb(H_2O)_9]^{3+}$	$[Tb(H_2O)_9]^{3+}$
Dysprosium(III)	$[Dy(H_2O)_8]^{3+}$, $[Dy(H_2O)_9]^{3+}$	$[Dy(H_2O)_9]^{3+}$
Holmium(III)	$[Ho(H_2O)_8]^{3+}$, $[Ho(H_2O)_{8.91}]^{3+}$, $[Ho(H_2O)_9]^{3+}$	$[Ho(H_2O)_{8.91}]^{3+}$
Erbium(III)	$[Er(H_2O)_8]^{3+}$, $[Er(H_2O)_{8.95}]^{3+}$, $[Er(H_2O)_9]^{3+}$	$[Er(H_2O)_{8.95}]^{3+}$
Thulium(III)	$[Tm(H_2O)_8]^{3+}$, $[Tm(H_2O)_{8.8}]^{3+}$, $[Tm(H_2O)_9]^{3+}$	$[Tm(H_2O)_{8.8}]^{3+}$
Ytterbium(II)		*$[Yb(H_2O)_8]^{2+}$*
Ytterbium(III)	$[Yb(H_2O)_8]^{3+}$, $[YbH_2O)_{8.7}]^{3+}$, $[Yb(H_2O)_9]^{3+}$	$[Yb(H_2O)_{8.7}]^{3+}$
Lutetium(III)	$[Lu(H_2O)_8]^{3+}$, $[Lu(H_2O)_{8.2}]^{3+}$, $[Lu(H_2O)_9]^{3+}$	$[Lu(H_2O)_{8.2}]^{3+}$
Actinium(III)		*$[Ac(H_2O)_{11}]^{3+}$*
Thorium(IV)	$[Th(H_2O)_8]^{4+}$, $[Th(H_2O)_9]^{4+}$, $[Th(H_2O)_{10}]^{4+}$	$[Th(H_2O)_9]^{4+}$
Uranium(III)	$[U(H_2O)_9]^{3+}$	$[U(H_2O)_9]^{3+}$
Uranium(IV)		*$[U(H_2O)_9]^{4+}$*
Uranium(VI)	$[UO_2(H_2O)_5]^{2+}$	$[UO_2(H_2O)_5]^{2+}$
Neptunium(III)	$[Np(H_2O)_9]^{3+}$	$[Np(H_2O)_9]^{3+}$
Neptunium(IV)		*$[Np(H_2O)_9]^{4+}$*
Neptunium(V)		*$[NpO_2(H_2O)_5]^{+}$*
Neptunium(VI)	$[NpO_2(H_2O)_5]^{2+}$	$[NpO_2(H_2O)_5]^{2+}$
Plutonium(III)	$[Pu(H_2O)_9]^{3+}$	$[Pu(H_2O)_9]^{3+}$
Plutonium(IV)		*$[Pu(H_2O)_9]^{4+}$*
Plutonium(V)		*$[PuO_2(H_2O)_5]^{+}$*
Plutonium(VI)	$[PuO_2(H_2O)_5]^{2+}$	$[PuO_2(H_2O)_5]^{2+}$
Americium(III)	$[Am(H_2O)_9]^{3+}$	$[Am(H_2O)_9]^{3+}$
Curium(III)	$[Cm(H_2O)_9]^{3+}$	$[Am(H_2O)_9]^{3+}$
Berkelium(III)		*$[Bk(H_2O)_9]^{3+}$*
Californium(III)	$[Cf(H_2O)_9]^{3+}$	$[Cf(H_2O)_9]^{3+}$
Einsteinium(III)		*$[Es(H_2O)_9]^{3+}$*
Fermium(III)		*$[Fm(H_2O)_9]^{3+}$*
Medelevium(III)		*$[Md(H_2O)_9]^{3+}$*

Table 1. *Cont.*

Metal Ion	Solid State	Aqueous Solution
Nobelium(III)		*$[No(H_2O)_9]^{3+}$*
Lawrentium(III)		*$[Lr(H_2O)_9]^{3+}$*
Titanium(III)	$[Ti(H_2O)_6]^{3+}$	$[Ti(H_2O)_6]^{3+}$
Titanium(IV)		*$[TiO(H_2O)_5]^{2+}$*
Zirconium(IV)		*$[Zr(H_2O)_8]^{4+}$*
Hafnium(IV)		*$[Hf(H_2O)_8]^{4+}$*
Vanadium(II)	$[V(H_2O)_6]^{2+}$	$[V(H_2O)_6]^{2+}$
Vanadium(III)	$[V(H_2O)_6]^{3+}$	$[V(H_2O)_6]^{3+}$
Vanadium(IV)	$[VO(H_2O)_6]^{2+}$	$[VO(H_2O)_5]^{2+}$
Vanadium(V)		*cis-$[VO_2(H_2O)_4]^{+}$*
Chromium(II)	$[Cr(H_2O)_6]^{2+}$	$[Cr(H_2O)_6]^{2+}$
Chromium(III)	$[Cr(H_2O)_6]^{3+}$	$[Cr(H_2O)_6]^{3+}$
Molybdenum(II)		*$[Mo_2(H_2O)_4]^{4+}$*
Molybdenum(III)	$[Mo(H_2O)_6]^{3+}$	$[Mo(H_2O)_6]^{3+}$
Manganese(II)	$[Mn(H_2O)_6]^{2+}$	$[Mn(H_2O)_6]^{2+}$
Manganese(III)	$[Mn(H_2O)_6]^{3+}$	$[Mn(H_2O)_6]^{3+}$
Iron(II)	$[Fe(H_2O)_6]^{2+}$	$[Fe(H_2O)_6]^{2+}$
Iron(III)	$[Fe(H_2O)_6]^{3+}$	$[Fe(H_2O)_6]^{3+}$
Ruthenium(II)	$[Ru(H_2O)_6]^{2+}$	$[Ru(H_2O)_6]^{2+}$
Ruthenium(III)	$[Ru(H_2O)_6]^{3+}$	$[Ru(H_2O)_6]^{3+}$
Cobalt(II)	$[Co(H_2O)_6]^{2+}$, $[Co_2(H_2O)_{10}]^{4+}$	$[Co(H_2O)_6]^{2+}$
Cobalt(III)	$[Co(H_2O)_6]^{3+}$	$[Co(H_2O)_6]^{3+}$
Rhodium(III)	$[Rh(H_2O)_6]^{3+}$	$[Rh(H_2O)_6]^{3+}$
Iridium(III)	$[Ir(H_2O)_6]^{3+}$	$[Ir(H_2O)_6]^{3+}$
Nickel(II)	$[Ni(H_2O)_6]^{2+}$, $[Ni_2(H_2O)_{10}]^{4+}$	$[Ni(H_2O)_6]^{2+}$
Palladium(II)	$[Pd(H_2O)_4]^{2+}$	*$[Pd(H_2O)_{5/6}]^{2+}$*
Platinum(II)	$[Pt(H_2O)_4]^{2+}$	*$[Pt(H_2O)_{5/6}]^{2+}$*
Copper(I)		*$[Cu(H_2O)_4]^{+}$*
Copper(II)	$[Cu(H_2O)_4]^{2+}$, $[Cu(H_2O)_5]^{2+}$, $[Cu(H_2O)_6]^{2+}$	$[Cu(H_2O)_6]^{2+}$
Silver(I)	$[Ag(H_2O)_2]^{+}$, $[Ag(H_2O)_3]^{+}$	$[Ag(H_2O)_2]^{+}$
Gold(III)	$[Au(H_2O)_4]^{3+}$	$[Au(H_2O)_4]^{3+}$
Zinc(II)	$[Zn(H_2O)_4]^{2+}$, $[Zn(H_2O)_5]^{2+}$, $[Zn(H_2O)_6]^{2+}$	$[Zn(H_2O)_6]^{2+}$
Cadmium(II)	$[Cd(H_2O)_6]^{2+}$, $[Cd_2(H_2O)_{10}]^{4+}$	$[Cd(H_2O)_6]^{2+}$
Mercury(I)	$[Hg_2(H_2O)_2]^{2+}$	*$[Hg_2(H_2O)_6]^{2+}$*
Mercury(II)	$[Hg(H_2O)_6]^{2+}$	$[Hg(H_2O)_6]^{2+}$
Aluminum(III)	$[Al(H_2O)_6]^{3+}$	$[Al(H_2O)_6]^{3+}$
Gallium(III)	$[Ga(H_2O)_6]^{3+}$	$[Ga(H_2O)_6]^{3+}$
Indium(III)	$[In(H_2O)_6]^{3+}$	$[In(H_2O)_6]^{3+}$
Thallium(I)		*$[Tl(H_2O)_n]^{+}$*

Table 1. *Cont.*

Metal Ion	Solid State	Aqueous Solution
Thallium(III)	$[Tl(H_2O)_6]^{3+}$	$[Tl(H_2O)_6]^{3+}$
Tin(II)	$[Sn(H_2O)_3]^{2+}$	$[Sn(H_2O)_3]^{2+}$
Lead(II)	$[Pb(H_2O)_3]^{2+}$, $[Pb(H_2O)_4]^{2+}$	$[Pb(H_2O)_{5–7}]^{2+}$
Bismuth(III)	$[Bi(H_2O)_9]^{3+}$	$[Bi(H_2O)_8]^{3+}$

Table 2. Summary of mean M–O bond distances in hydrated metal ions in solid state and solution determined experimentally. The following abbreviations are used in this table, coordination, and configuration: two—linear, $D_{\infty h}$, three: triangular, $D_{3h}(3)$, trigonal pyramidal, C_{3v}, four—tetrahedral, T_d, and square-planar, $D_{4h}(4)$, five: square pyramidal, C_{4v}, and trigonal bipyramidal, $D_{3h}(5)$, six: octahedral O_h, and tetragonally elongated octahedral (Jahn–Teller distorted octahedral), C_{4v}, eight: square antiprism, $D_{4h}(8)$, nine: tricapped prismatic, $D_{3h}(9)$, and compounds without a well-defined coordination geometry, the coordination number is given by a figure.

Metal Ion	Solid State	Aqueous Solution
$Li(H_2O)_4^+$	1.941–T_d (Table S1a)	1.95–T_d [10–16]
$Na(H_2O)_6^+$	2.417–O_h (Table S1b)	2.43–O_h [10,21–25]
$Na(H_2O)_5^{3+}$	2.341–D_{3h} (Table S1b)	
$K(H_2O)_7^+$		2.81–7 [10,28–30,32]
$Rb(H_2O)_8^+$		2.98–D_{4h}(8) [33,34]
$Cs(H_2O)_8^+$		3.10–D_{4h}(8) [10,33,36,37]
$Be(H_2O)_4^{2+}$	1.613–T_d (Table S2a)	1.62–T_d [a]
$Mg(H_2O)_6^{2+}$	2.066–O_h (Table S2b)	2.07–O_h [28,29]
$Ca(H_2O)_8^{2+}$	2.476–D_{4h}(8) (Table S2c)	2.48–D_{4h}(8) [28,29,43–52]
$Ca(H_2O)_7^{3+}$	2.401–C_{3v}(7) (Table S2c)	
$Ca(H_2O)_6^{3+}$	2.324–O_h (Table S2c)	
$Sr(H_2O)_8^{2+}$	2.613–D_{4h} (Table S2d)	2.61–D_{4h}(8) [58–67]
$Ba(H_2O)_8^{2+}$	2.80–D_{4h} (Table S2e)	2.81–D_{4h}(8) [58,69,74–76]
$Ra(H_2O)_{10}^{2+}$		2.93–10 [74]
$Sc(H_2O)_8^{3+}$	2.170 + 2.459–D_{3h}(9) (Table S3a)	2.17 + 2.33–D_{3h}(9) [81]
$Y(H_2O)_8^{3+}$	2.353–D_{4h}(8) (Table S3b)	2.367–D_{4h}(8) [82]
$La(H_2O)_9^{3+}$	2.528 + 2.611–D_{3h}(9) (Table S4a)	2.53 + 2.60–D_{3h}(9) [88]
$La(H_2O)_8^{3+}$	2.495–D_{4h}(8) (Table S4a)	
$Ce(H_2O)_9^{3+}$	2.491 + 2.597–D_{3h}(9) (Table S4b)	2.49 + 2.62–D_{3h}(9) [88]
$Ce(H_2O)_8^{3+}$	2.485–D_{4h}(8) (Table S4b)	
$Pr(H_2O)_9^{3+}$	2.475 + 2.568–D_{3h}(9) (Table S4c)	2.48 + 2.58–D_{3h}(9) [88]
$Pr(H_2O)_8^{3+}$	2.476–D_{4h}(8) (Table S4c)	
$Nd(H_2O)_9^{3+}$	2.458 + 2.565–D_{3h}(9) (Table S4d)	2.47 + 2.60–D_{3h}(9) [88]
$Nd(H_2O)_8^{3+}$	2.439–D_{4h}(8) (Table S4d)	
$Sm(H_2O)_9^{3+}$	2.436 + 2.545–D_{3h}(9) (Table S4e)	2.43 + 2.53–D_{3h}(9) [88]
$Sm(H_2O)_8^{3+}$	2.430–D_{4h}(8) (Table S4e)	

Table 2. *Cont.*

Metal Ion	Solid State	Aqueous Solution
$Eu(H_2O)_8{}^{2+}$		2.584–D_{4h}(8) [66]
$Eu(H_2O)_9{}^{3+}$	2.414 + 2.535–D_{3h}(9) (Table S4f)	2.42 + 2.57–D_{3h}(9) [88]
$Eu(H_2O)_8{}^{3+}$	2.416–D_{4h}(8) (Table S4f)	
$Gd(H_2O)_9{}^{3+}$	2.404 + 2.542–D_{3h}(9) (Table S4g)	2.41 + 2.53–D_{3h}(9) [88]
$Gd(H_2O)_8{}^{3+}$	2.403–D_{4h}(8) (Table S4g)	
$Tb(H_2O)_9{}^{3+}$	2.388 + 2.514–D_{3h}(9) (Table S4h)	2.40 + 2.53–D_{3h}(9) [88]
$Tb(H_2O)_8{}^{3+}$	2.381–D_{4h}(8) (Table S4h)	
$Dy(H_2O)_9{}^{3+}$	2.368 + 2.519–D_{3h}(9) (Table S4i)	2.38 + 2.50–D_{3h}(9) [88]
$Dy(H_2O)_8{}^{3+}$	2.371–D_{4h}(8) (Table S4i)	
$Ho(H_2O)_9{}^{3+}$	2.375 + 2.494–D_{3h}(9) (Table S4j)	2.37 + 2.50–D_{3h}(9) [88]
$Ho(H_2O)_8{}^{3+}$	2.356–D_{4h}(8) (Table S4j)	
$Er(H_2O)_{8.9}{}^{3+}$	2.351 + 2.515–D_{3h}(9) (Table S4k)	2.36 + 2.48–D_{3h}(9) [88]
$Er(H_2O)_8{}^{3+}$	2.345–D_{4h}(8) (Table S4k)	
$Tm(H_2O)_{8.9}{}^{3+}$	2.331 + 2.513–D_{3h}(9) (Table S4l)	2.35 + 2.47–D_{3h}(9) [88]
$Tm(H_2O)_8{}^{3+}$	2.339–D_{4h}(8) (Table S4l)	
$Yb(H_2O)_{8.7}{}^{3+}$	2.314 + 2.501–D_{3h}(9) (Table S4m)	2.32 + 2.43–D_{3h}(9) [88]
$Yb(H_2O)_8{}^{3+}$	2.326–D_{4h}(8) (Table S4m)	
$Lu(H_2O)_{8.2}{}^{3+}$	2.296 + 2.506–D_{3h}(9) (Table S4n)	2.28 + 2.37–D_{3h}(9) [88]
$Lu(H_2O)_8{}^{3+}$	2.318–D_{4h}(8) (Table S4n)	
$Ac(H_2O)_{11}{}^{3+}$		2.63–11 [116]
$Th(H_2O)_{10}{}^{4+}$	2.498–10 2.345–D_{4h}(8) (Table S4k)	
$Th(H_2O)_9{}^{4+}$	2.448–9 (Table S5a)	2.45–9 [119,122]
$U(H_2O)_9{}^{3+}$	2.509 + 2.596–D_{3h}(9) (Table S5b)	2.50 + 2.58–D_{3h}(9) [110]
$U(H_2O)_9{}^{4+}$	2.41–9 (Table S5b)	2.42–9 [123,124]
$U(H_2O)_8{}^{4+}$	2.395–D_{4h}(8) (Table S5b)	
$UO_2(H_2O)_5{}^{2+}$	1.760 + 2.444–D_{5h} (Table S5b)	1.77 + 2.42–D_{5h} [111,122]
$Np(H_2O)_9{}^{3+}$	2.491 + 2.571–D_{3h} (Table S5c)	2.48 + 2.56–D_{3h} [111]
$Np(H_2O)_9{}^{4+}$		2.40–9 [110]
$Np(H_2O)_8{}^{4+}$	2.335–D_{4h}(8) (Table S5c)	
$NpO_2(H_2O)_5{}^{2+}$	1.744 + 2.416–D_{5h} (Table S5c)	1.83 + 2.48–D_{5h} [111,126]
$Pu(H_2O)^{3+}$	2.475 + 2.571–D_{3h} (Table S5d)	2.45 + 2.55–D_{3h}(9) [110–112,114,115]
$Pu(H_2O)_8{}^{4+}$	2.39–9 (Table S5d)	2.395–D_{4h}(8) [114,115]
$PuO_2(H_2O)_5{}^{+}$		1.81 + 2.47–D_{5h} [114,115]
$PuO_2(H_2O)_5{}^{2+}$	1.732 + 2.409–D_{3h} (Table S5d)	1.75 + 2.41–D_{5h} [114,115]
$Am(H_2O)_9{}^{3+}$	2.466 + 2.578–D_{3h}(9) (Table S5e)	2.43 + 2.54–D_{3h}(9) [108,110,113]
$Cm(H_2O)_9{}^{3+}$	2.453 + 2.555–D_{3h}(9) (Table S5e)	2.40 + 2.53–D_{3h}(9) [108,110,113]
$Bk(H_2O)_9{}^{3+}$		2.38 + 2.50–D_{3h}(9) [110]
$Cf(H_2O)_9{}^{3+}$	2.423 + 2.550–D_{3h}(9) (Table S5e)	2.37 + 2.49–D_{3h}(9) [110]

Table 2. *Cont.*

Metal Ion	Solid State	Aqueous Solution
$Ti(H_2O)_6^{3+}$	2.02–O_h (Table S6a)	2.03–O_h [a]
$TiO(H_2O)_5^{2+}$	1.644 + 2.034 + 2.233 [a] (Table S6a)	
$Zr(H_2O)_8^{4+}$		2.19–D_{4h}(8) [132]
$Hf(H_2O)_8^{4+}$		2.16–D_{4h}(8) [132]
$V(H_2O)_6^{2+}$	2.131–O_h (Table S7)	2.13–O_h [a]
$V(H_2O)_6^{3+}$	1.994–O_h (Table S7)	1.99–O_h [133]
$VO(H_2O)_5^{2+}$	1.580 + 2.026 + 2.187 (Table S7)	1.60 + 2.03 + 2.20 [134]
$VO_2(H_2O)_4^{+}$		1.62 + 1.99 + 2.22 [134]
$Cr(H_2O)_6^{2+}$	2.083 + 2.333–C_{4v} (Table S8)	2.08 + ~2.3–C_{4v} [133]
$Cr(H_2O)_6^{3+}$	1.965–O_h (Table S8)	1.97–O_h [28,29,135]
$Mo(H_2O)_6^{3+}$	2.089–O_h (Table S8)	2.09–O_h [138]
$Mn(H_2O)_6^{2+}$	2.174–O_h (Table S9)	2.17–O_h [28,29,139,140]
$Mn(H_2O)_6^{3+}$	2.088–O_h (Table S9)	
$Fe(H_2O)_6^{2+}$	2.120–O_h (Table S10a)	2.12–O_h [28,29,145,146]
$Fe(H_2O)_6^{3+}$	1.995–O_h (Table S10a)	2.00–O_h [28,29,145]
$Ru(H_2O)_6^{2+}$	2.111–O_h (Table S10b)	2.11–O_h [a]
$Ru(H_2O)_6^{3+}$	2.021–O_h (Table S10b)	2.03–O_h [a]
$Co(H_2O)_6^{2+}$	2.087–O_h (Table S11a)	2.09–O_h [28,29,147–150]
$Co(H_2O)_6^{3+}$	1.873–O_h (Table S11a)	
$Rh(H_2O)_6^{3+}$	2.014–O_h (Table S11b)	2.04–O_h [152–154]
$Ir(H_2O)_6^{3+}$	2.04–O_h (Table S11b)	
$Ni(H_2O)_6^{2+}$	2.055–O_h (Table S12a)	2.055–O_h [28,29,155]
$Pd(H_2O)_6^{2+}$		2.01 + 2.8–D_{4h}(6) [156–160]
$Pd(H_2O)_4^{2+}$	2.029–D_{4h}(4) (Table S12b)	
$Pt(H_2O)_6^{2+}$		2.02 + 2.8–D_{4h}(6) [156–160]
$Pt(H_2O)_4^{2+}$	2.012–D_{4h}(4) (Table S12b)	
$Cu(H_2O)_4^{+}$		2.14–T_d [b] [161]
$Cu(H_2O)_6^{2+}$	1.980 + 2.322–C_{4v}(6) (Table S13a)	1.96 + 2.32–D_{4h}(6) [163,167]
$Cu(H_2O)_6^{2+}$	1.944 + 2.248–C_{4v}(5) (Table S13a)	
$Ag(H_2O)_4^{+}$		2.32 + 2.54–2 + 2 [181]
$Ag(H_2O)_2^{+}$	2.129–$D_{\infty h}$ (Table S13b)	2.34–2 [182]
$Au((H_2O)_4^{3+}$	2.155–D_{4h}(4) (Table S13c)	
$Zn(H_2O)_6^{2+}$	2.088–O_h (Table S14a)	2.09–O_h [28,29]
$Zn(H_2O)_5^{2+}$	2.017–D_{3h}(5) (Table S14a)	
$Cd(H_2O)_6^{2+}$	2.266–O_h (Table S14b)	2.27–O_h [184]
$Hg_2(H_2O)_6^{2+}$		2.26–T_d [185]
$Hg_2(H_2O)_2^{2+}$	2.14–T_d (Table S14c)	
$Hg(H_2O)_7^{2+}$		2.32–7 [188]
$Hg(H_2O)_6^{2+}$	2.342–O_h (Table S14c)	2.34–O_h [187]

Table 2. *Cont.*

Metal Ion	Solid State	Aqueous Solution
$Al(H_2O)_6{}^{3+}$	1.883–O_h (Table S15a)	1.89–O_h [28,29]
$Ga(H_2O)_6{}^{3+}$	1.946–O_h (Table S15b)	1.96–O_h [28,29,135]
$In(H_2O)_6{}^{3+}$	2.124–O_h (Table S15c)	2.14–O_h [135]
$Ti(H_2O)_n{}^{+}$		2.73 + 3.18–2 + 2 [202]
$Tl(H_2O)_6{}^{3+}$	2.23–O_h (Table S15d)	2.23–O_h [193,194]
$Sn(H_2O)_3{}^{2+}$	2.204–C_{3v} (Table S16a)	2.21–C_{3v} [203]
$Pb(H_2O)_6{}^{2+}$		2.53–6 [204,205]
$Bi(H_2O)_9{}^{3+}$	2.440 + 2.582–D_{3h}(9) (Table S16c)	
$Bi(H_2O)_8{}^{3+}$		2.41–D_{4h}(8) [207]

[a] Extrapolated from solid state structures; [b] Extrapolated from physico-chemical parameters.

Supplementary Materials: The following supporting information can be downloaded at: https://www.mdpi.com/article/10.3390/liquids2030014/s1.

Funding: This research received no external funding.

Acknowledgments: I acknowledge the long-time support from the Swedish Research Council of my projects dealing with hydrated and solvated ions and all fruitful and enjoyable collaborations with former Ph.D. students and post-docs.

Conflicts of Interest: The author declares no conflict of interest.

References

1. Bunker, G. *Introduction XAFS: A Pratical Guide to X-ray Absorption Fine Structure Spectroscopy*; Cambridge University Press: Cambridge, UK, 2010; ISBN 978-0-521-76775-0.
2. Rehr, J.J.; Albers, R.C. Theoretical approaches to X-ray absorption fine structure. *Rev. Mod. Phys.* **2000**, *72*, 621–654. [CrossRef]
3. Stern, E.A.; Sayers, D.E.; Lytle, F.W. Extended X-ray Absorption Fine-Structure Technique. III. Determination of Physical Parameters. *Phys. Rev. B* **1975**, *11*, 4836–4846. [CrossRef]
4. Van Bokhaven, J. *X-ray Absorption and X-ray Emission Spectroscopy—Theory and Applications*; John Wiley & Sons Inc.: Hoboken, NJ, USA, 2016; ISBN 9781118844236.
5. Levy, H.A.; Danford, M.D.; Narten, A.H. *Data Collection and Evaluation with an X-ray Diffractometer Designed for the Study of Liquid Structure*; Technical Report ORNL-3960; Oak Ridge National Laboratory: Oak Ridge, TN, USA, 1966. Available online: https://www.osti.gov/servlets/purl/4524253 (accessed on 27 August 2022).
6. Magini, M.; Licheri, G.; Paschina, G.; Piccaluga, G.; Pinna, G. *X-ray Diffraction of Ions in Aqueous Solutions: Hydration and Complex Formation*; CRC Press: Boca Raton, FL, USA, 2018; Chapter 1, ISBN 978-1-351-08622-6.
7. Neilson, G.W. Diffraction studies of aqueous electrolyte solutions. *Pure Appl. Chem.* **1988**, *60*, 1797–1806. [CrossRef]
8. Linert, W.; Bridge, M.; Ohtaki, H. *Comprehensive Coordination Chemistry II*; Elsevier Science: Amsterdam, Netherlands, 2003; Chapter 2.46 Solvation.
9. Stangret, J.; Gampe, T. Hydration sphere of tetrabutylammonium cation. FTIR studies of HDO spectra. *J. Phys. Chem. B* **1999**, *103*, 3778–3783. [CrossRef]
10. Mähler, J.; Persson, I. A study of the hydration of the alkali metal ions in aqueous solution. *Inorg. Chem.* **2012**, *51*, 425–438. [CrossRef]
11. Smirnov, P.R.; Trostin, V.N. Structure of the nearest surrounding of the Li^+ ion in aqueous solutions of its salts. *Russ. J. Gen. Chem.* **2006**, *76*, 175–182. [CrossRef]
12. Marcus, Y. Effect of ions on the structure of water: Structure making and breaking. *Chem. Rev.* **2009**, *109*, 1346–1370. [CrossRef]
13. Persson, I. Hydrated metal ions in aqueous solution: How regular are their structures? *Pure Appl. Chem.* **2010**, *82*, 1901–1917. [CrossRef]
14. Vinogradov, E.V.; Smirnov, P.R.; Trostin, V.N. Structure of hydrated complexes formed by metal ions of Groups I—III of the Periodic Table in aqueous electrolyte solutions under ambient conditions. *Russ. Chem. Bull.* **2003**, *52*, 1253–1271. [CrossRef]
15. Marcus, Y. Ionic radii in aqueous solutions. *Chem. Rev.* **1988**, *88*, 1475–1498. [CrossRef]
16. Mason, P.E.; Ansell, S.; Neilson, G.W.; Rempe, S.B. Neutron Scattering Studies of the Hydration Structure of Li^+. *J. Phys. Chem. B* **2015**, *119*, 2003–2009. [CrossRef]
17. Loeffler, H.; Rode, B. The hydration structure of the lithium ion. *J. Chem. Phys.* **2002**, *117*, 110–117. [CrossRef]

18. Bouazizi, S.; Nasr, S. Local order in aqueous lithium chloride solutions as studied by X-ray scattering and molecular dynamics simulations. *J. Mol. Struct.* **2007**, *837*, 206–213. [CrossRef]
19. Rudolph, W.; Brooker, M.; Pye, C. Hydration of lithium ion in aqueous solutions. *J. Phys. Chem.* **1995**, *99*, 3793–3797. [CrossRef]
20. Du, H.; Rasaiah, J.C.; Miller, J.D. Structural and dynamic properties of concentrated alkali halide solutions: A molecular dynamics simulation study. *J. Phys. Chem. B* **2007**, *111*, 209–217. [CrossRef]
21. Smirnov, P.R.; Trostin, V.N. Structure of the nearest surrounding of the Na^+ ion in aqueous solutions of its salts. *Russ. J. Gen. Chem.* **2007**, *77*, 844–850. [CrossRef]
22. Smirnov, P.R. Structure of the nearest environment of Na^+, K^+, Rb^+, and Cs^+ ions in oxygen-containing solvents. *Russ. J. Gen. Chem.* **2020**, *90*, 1693–1702. [CrossRef]
23. Azam, S.S.; Hofer, T.S.; Randolf, B.R.; Rode, B.M. Hydration of sodium(I) and potassium(I) revisited: A comparative QM/MM and QMCF MD simulation study of weakly hydrated ions. *J. Phys. Chem. A* **2009**, *113*, 1827–1834. [CrossRef]
24. Rowley, N.C.; Roux, B. The solvation structure of Na^+ and K^+ in liquid water determined from high level ab initio molecular dynamics simulations. *J. Chem. Theor. Comput.* **2012**, *8*, 3526–3535. [CrossRef]
25. Bankura, A.; Carnevale, V.; Klein, M.L. Structure and spectroscopy of hydrated sodium ions at different temperatures and the cluster stability rules. *J. Chem. Phys.* **2013**, *138*, 014501. [CrossRef]
26. Shannon, R.D. Revised effective ionic radii and systematic studies of interatomic distances in halides and chalcogenides. *Acta Crystallogr. Sect. A* **1976**, *32*, 751–767. [CrossRef]
27. Beattie, J.K.; Best, S.P.; Skelton, B.W.; White, A.H. Structural studies on the caesium alums, $CsM^{III}[SO_4]_2{\cdot}12H_2O$. *J. Chem. Soc. Dalton Trans.* **1981**, 2105–2111. [CrossRef]
28. Ohtaki, H.; Radnai, T. Structure and dynamics of hydrated ions. *Chem. Rev.* **1993**, *93*, 1157–1204. [CrossRef]
29. Johansson, G. Structures of complexes in solution derived from X-ray diffraction measurements. *Adv. Inorg. Chem.* **1992**, *39*, 159–232.
30. Smirnov, P.R.; Trostin, V.N. Structures of the nearest surroundings of the K^+, Rb^+, and Cs^+ ions in aqueous solutions of their salt. *Russ. J. Gen. Chem.* **2007**, *77*, 2101–2107. [CrossRef]
31. Tongaar, A.; Leidl, K.R.; Rode, B.M. Born–Oppenheimer ab initio QM/MM dynamics simulations of Na^+ and K^+ in water: From structure making to structure breaking effects. *J. Phys. Chem. A* **1998**, *103*, 10340–10347. [CrossRef]
32. Glezakou, V.-A.; Chen, Y.; Fulton, J.L.; Schenter, G.K.; Dang, L.X. Electronic structure, statistical mechanical simulations, and EXAFS spectroscopy of aqueous potassium. *Theor. Chem. Acc.* **2006**, *115*, 86–99. [CrossRef]
33. Caralampio, D.Z.; Martínez, J.M.; Pappalardo, R.R.; Sánchez Marcos, E. The hydration structure of the heavy-alkalines Rb^+ and Cs^+ through molecular dynamics and X-ray absorption spectroscopy: Surface clusters and eccentricity. *Phys. Chem. Chem. Phys.* **2017**, *19*, 28993–29004. [CrossRef]
34. D'Angelo, P.; Persson, I. Structure of the hydrated and dimethyl sulfoxide solvated rubidium ions in solution. *Inorg. Chem.* **2004**, *43*, 3543–3549. [CrossRef]
35. Audebrand, N.; Jeanneau, E.; Bataille, T.; Raite, S.; Louër, D. A family of microporous mixed oxalates with isotypic-framework structures based on eight-coordinate metals. *Solid State Sci.* **2004**, *6*, 579–591. [CrossRef]
36. Pálinkás, G.; Radnai, T.; Hajdu, H. Ion-solvent and solvent-solvent interactions. X-ray study of aqueous alkali chloride solutions. *Z. Naturforsch. Ser. A* **1980**, *35*, 107–114. [CrossRef]
37. Bertagnolli, H.; Weidner, J.-U.; Zimmermann, H. Röntgenstrukturuntersuchung wäßriger Caesiumfluorid-Lösungen. *Ber. Bunsen-Ges. Phys. Chem.* **1974**, *78*, 2–19.
38. Yamaguchi, T.; Ohtaki, H.; Spohr, E.; Pálinkás, G.; Heinzinger, K.; Probst, M.M. Molecular dynamics and X-ray diffraction study of aqueous beryllium(II) chloride solutions. *Z. Naturforsch. Ser. A* **1986**, *41*, 1175–1185. [CrossRef]
39. Di Tommaso, S.; de Leeuw, N.H. Structure and dynamics of the hydrated magnesium ion and of the solvated magnesium carbonates: Insights from first principles simulations. *Phys. Chem. Chem. Phys.* **2010**, *12*, 894–901. [CrossRef] [PubMed]
40. Ikeda, T.; Boero, M.; Terakura, K. Hydration properties of magnesium and calcium ions from constrained first principles molecular dynamics. *J. Chem. Phys.* **2007**, *127*, 074503. [CrossRef] [PubMed]
41. Duboué-Dijon, E.; Mason, P.; Fischer, H.; Jungwirth, P. Hydration and ion pairing in Aqueous Mg^{2+} and Zn^{2+} solutions: Force-field description aided by neutron scattering experiments and ab Initio molecular dynamics simulations. *J. Phys. Chem. B* **2017**, *122*, 3296–3306. [CrossRef] [PubMed]
42. Delgado, A.A.A.; Sethio, D.; Kraka, E. Assessing the intrinsic strengths of ion–solvent and solvent–solvent interactions for hydrated Mg^{2+} clusters. *Inorganics* **2021**, *9*, 31. [CrossRef]
43. Smirnov, P.; Yamagami, M.; Wakita, H.; Yamaguchi, T. An X-ray diffraction study on concentrated aqueous calcium nitrate solutions at subzero semperatures. *J. Mol. Liq.* **1997**, *73–74*, 305–316. [CrossRef]
44. Jalilehvand, F.; Spångberg, D.; Lindqvist-Reis, P.; Hermansson, K.; Persson, I.; Sandström, M. Hydration of the calcium ion. An EXAFS, large-angle X-ray scattering, and molecular dynamics simulation study. *J. Am. Chem. Soc.* **2001**, *123*, 431–441. [CrossRef]
45. Fulton, J.L.; Heald, S.M.; Badyal, Y.S.; Simonson, J.M. Understanding the effects of concentration on the solvation structure of Ca^{2+} in aqueous solution. I: The perspective on local structure from EXAFS and XANES. *J. Phys. Chem. A* **2003**, *107*, 4688–4696. [CrossRef]

46. Badyal, Y.S.; Barnes, A.C.; Cuello, G.J.; Simonson, J.M. Understanding the effects of concentration on the solvation structure of Ca^{2+} in aqueous solutions. II: Insights into longer range order from neutron diffraction isotope substitution. *J. Phys. Chem. A* **2004**, *108*, 11819–11827. [CrossRef]
47. D'Angelo, P.; Petit, P.E.; Pavel, N.V. Double-electron excitation channels at the Ca^{2+} K-edge of hydrated calcium ion. *J. Phys. Chem. B* **2004**, *108*, 11857–11865. [CrossRef]
48. Gaspar, A.M.; Marques, M.A.; Cabaco, M.I.; Marques, M.I.D.; Buslaps, T.; Honkimaki, V.; European Mol Liquid Grp; Japenese Mol Liquid Grp. X-ray diffraction investigations of concentrated aqueous solutions of calcium halides. *J. Mol. Liq.* **2004**, *110*, 15–22. [CrossRef]
49. Fulton, J.L.; Chen, Y.; Heald, S.M.; Balasubramanian, M. High-pressure, High-temperature X-ray absorption fine structure transmission cell for the study of aqueous ions with low absorption-edge energies. *Rev. Sci. Instrum.* **2004**, *75*, 5228–5231. [CrossRef]
50. Megyes, T.; Grosz, T.; Radnai, T.; Bako, I.; Palinkas, G. Solvation of calcium ion in polar solvents: An X-ray diffraction and ab initio study. *J. Phys. Chem. A* **2004**, *108*, 7261–7271. [CrossRef]
51. Megyes, T.; Bako, I.; Balint, S.; Grosz, T.; Radnai, T. Ion pairing in aqueous calcium chloride solution: Molecular dynamics simulation and diffraction studies. *J. Mol. Liq.* **2006**, *129*, 63–74. [CrossRef]
52. Bucher, D.; Kuyucak, S. Polarization of water in the first hydration shell of K^+ and Ca^{2+} ions. *J. Phys. Chem. B* **2008**, *112*, 10786–10790. [CrossRef]
53. Pham, V.-T.; Fulton, J.L. Ion-pairing in Aqueous $CaCl_2$ and RbBr Solutions: Simultaneous Structural Refinement of XAFS and XRD Data. *J. Chem. Phys.* **2013**, *138*, 044201. [CrossRef]
54. Rudolph, W.; Irmer, G. Hydration of calcium(II) ion in aqueous solution of common anions (ClO_4^-, Cl^-, Br^- and NO_3^-). *Dalton Trans.* **2013**, *37*, 3919–3935. [CrossRef]
55. Yoo, J.; Wilson, J.; Aksimentiev, A. Improved model of hydrated calcium ion for molecular dynamics simulations using classical biomolecular force fields. *Biopolymers* **2016**, *105*, 752–763. [CrossRef]
56. Delgado, A.A.A.; Sethio, D.; Munar, I.; Aviyente, V.; Kraka, E. Local vibrational mode analysis of ion-solvent and solvent-solvent interactions for hydrated Ca^{2+} clusters. *J. Chem. Phys.* **2020**, *153*, 224303. [CrossRef] [PubMed]
57. Shi, R.; Zhao, Z.; Huang, X.; Wang, P.; Su, Y.; Sai, L.; Liang, X.; Han, H.; Zhao, J. Ground-state structures of hydrated calcium ion clusters from comprehensive genetic algorithm search. *Front. Chem.* **2021**, *9*, 637750. [CrossRef] [PubMed]
58. Persson, I.; Sandström, M.; Yokoyama, H.; Chaudhry, M. Structures of the solvated barium and strontium ions in aqueous, dimethyl sulfoxide and pyridine solution, and crystal structure of strontium and barium hydroxide octahydrate. *Z. Naturforsch. Sect. B.* **1995**, *50*, 21–37. [CrossRef]
59. D'Angelo, P.; Migliorati, V.; Sessa, F.; Mancini, G.; Persson, I. XANES reveals the flexible nature of hydrated strontium in aqueous solution. *J. Phys. Chem. B* **2016**, *120*, 4114–4124. [CrossRef]
60. O'Day, P.A.; Newvillre, M.; Neuhoff, P.S.; Sahai, N.; Carroll, S.A. X-ray absorption spectroscopy of strontium (II) coordination: I. Static and thermal disorder in crystalline, hydrated, and precipitated solids and in aqueous solution. *J. Colloid Interface Sci.* **2000**, *222*, 184–197.
61. Axe, L.; Bunker, G.B.; Anderson, P.R.; Tyson, T.A. An XAFS analysis of strontium at the hydrous ferric oxide surface. *J. Colloid Interface Sci.* **1998**, *199*, 44–52. [CrossRef]
62. Caminiti, R.; Musinu, A.; Paschina, G.; Pinna, G. X-ray diffraction study of aqueous $SrCl_2$ solutions. *J. Appl. Crystallogr.* **1982**, *15*, 482–487. [CrossRef]
63. Albright, J.N. X-ray diffraction studies of aqueous alkaline-earth chloride solutions. *J. Chem. Phys.* **1972**, *56*, 3783–3786. [CrossRef]
64. Ramos, S.; Neilson, G.W.; Barnes, A.C.; Capitán, M.J. Anomalous X-ray diffraction studies of Sr^{2+} hydration in aqueous solution. *J. Chem. Phys.* **2003**, *118*, 5542–5546. [CrossRef]
65. Pfund, D.M.; Darab, J.G.; Fulton, J.L.; Ma, Y. An XAFS study of strontium ions and krypton in supercritical water. *J. Phys. Chem.* **1994**, *98*, 13102–13107. [CrossRef]
66. Moreau, G.; Helm, L.; Purans, J.; Merbach, A.E. Structural Investigation of the Aqueous Eu^{2+} Ion: Comparison with Sr^{2+} Using the XAFS Technique. *J. Phys. Chem. A* **2002**, *106*, 3034–3043. [CrossRef]
67. Steward, T.M.; Henderson, C.M.B.; Charnock, J.M.; Driesner, T. An EXAFS study of solvation and ion pairing in aqueous strontium solutions to 300 °C. *Geochim. Cosmochim. Acta* **1999**, *63*, 2409–2418. [CrossRef]
68. Harris, D.J.; Brodholt, J.P.; Sherman, D.M. Hydration of Sr^{2+} in Hydrothermal Solutions from ab Initio Molecular Dynamics. *J. Phys. Chem. B* **2003**, *107*, 9056–9058. [CrossRef]
69. Chaudhari, M.I.; Rempee, S.B. Strontium and barium in aqueous solution and a potassium channel binding site. *J. Chem. Phys.* **2018**, *148*, 222831. [CrossRef] [PubMed]
70. Spohr, E.; Pálinkás, G.; Heinzinger, K.; Bopp, B.; Probst, M.M. Molecular dynamics study of an aqueous strontium chloride solution. *J. Phys. Chem.* **1988**, *92*, 6754–6761. [CrossRef]
71. Palmer, B.J.; Pfund, D.M.; Fulton, J.L. Direct modeling of EXAFS spectra from molecular dynamics simulations. *J. Phys. Chem.* **1996**, *100*, 13393–13398. [CrossRef]
72. Boda, A.; De, S.; Ali, S.M.; Tulishetti, S.; Khan, S.; Singh, J.K. From microhydration to bulk hydration of Sr^{2+} metal ion: DFT, MP2 and molecular dynamics study. *J. Mol. Liq.* **2012**, *172*, 110–118. [CrossRef]

73. Hofer, T.S.; Randolf, B.R.; Rode, B.M. Sr(II) in water: A labile hydrate with a highly mobile structure. *J. Phys. Chem. B* **2006**, *110*, 20409–20417. [CrossRef]
74. Pappalardo, R.R.; Caralampio, D.Z.; Martínez, J.M.; Sánchez Marcos, E. Hydration of heavy alkaline-earth cations studied by molecular dynamics simulations and X-ray absorption spectroscopy. *Inorg. Chem.* **2021**, *60*, 13578–13587. [CrossRef] [PubMed]
75. Hofer, T.S.; Rode, B.M.; Randolf, B.R. Structure and dynamics of solvated Ba(II) in dilute aqueous solution—An ab initio QM/MM MD approach. *Chem. Phys.* **2005**, *312*, 81–88. [CrossRef]
76. Stack, A.G.; Rustad, J.R. Structure and dynamics of water on aqueous barium ion and the {001} barite surface. *J. Phys. Chem. C* **2007**, *111*, 16387–16391. [CrossRef]
77. Hoogerstraete, T.V.; Brooks, N.R.; Onghena, B.; Van Meervelt, L.; Binnemans, K. Crystal structures of hydrated rare-earth bis(trifluoromethylsulfonyl)imide salts. *CrystEngComm* **2015**, *17*, 7142–7149. [CrossRef]
78. Waller, F.J.; Barrett, A.G.M.; Braddock, D.C.; Ramprasad, D.; McKinnell, R.M.; White, A.J.P.; Williams, D.J.; Ducray, R. Tris(trifluoromethanesulfonyl)methide ("Triflide") anion: Convenient preparation, X-ray crystal structures, and exceptional catalytic activity as a counterion with ytterbium(III) and scandium(III). *J. Org. Chem.* **1999**, *64*, 2910–2913. [CrossRef] [PubMed]
79. Abbasi, A.; Lindqvist-Reis, P.; Eriksson, L.; Sandström, D.; Lidin, S.; Persson, I.; Sandström, M. Highly Hydrated cations: Deficiency, mobility, and coordination of water in crystalline nonahydrated scandium(III), yttrium(III) and lanthanoid(III) trifluoromethanesulfonates. *Chem. Eur. J.* **2005**, *11*, 4065–4077. [CrossRef] [PubMed]
80. Castellani, C.B.; Carugo, O.; Giusti, M.; Sardone, N. Crystal structure of scandium(III) triflate enneahydrated. *Eur. J. Solid State Inorg. Chem.* **1995**, *32*, 1089–1099.
81. Lindqvist-Reis, P.; Persson, I.; Sandström, M. The hydration of the scandium(III) ion in aqueous solution and crystalline hydrates studied by xafs spectroscopy, large angle X-ray scattering and crystallography. *Dalton Trans.* **2006**, 3868–3878. [CrossRef]
82. Lindqvist-Reis, P.; Lamble, K.; Pattanaik, S.; Sandström, M.; Persson, I. Hydration of the yttrium(III) ion in aqueous solution. an X-ray diffraction and XAFS structural study. *J. Phys. Chem. Sect. B* **2000**, *104*, 402–408. [CrossRef]
83. Van de Voorde, M.; Geboes, B.; Tom Vander Hoogerstraete, T.; Van Hecke, K.; Cardinaels, T.; Binnemans, K. Stability of europium(II) in aqueous nitrate solutions. *Dalton Trans.* **2019**, *48*, 14758–14768. [CrossRef]
84. Szostak, M.; Spain, M.; Procter, D.J. Preparation of samarium(II) iodide: Quantitative evaluation of the effect of water, oxygen, and peroxide content, preparative methods, and the activation of samarium metal. *J. Org. Chem.* **2012**, *77*, 3049–3059. [CrossRef]
85. Tsuneo, I.; Yasuhiro, K.; Susumu, H. Cerium(IV) trifluoromethanesulfonate as a strong oxidizing agent. *Chem. Lett.* **1990**, *19*, 1445–1446.
86. Marsac, R.; Réal, F.; Lal Banik, N.; Pédrot, M.; Pourret, O.; Vallet, V. Aqueous chemistry of Ce(IV): Estimations using actinide analogues. *Dalton Trans.* **2017**, *46*, 13553–13561. [CrossRef] [PubMed]
87. Ikeda-Ohno, A.; Tsushima, S.; Hennig, C.; Yaitab, T.; Bernharda, G. Dinuclear complexes of tetravalent cerium in an aqueous perchloric acid solution. *Dalton Trans.* **2012**, *41*, 7190–7192. [CrossRef] [PubMed]
88. Persson, I.; D'Angelo, P.; De Panfilis, S.; Sandström, M.; Eriksson, L. Hydration of the lanthanoid(III) ions in aqueous solution and crystalline hydrates studied by EXAFS spectroscopy and crystallography. The myth of the "Gadolinium Break". *Chem. Eur. J.* **2008**, *14*, 3056–3066. [CrossRef] [PubMed]
89. Lundberg, D.; Persson, I.; Eriksson, L.; D'Angelo, P.; de Panfilis, S. Structural study of the *N,N'*-dimethylpropyleneurea solvated lanthanoid(III) ions in solution and solid state with an analysis of the ionic radii of lanthanoid(III) Ions. *Inorg. Chem.* **2010**, *49*, 4420–4432. [CrossRef]
90. Spedding, F.H.; Pikal, M.J.; Ayers, B.O. Apparent molal volumes of some aqueous rare earth chloride and nitrate solutions at 25°. *J. Phys. Chem.* **1966**, *70*, 2440–2449. [CrossRef]
91. Spedding, F.H.; Jones, K.C. Heat capacities of aqueous rare earth chloride solutions at 25 °C. *J. Phys. Chem.* **1966**, *70*, 2450–2455. [CrossRef]
92. Spedding, F.H.; Pikal, M.J. Relative viscosities of some aqueous rare earth chloride solutions at 25 °C. *J. Phys. Chem.* **1966**, *70*, 2430–2440. [CrossRef]
93. Marsh, K. 125. The dimethyl phosphates of the rare-earth metals. *J. Chem. Soc.* **1939**, *61*, 554–558. [CrossRef]
94. Templeton, D.H.; Dauben, C.H. Lattice parameters of some rare earth compounds and a set of crystal radii. *J. Am. Chem. Soc.* **1954**, *76*, 5237–5239. [CrossRef]
95. Cossy, C.; Helm, L.; Merbach, A.E. Oxygen-17 nuclear magnetic resonance kinetic study of water exchange on the lanthanide(III) aqua ions. *Inorg. Chem.* **1988**, *27*, 1973–1979. [CrossRef]
96. Cossy, C.; Helm, L.; Merbach, A.E. High-pressure NMR study. 38. Water-exchange mechanisms on the terbium to thulium octaaqualanthanide (III) ions: A variable-pressure oxygen-17 NMR study. *Inorg. Chem.* **1989**, *28*, 2699–2703. [CrossRef]
97. Micskei, K.; Powell, D.H.; Helm, L.; Brucher, E.; Merbach, A.E. Water exchange on $Gd(H_2O)_8]^{3+}$ and $[Gd(PDTA)(H_2O)_2]^-$ in aqueous solution: A variable-pressure, -temperature and -magnetic field ^{17}O NMR study. *Magn. Reson. Chem.* **1993**, *31*, 1011–1020. [CrossRef]
98. Powell, D.H.; Merbach, A.E. Water exchange on the light lanthanide aqua ions $[Pr(H_2O)_9]^{3+}$ and $[Nd(H_2O)_9]^{3+}$: A variable temperature and magnetic field ^{17}O NMR study. *Magn. Reson. Chem.* **1994**, *32*, 739–745. [CrossRef]
99. Fay, D.; Litchinski, D.; Purdie, N. Ultrasonic absorption in aqueous salts of the lanthanides. *J. Phys. Chem.* **1969**, *73*, 544–552. [CrossRef]

100. Garza, V.L.; Purdie, N. Ultrasonic absorption in aqueous salts of the lanthanides. II. Acetates. *J. Phys. Chem.* **1970**, *74*, 275–280. [CrossRef]
101. Peppard, D.F.; Mason, G.W.; Lewey, S. A tetrad effect in the liquid-liquid extraction ordering of lanthanides(III). *J. Inorg. Nucl. Chem.* **1969**, *31*, 2271–2272. [CrossRef]
102. Schwarzenbach, G.; Gut, R.; Anderegg, G. Komplexone XXV. Die polarographische Untersuchung von Austauschgleichgewichten. Neue Daten der Bildungskonstanten von Metallkomplexen der Äthylendiamin-tetraessigsäure und der 1, 2-Diaminocyclohexan-tetraessigsäure. *Helv. Chim. Acta* **1954**, *37*, 937–957. [CrossRef]
103. Wheelwright, E.; Spedding, F.H.; Schwarzenbach, G. The stability of the rare earth complexes with ethylenediaminetetraacetic acid. *J. Am. Chem. Soc.* **1953**, *75*, 4196–4201. [CrossRef]
104. Peppard, D.F.; Bloomquist, C.A.A.; Horwitz, E.P.; Lewey, S.; Mason, G.W. Analogous actinide (III) and lanthanide (III) tetrad effects. *J. Inorg. Nucl. Chem.* **1970**, *32*, 339–343. [CrossRef]
105. Lundberg, D.; Persson, I. The size of actinoid(III) ions—Structural analysis vs. common misinterpretations. *Coord. Chem. Rev.* **2016**, *318*, 131–134. [CrossRef]
106. Apostolidis, C.; Schimmelpfennig, B.; Magnani, N.; Lindqvist-Reis, P.; Walter, O.; Sykora, R.; Morgenstern, A.; Colineau, E.; Caciuffo, R.; Klenze, R.; et al. $[An(H_2O)_9](CF_3SO_3)_3$ (An = U–Cm, Cf): Exploring their stability, structural chemistry, and magnetic behavior by experiment and theory. *Angew. Chem. Int. Ed.* **2010**, *49*, 6343–6347. [CrossRef] [PubMed]
107. Matonic, J.H.; Scott, B.L.; Neu, M.P. High-yield synthesis and single-crystal X-ray structure of a plutonium(III) aquo complex: $[Pu(H_2O)_9][CF_3SO_3]_3$. *Inorg. Chem.* **2001**, *40*, 2638–2639. [CrossRef] [PubMed]
108. Lindqvist-Reis, P.; Apostolidis, C.; Rebizant, J.; Morgenstern, A.; Klenze, R.; Walter, O.; Fanghänel, T.; Haire, R.G. The structures and optical spectra of hydrated transplutonium ions in the solid state and in solution. *Angew. Chem. Int. Ed.* **2007**, *46*, 919–922. [CrossRef] [PubMed]
109. Skanthakumar, S.; Antonio, M.R.; Wilson, R.E.; Soderholm, L. The curium aqua ion. *Inorg. Chem.* **2007**, *46*, 3485–3491. [CrossRef]
110. D'Angelo, P.; Martelli, F.; Spezia, R.; Filipponi, A.; Denecke, M.A. Hydration properties and ionic radii of actinide(III) ions in aqueous solution. *Inorg. Chem.* **2013**, *52*, 10318–10324. [CrossRef]
111. Allen, P.G.; Bucher, J.J.; Shuh, D.K.; Edelstein, N.M.; Reich, T. Investigation of aquo and chloro complexes of UO_2^{2+}, NpO_2^{+}, Np^{4+}, and Pu^{3+} by X-ray absorption fine structure spectroscopy. *Inorg. Chem.* **1997**, *36*, 4676–4783. [CrossRef]
112. Wilson, R.E.; Almond, P.M.; Burns, P.C.; Soderholm, L. The structure and synthesis of plutonium(III) chlorides from aqueous solution. *Inorg. Chem.* **2006**, *45*, 8483–8485. [CrossRef]
113. Allen, P.G.; Bucher, J.J.; Shuh, D.K.; Edelstein, N.M.; Craig, I. Coordination chemistry of trivalent lanthanide and actinide ions in dilute and concentrated chloride solutions. *Inorg. Chem.* **2000**, *39*, 595–601. [CrossRef]
114. Conradson, S.D.; Abney, K.D.; Begg, B.D.; Brady, E.D.; Clark, D.L.; den Auwer, C.; Ding, M.; Dorhout, P.K.; Espinosa-Faller, F.J.; Gordon, P.L.; et al. Higher order speciation effects on plutonium L_3 X-ray absorption near edge spectra. *Inorg. Chem.* **2004**, *43*, 116–131. [CrossRef]
115. Conradson, S.D.; Clark, D.L.; Neu, M.P.; Runde, W.H.; Tait, C.D. Characterizing the plutonium aquo ions by XAFS spectroscopy. *Los Alamos Sci.* **2000**, *26*, 418–421.
116. Ferrier, M.G.; Stein, B.W.; Batista, E.R.; Berg, J.M.; Birnbaum, E.R.; Engle, J.W.; John, K.D.; Kozimor, S.A.; Lezama Pacheco, J.S.; Redman, L.N. Synthesis and characterization of the actinium aquo ion. *ACS Central Sci.* **2017**, *3*, 176–185. [CrossRef]
117. Deblonde, G.J.-P.; Zavarin, M.; Kersting, A.B. The coordination properties and ionic radius of actinium: A 120-year old enigma. *Coord. Chem. Rev.* **2021**, *446*, 214130. [CrossRef]
118. Wilson, R.E.; Skanthakumar, S.; Burns, P.C.; Soderholm, L. Structure of the homoleptic thorium(IV) aqua ion $[Th(H_2O)_{10}]Br_4$. *Angew. Chem, Int. Ed.* **2007**, *46*, 8043–8045. [CrossRef]
119. Torapava, N.; Persson, I.; Eriksson, L.; Lundberg, D. Hydration and hydrolysis of thorium(IV) in aqueous solution and the structures of two crystalline thorium(IV) hydrates. *Inorg. Chem.* **2009**, *48*, 11712–11723. [CrossRef]
120. Pohl, R.W.H.; Wiebke, J.; Klein, A.; Dolg, M.; Maggiarosa, N. New 5,5′-bitetrazole thorium(IV) compound: Synthesis, crystal structure and quantum chemical investigation. *Eur. J. Inorg. Chem.* **2009**, *2009*, 2472–2476. [CrossRef]
121. Thuery, P. Second-sphere complexation of thorium(IV) by cucurbit[6]uril with included perrhenate counterions—Crystal structure and Hirshfeld surface analysis. *Eur. J. Inorg. Chem.* **2015**, *2015*, 2037–2040. [CrossRef]
122. Hennig, C.; Schmeide, K.; Brendler, V.; Moll, H.; Tsushima, S.; Scheinost, A.C. EXAFS investigation of U(VI), U(IV), and Th(IV) sulfato complexes in aqueous solution. *Inorg. Chem.* **2007**, *46*, 5882–5892. [CrossRef]
123. Moll, H.; Denecke, M.A.; Jalilehvand, F.; Sandström, M.; Grenthe, I. Structure of the aqua ions and fluoride complexes of uranium (IV) and thorium (IV) in aqueous solution an EXAFS study. *Inorg. Chem.* **1999**, *38*, 1795–1799. [CrossRef]
124. Ikeda-Ohno, A.; Hennig, C.; Tsushima, S.; Scheinost, A.C.; Bernhard, G.; Yaita, T. Speciation and structural study of U(IV) and-(VI) in perchloric and nitric acid solutions. *Inorg. Chem.* **2009**, *48*, 7201–7210. [CrossRef]
125. Ikeda-Ohno, A.; Hennig, C.; Rossberg, A.; Fune, H.; Scheinost, A.C.; Bernhard, G.; Yaita, T. Electrochemical and complexation behavior of neptunium in aqueous perchlorate and nitrate solutions. *Inorg. Chem.* **2008**, *47*, 8294–8305. [CrossRef]
126. Skanthakumar, S.; Antonio, M.R.; Soderholm, L. A comparison of neptunyl(V) and neptunyl(VI) solution coordination: The stability of cation-cation interactions. *Inorg. Chem.* **2008**, *47*, 4591–4595. [CrossRef]
127. Miyanaga, T.; Watanabe, I.; Ikeda, S. Structures of hydrated titanium and vanadium ions in aqueous-solutions studied by X-ray absorption-spectroscopy. *Bull. Chem. Soc. Jpn.* **1990**, *63*, 3282–3287. [CrossRef]

128. Tachikawa, H.; Ichikawa, T.; Yoshida, H. Hydration structure of titanium(III) ion. Electron spin resonance and electron spin-echo study. *J. Am. Chem. Soc.* **1990**, *112*, 977–982. [CrossRef]
129. Persson, I.; Lundberg, D. Structure of a hydrated sulfonatotitanyl(IV) complex in aqueous solution and the dimethylsulfoxide solvated titanyl(IV) ion in solution and solid state. *J. Solution Chem.* **2017**, *46*, 476–487.
130. Rabe, S.; Muller, U. Oxotitan-Addukte mit Dimethylsulfoxid: $[TiO(OSMe_2)_5]Cl_2$ und $[Ti_4O_6(OSMe_2)_{12}]Cl_4{\cdot}5Me_2SO{\cdot}\frac{1}{2}H_2O$/Oxotitanium compounds with dimethylsulfoxide: $[TiO(OSMe_2)_5]Cl_2$ and $[Ti_4O_6(OSMe_2)_{12}]Cl_4{\cdot}5Me_2SO{\cdot}\frac{1}{2}H_2O$. *Z. Naturforsch. Teil B* **1997**, *52*, 1291–1295. [CrossRef]
131. Enders, M.; Rudolph, R.; Pritzkow, H. Synthese und kristallstruktur von pentakis(dimethylsulfoxid)-oxo-titan(IV)chlorid/synthesis and crystal structure of pentakis(dimethylsulfoxide)-oxo-titanium(IV) chloride. *Z. Naturforsch. Teil B* **1997**, *52*, 496–499. [CrossRef]
132. Hagfeldt, C.; Kessler, V.; Persson, I. Structure of the hydrated, hydrolysed and solvated zirconium(IV) and hafnium(IV) ions in water and aprotic oxygen donor solvents. A crystallographic, EXAFS spectroscopic and large angle X-ray scattering study. *Dalton Trans.* **2004**, 2142–2151. [CrossRef] [PubMed]
133. Miyanaga, T.; Watanabe, I.; Ikeda, S. Amplitude in EXAFS and ligand-exchange reaction of hydrated 3D transition-metal complexes. *Chem. Lett.* **1988**, *17*, 1073–1076. [CrossRef]
134. Krakowiak, J.; Lundberg, D.; Persson, I. A coordination chemistry study of hydrated and solvated cationic vanadium ions in oxidation states +III, +IV, and +V in solution and solid state. *Inorg. Chem.* **2012**, *51*, 9598–9609. [CrossRef]
135. Lindqvist-Reis, P.; Díaz-Moreno, S.; Munoz-Páez, A.; Pattanaik, S.; Persson, I.; Sandström, M. On the structure of hydrates gallium, indium and chromium(III) ions in aqueous solutions. A large angle X-ray scattering and EXAFS study. *Inorg. Chem.* **1998**, *37*, 6675–6683. [CrossRef]
136. Torapava, N.; Radkevich, A.; Davydov, D.; Titov, I.; Persson, I. Composition and structure of polynuclear chromium(III) hydroxo complexes. *Inorg. Chem.* **2009**, *48*, 10383–10388. [CrossRef]
137. Cramer, S.P.; Eidem, P.K.; Pafett, M.T.; Winkler, J.R.; Dori, Z.; Gray, H.B. X-ray absorption edge and EXAFS spectroscopic studies of molybdenum ions in aqueous solution. *J. Am. Chem. Soc.* **1983**, *105*, 799–802. [CrossRef]
138. Brorson, M.; Gajhede, M. Crystal structure of cesium molybdenum alum, $Cs[Mo(H_2O)_6](SO_4)_2{\cdot}6H_2O$, at 110 K. *Inorg. Chem.* **1987**, *26*, 2109–2112. [CrossRef]
139. Konieczna, H.; Lundberg, D.; Persson, I. Solvation and coordination chemistry of manganese(II) ion in solvents with solvation properties. A transfer thermodynamic, complex formation, EXAFS spectroscopic and crystallographic study. *Polyhedron* **2021**, *195*, 114961. [CrossRef]
140. Chen, Y.; Fulton, J.L.; Partenheimer, W. A XANES and EXAFS study of hydration and ion pairing in ambient aqueous $MnBr_2$ solutions. *J. Solut. Chem.* **2005**, *34*, 993–1007. [CrossRef]
141. Caminiti, R.; Cucca, G.; Pintori, G. Hydration and ion-pairing in concentrated aqueous $Mn(NO_3)_2$ solutions. An X-ray and Raman spectroscopy study. *Chem. Phys.* **1984**, *88*, 155–161. [CrossRef]
142. Chukanov, N.V.; Zubkova, N.V.; Pautov, L.A.; Göttlicher, J.; Kasatkin, A.V.; Van, K.V.; Ksenofontov, D.A.; Pekov, I.V.; Vozchikova, S.A.; Pushcharovsky, D.Y. Jouravskite: Refined data on the crystal structure, chemical composition and spectroscopic properties. *Phys. Chem. Miner.* **2019**, *46*, 417–4257. [CrossRef]
143. Allen, F.H. The Cambridge Structural Database: A quarter of a million crystal structures and rising. *Acta Crystallogr. Sect. B* **2002**, *58*, 380–388, Cambridge Structure Database (Conquest 2022.1.0). [CrossRef]
144. *Inorganic Crystal Structure Database 1.4.6, Release 2021-2*; FIZ Karlsruhe: Eggenstein-Leopoldshafen, Germany, 2021.
145. Lundberg, D.; Ullström, A.-S.; D'Angelo, P.; Persson, I. A structural study of the hydrated and the dimethylsulfoxide, *N,N'*-dimethylpropyleneurea, and *N,N*-dimethylthioformamide solvated iron(II) and iron(III) ions in solution and solid state. *Inorg. Chim. Acta* **2007**, *360*, 1809–1818. [CrossRef]
146. Kálmán, E.; Radnai, T.; Pálinkás, G.; Hajdus, F.; Vertes, A. Hydration of iron(II) ion in aqueous solutions. Hydration of iron(II) ion in aqueous solution. *Electrochim. Acta* **1988**, *33*, 1223–1228. [CrossRef]
147. D'Angelo, P.; Barone, V.; Chillemi, G.; Sanna, N.; Meyer-Klauche, W.; Pavel, N.V. Hydrogen and higher shell contributions in Zn^{2+}, Ni^{2+}, and Co^{2+} aqueous solutions: An X-ray absorption fine structure and molecular dynamics study. *J. Am. Chem. Soc.* **2002**, *124*, 1958–1967. [CrossRef] [PubMed]
148. Chen, C.; Hayes, K. X-ray absorption spectroscopy investigation of aqueous Co(II) and Sr(II) sorption at clay–water interfaces. *Geochim. Cosmochim. Acta* **1999**, *63*, 3205–3215. [CrossRef]
149. Spezia, R.; Duvail, M.; Vitorge, P.; Cartailler, T.; Tortajada, J.; Chillemi, G.; D'Angelo, P. A coupled Car-Parrinello molecular dynamics and EXAFS data analysis investigation of aqueous Co^{2+}. *J. Phys. Chem. A* **2006**, *110*, 13081–13088. [CrossRef]
150. Caralampio, D.Z.; Reeves, B.; Beccia, M.R.; Martínez, J.M.; Pappalardo, R.R.; Christophe den Auwer, C.; Sánchez Marcos, E. Revisiting the cobalt(II) hydration from molecular dynamics and X-ray absorption spectroscopy. *Mol. Phys.* **2019**, *117*, 3320–3328. [CrossRef]
151. Magini, M.; Giubileo, G. Solute Structuring in Concentrated A Co(II) and Ni(II) Perchlorate Solutions. *Gazz. Chim. Ital.* **1981**, *111*, 449–454.
152. Caminiti, R.; Atzei, D.; Cucca, P.; Anedda, A.; Bongiovanni, G. Structure of Rhodium(III) Nitrate Aqueous Solutions. *An Investigation by X-ray Diffraction and Raman Spectroscopy. J. Phys. Chem.* **1986**, *90*, 238–243.
153. Caminiti, R.; Cucca, P. An X-ray diffraction study of Rh(III) coordination in a dilute aqueous-solution of $Rh(ClO_4)_3$. *Phys. Chem. Lett.* **1984**, *108*, 51–57. [CrossRef]

154. Read, M.C.; Sandström, M. 2nd-sphere hydration of rhodium(III) and chromium(III) in aqueous-solution—A large-angle X-ray-scattering and EXAFS study. *Acta Chem. Scand.* **1992**, *46*, 1177–1182. [CrossRef]
155. Kristiansson, O.; Persson, I.; Bobicz, D.; Xu, D. A structural study of the hydrated and the dimethylsulfoxide, *N,N'*-dimethylpropyleneurea, acetonitrile, pyridine and *N,N*-dimethylthioformamide solvated nickel(II) ion solution and solid state. *Inorg. Chim. Acta* **2003**, *344*, 15–27. [CrossRef]
156. Purans, J.; Fourest, B.; Cannes, C.; Sladkov, V.; David, F.; Venault, L.; Lecomte, M. Structural investigation of Pd(II) in concentrated nitric and perchloric acid solutions by XAFS. *J. Chem. Phys. B* **2005**, *109*, 11074–11082. [CrossRef]
157. Hofer, T.S.; Randolf, B.R.; Ali Shah, A.; Rode, B.M.; Persson, I. Structure and dynamics of the hydrated palladium(II) ion in aqueous solution. A QMCF MD simulation and EXAFS spectroscopic study. *Chem. Phys. Lett.* **2007**, *445*, 193–197. [CrossRef]
158. Beret, E.C.; Pappalardo, R.R.; Doltsinis, N.L.; Marx, D.; Marcos, E.S. Aqueous Pd^{II} and Pt^{II}: Anionic hydration revealed by Car–Parrinello simulations. *ChemPhysChem* **2008**, *9*, 237–240. [CrossRef]
159. Jalilehvand, F.; Laffin, L.J. Structure of the hydrated platinum(II) ion and the *cis*-diammineplatinum(II) complex in acidic aqueous solution: An EXAFS study. *Inorg. Chem.* **2008**, *47*, 3248–3254. [CrossRef]
160. Hofer, T.S.; Randolf, B.R.; Rode, B.M.; Persson, I. The hydrated platinum(II) ion in aqueous solution—A combined theoretical and EXAFS spectroscopic study. *Dalton Trans.* **2009**, 1512–1515. [CrossRef]
161. Persson, I.; Penner-Hahn, J.E.; Hodgson, K.O. An EXAFS spectroscopic study of solvates of copper(I) and copper(II) in acetonitrile, dimethyl sulfoxide, pyridine and tetrahydrothiophene solution and a large angle X-ray scattering study of the copper(II) acetonitrile solvate in solution. *Inorg. Chem.* **1993**, *32*, 2497–2501. [CrossRef]
162. Persson, I.; Persson, P.; Sandström, M.; Ullström, A.-S. Structure of Jahn-Teller distorted solvated copper(II) ions in solution, and in solids with apparently regular octahedral coordination geometry. *J. Chem. Soc. Dalton Trans.* **2002**, 1256–1265. [CrossRef]
163. Persson, I.; Lundberg, D.; Bajnoczi, E.G.; Klementiev, K.; Just, J.; Sigfridsson Clauss, K.G.V. EXAFS study on the coordination chemistry of the solvated copper(II) ion in a series of oxygen donor solvents. *Inorg. Chem.* **2020**, *59*, 9538–9550. [CrossRef]
164. Peisach, J.; Mims, W.B. Deviations from centrosymmetry in some simple Cu^{2+} complexes. *Chem. Phys. Lett.* **1976**, *37*, 307–310. [CrossRef]
165. Peisach, J.; Mims, W.B. Linear electric-field effect in stellacyanin, azurin and in some simple-model compounds. *Eur. J. Biochem.* **1978**, *84*, 207–214. [CrossRef]
166. Pasquarello, A.; Petri, I.; Salmon, P.S.; Parisel, O.; Car, R.; Toth, E.; Powell, D.H.; Fischer, H.E.; Helm, L.; Merbach, A.E. First solvation shell of the Cu(II) aqua ion: Evidence for fivefold coordination. *Science* **2001**, *291*, 856–859. [CrossRef]
167. Benfatto, M.; D'Angelo, P.; Della Longa, S.; Pavel, N.V. Evidence of distorted fivefold coordination of the Cu^{2+} aqua ion from an X-ray-absorption spectroscopy quantitative analysis. *Phys. Rev. B* **2002**, *65*, 174205. [CrossRef]
168. Franck, P.; Benfatto, M.; Szilagyi, R.K.; D'Angelo, P.; Della Longa, S.; Hodgson, K.O. The solution structure of $[Cu(aq)]^{2+}$ and its implications for rack-induced bonding in blue copper protein active sites. *Inorg. Chem.* **2005**, *44*, 1922–1933. [CrossRef] [PubMed]
169. Chaboy, J.; Munoz-Paez, A.; Marcos, E.S. The interplay of the $3d^9$ and $3d^{10}$ L electronic configurations in the copper *K*-edge XANES spectra of Cu(II) compounds. *J. Synchrotron Rad.* **2006**, *13*, 471–476. [CrossRef]
170. Bryantsev, V.S.; Diallo, M.S.; van Duin, A.C.T.; Goddard III, W.A. Hydration of copper(II): New insights from density functional theory and the COSMO solvation model. *J. Phys. Chem. A* **2008**, *112*, 9104–9112. [CrossRef]
171. De Almeida, K.J.; Murugan, N.A.; Rinkevicius, Z.Z.; Hugosson, H.-W.; Vahtras, O.; Ågren, H.; Cesar, A. Conformations, structural transitions and visible near-infrared absorption spectra of four-, five- and six-coordinated Cu(II) aqua complexes. *Phys. Chem. Chem. Phys.* **2009**, *11*, 508–519. [CrossRef]
172. Liu, X.; Lu, X.; Meijer, E.J.; Wang, R. Hydration mechanisms of Cu^{2+}: Tetra-, penta- or hexa-coordinated? *Phys. Chem. Chem. Phys.* **2010**, *12*, 10801–10804. [CrossRef]
173. Gomez-Salces, S.; Aguado, F.; Valiente, R.; Rodrigues, F. Unraveling the coordination geometry of copper(II) ions in aqueous solution through absorption intensity. *Angew. Chem. Int. Ed.* **2012**, *51*, 9335–9338. [CrossRef]
174. Bowron, D.T.; Amboage, M.; Boada, R.; Freeman, A.; Hayama, S.; Diaz-Moreno, S. The hydration structure of Cu^{2+}: More tetrahedral than octahedral? *RSC Adv.* **2013**, *3*, 17803–17812. [CrossRef]
175. Moin, S.T.; Hofer, T.S.; Weiss, A.K.H.; Rode, B.M. Dynamics of ligand exchange mechanism at Cu(II) in water: An ab initio quantum mechanical charge field molecular dynamics study with extended quantum mechanical region. *J. Chem. Phys.* **2013**, *139*, 014503. [CrossRef]
176. Bowron, D.T.; Diaz-Moreno, S. The structure of water in solutions containing di- and trivalent cations by empirical potential structure refinement. *J. Phys. Condens. Matter* **2013**, *25*, 454213. [CrossRef]
177. Frank, P.; Benfatto, M.; Qayyam, M.; Hedman, B.; Hogdson, K.O. A high-resolution XAS study of aqueous Cu(II) in liquid and frozen solutions: Pyramidal, polymorphic, and non-centrosymmetric. *J. Chem. Phys.* **2015**, *142*, 084310. [CrossRef]
178. Chillemi, G.; Pace, E.; D'Abramo, M.; Benfatto, M. Equilibrium between 5- and 6-fold coordination in the first hydration shell of Cu(II). *J. Phys. Chem. A* **2016**, *120*, 3958–3965. [CrossRef]
179. Galván-Garcáa, E.A.; Agacino-Valdés, E.; Franco-Pérez, M.; Gómez-Balderas, R. $[Cu(H_2O)_n]^{2+}$ (n = 1–6) complexes in solution phase: A DFT hierarchical study. *Theo. Chem. Acc.* **2017**, *136*, 29. [CrossRef]
180. Frank, P.; Benfatto, M.; Qayyum, M. $[Cu(aq)]^{2+}$ is structurally plastic and the axially elongated octahedron goes missing. *J. Chem. Phys.* **2018**, *148*, 204302. [CrossRef]

181. Persson, I.; Nilsson, K.B. Coordination chemistry of the solvated silver(I) ion in oxygen donor solvents water, dimethyl sulfoxide and *N,N'*-dimethylpropyleneurea. *Inorg. Chem.* **2006**, *45*, 7428–7434. [CrossRef]
182. Busato, M.; Melchior, A.; Migliorati, V.; Colella, A.; Persson, I.; Mancini, G.; Veclaniz, D.; D'Angelo, P. The elusive coordination of the Ag^+ ion in aqueous solution: Evidence for a linear structure. *Inorg. Chem.* **2020**, *59*, 10291–10302. [CrossRef]
183. Lin, R.-L.; Dong, Y.-P.; Tang, M.; Liu, Z.; Tao, Z.; Liu, J.-X. Selective Recovery and Detection of Gold with Cucurbit[*n*]urils (n = 5–7). *Inorg. Chem.* **2020**, *59*, 3850–3855. [CrossRef] [PubMed]
184. D'Angelo, P.; Chillemi, G.; Barone, V.; Mancini, G.; Sanna, N.; Persson, I. Experimental evidence for a variable first coordination shell of the cadmium(II) ion in aqueous, dimethyl sulfoxide and *N,N'*-dimethylpropyleneurea solution. *J. Phys. Chem. B* **2005**, *109*, 9178–9185. [CrossRef]
185. Rosdahl, J.; Persson, I.; Kloo, L.; Ståhl, K. On the solvation of the mercury(I) ion. A Structural, vibration spectroscopic and quantum chemical study. *Inorg. Chim. Acta* **2004**, *357*, 2624–2634. [CrossRef]
186. Hofer, T.S.; Randolf, B.R.; Rode, B.M. The hydration of the mercury(I)-dimer—A quantum mechanical charge field molecular dynamics study. *Chem. Phys.* **2008**, *349*, 210–218. [CrossRef]
187. Persson, I.; Eriksson, L.; Lindqvist-Reis, P.; Persson, P.; Sandström, M. An EXAFS spectroscopic, large-angle X-ray scattering, and crystallographic study of hexahydrated, dimethyl sulfoxide and pyridine 1-oxide hexasolvated mercury(II) Ions. *Chem. Eur. J.* **2008**, *14*, 6687–6696. [CrossRef] [PubMed]
188. Chillemi, G.; Mancini, G.; Sanna, N.; Barone, V.; Della Longa, S.; Benfatto, M.; Pavel, N.V.; D'Angelo, P. Evidence for sevenfold coordination in the first solvation shell of Hg(II) aqua ion. *J. Am. Chem. Soc.* **2007**, *129*, 5430–5436. [CrossRef] [PubMed]
189. Sandström, M.; Persson, I. Crystal and molecular structure of hexakis(dimethylsulfoxide)mercury(II) perchlorate, $[Hg((CH_3)_2SO)_6](ClO_4)_2$. *Acta Chem. Scand. Ser. A* **1978**, *32*, 95–100. [CrossRef]
190. Strömberg, D.; Sandström, M.; Wahlgren, U. Theoretical calculations on the structure of the hexahydrated divalent zinc, cadmium and mercury ions. *Chem. Phys. Lett.* **1990**, *172*, 49–54. [CrossRef]
191. Bersuker, I.B. *The Jahn–Teller Effect and Vibronic Interactions in Modern Chemistry*; Plenum Press: New York, NY, USA, 1984; Chapters 1–5.
192. Bersuker, I.B.; Polinger, V.Z. *Vibronic Interactions in Molecules and Crystals*; Springer: Berlin, Germany, 1989; Chapters 3 and 4.
193. Glaser, J.; Johansson, G.; Jennische, P.; Wahlberg, A.; Bastiansen, O.; Braathen, G.; Fernholt, L.; Gundersen, G.; Nielsen, C.J.; Cyvin, B.N.; et al. On the Structures of the thallium(III) ion and Its bromide complexes in aqueous solution. *Acta Chem. Scand. A* **1982**, *36*, 125–135. [CrossRef]
194. Blixt, J.; Glaser, J.; Mink, J.; Persson, I.; Persson, P.; Sandström, M. On the structure of thallium(III) chloride, bromide and cyanide complexes in aqueous solution. *J. Am. Chem. Soc.* **1995**, *117*, 5089–5104. [CrossRef]
195. Shimoni-Livny, L.; Glusker, J.P.; Bock, C.W. Lone pair functionality in divalent lead compounds. *Inorg. Chem.* **1998**, *37*, 1853–1867. [CrossRef]
196. Gillespie, R.J.; Nyholm, R.S. Inorganic stereochemistry. *Q. Rev. London* **1957**, *11*, 339–380. [CrossRef]
197. Gillespie, R.J.; Hargittai, I. *The VSEPR Model of Molecular Geometry*; Allyn and Bacon: Boston, MA, USA, 1991; ISBN 978-0-205-12369-8.
198. Mudring, A.-V.; Rieger, F. Lone Pair Effect in Thallium(I) macrocyclic compounds. *Inorg. Chem.* **2005**, *44*, 6240–6243 and references therein. [CrossRef]
199. Walsh, G.; Watson, G.A. Influence of the anion on lone pair formation in Sn(II) monochalcogenides: A DFT study. *J. Phys. Chem. B* **2005**, *109*, 18868–18875. [CrossRef]
200. Stoltzfus, M.W.; Woodward, P.M.; Seshadri, R.; Klepeis, J.H.; Bursten, B. Structure and bonding in $SnWO_4$, $PbWO_4$, and $BiVO_4$: Lone pairs vs inert pairs. *Inorg. Chem.* **2007**, *46*, 3839–3850. [CrossRef]
201. Walsh, A.; Payne, D.J.; Edgell, R.G.; Watson, G.A. Stereochemistry of post-transition metal oxides: Revision of the classical lone pair model. *Chem. Soc. Rev.* **2011**, *40*, 4455–4463. [CrossRef]
202. Persson, I.; Jalilehvand, F.; Sandström, M. Structure of the solvated thallium(I) ion in aqueous, dimethyl sulfoxide, *N,N'*-dimethylpopyleneurea and *N,N*-dimethylthioformamide solution. *Inorg. Chem.* **2002**, *41*, 192–197. [CrossRef]
203. Persson, I.; D'Angelo, P.; Lundberg, D. Hydrated and solvated tin(II) ions in solution and the solid state, and a coordination chemistry overview of the $d^{10}s^2$. *Chem. Eur. J.* **2016**, *22*, 18583–18592. [CrossRef]
204. Swift, T.J.; Sayre, W.G. Determination of hydration numbers of cations in aqueous solution by means of proton NMR. *J. Chem. Phys.* **1966**, *44*, 3567–3574. [CrossRef]
205. Bargar, J.R.; Brown, G.E., Jr.; Parks, G.A. Surface complexation of Pb(II) at oxide-water interfaces: I. XAFS and bond-valence determination of mononuclear and polynuclear Pb(II) sorption products on aluminum oxides. *Geochim. Cosmochim. Acta* **1997**, *61*, 2617–2637. [CrossRef]
206. Persson, I.; Lyczko, K.; Lundberg, D.; Eriksson, L.; Płaczek, A. A Coordination chemistry study of hydrated and solvated lead(II) ions in solution and solid state. *Inorg. Chem.* **2011**, *50*, 1058–1072. [CrossRef]
207. Näslund, J.; Persson, I.; Sandström, M. Solvation of the bismuth(III) ion by water, dimethyl sulfoxide, *N,N'*-dimethylpropyleneurea, and *N,N*-dimethylthioformamide. An EXAFS, large-angle X-ray scattering, and crystallographic structural study. *Inorg. Chem.* **2000**, *39*, 4012–4021. [CrossRef]

liquids

MDPI

Review

Vibrational Raman Spectroscopy of the Hydration Shell of Ions

Nishith Ghosh [1], Subhadip Roy [2], Anisha Bandyopadhyay [2] and Jahur Alam Mondal [2,*]

1 Industrial Hygiene & Safety Section, Bhabha Atomic Research Center, Homi Bhabha National Institute, Trombay, Mumbai 400085, India

2 Radiation & Photochemistry Division, Bhabha Atomic Research Centre, Homi Bhabha National Institute, Trombay, Mumbai 400085, India

* Correspondence: mondal@barc.gov.in

Abstract: Ionic perturbation of water has important implications in various chemical, biological and environmental processes. Previous studies revealed the structural and dynamical perturbation of water in the presence of ions, mainly with concentrated electrolyte solutions having significant interionic interactions. These investigations highlighted the need of selective extraction of the hydration shell water from a dilute electrolyte solution that is largely free from interionic interactions. Double-difference infrared (DDIR) and Raman multivariate curve resolution (Raman-MCR), as well as MD simulation, provided valuable insight in this direction, suggesting that the perturbed water mainly resides in the immediate vicinity of the ion, called the hydration shell. Recently, we have introduced Raman difference spectroscopy with simultaneous curve fitting (Raman-DS-SCF) analysis that can quantitatively extract the vibrational response of the perturbed water pertaining to the hydration shell of fully hydrated ions/solute. The DS-SCF analysis revealed novel hydrogen-bond (H-bond) structural features of hydration water, such as the existence of extremely weakly interacting water–OH (ν_{max} ~ 3600 cm^{-1}) in the hydration shell of high-charge-density metal ions (Mg^{2+}, Dy^{3+}). In addition, Raman-DS-SCF retrieves the vibrational response of the shared water in the water–shared-ion pair (WSIP), which is different from the hydration shell water of either the interacting cation and anion. Herein, we discuss the perturbation of water H-bonding in the immediate vicinity of cation, anion, zwitterion and hydrophobes and also the inter-ionic interactions, with a focus on the recent results from our laboratory using Raman-DS-SCF spectroscopy.

Keywords: hydration shell; ion–water interaction; Raman difference spectroscopy

1. Introduction

Water plays pivotal roles in atmospheric, chemical and biological processes [1–4]. In the real world, water rarely exists in its pure form; ions and amphiphiles are integral parts of natural water. Water as a liquid is highly structured due to its extensive hydrogen (H)-bond network, which is easily affected by external stimuli, including temperature, pressure, charge (ions) and other solutes with varying H-bonding abilities [5–8]. The majority of molecular interactions that dictate cellular processes in living organisms occur in the ion and amphiphile containing aqueous media known as the intra- and extra-cellular fluids. The water molecules in the immediate vicinity of ions and biomolecules not only act as a solvent but also as a 'reactant'. The H-bonding of water is affected in the hydration shell of the ions and amphiphilic/hydrophobic biomolecules, which in turn influences the properties of these ions/molecules that play pivotal roles in biophysical/chemical processes, including the stability of secondary, tertiary and quaternary structures of proteins; selectivity of ion channel; cell-signaling; and the formation and flexibility of lipid membranes [9]. Being strongly associated, water exhibits the rapid delocalization of energy via its intra- and intermolecular vibrational modes; the structural perturbation of water in the ion hydration shell also affects the energy delocalization and reorganization dynamics in the hydration shell [6,10–15].

Early investigations categorized the charged groups/ions as the 'kosmotrope', i.e., the water structure maker, and the 'chaotrope', i.e., the water structure breaker, depending on their effect on the macroscopic properties of water such as viscosity, entropy and the salting in/out of proteins [6,16–22]. These descriptions, instead of the individual hydration of the ions in water, represent the cumulative effect of the cation and anion (produced by the electrolyte) on water H-bonding. As a result, the molecular-level description of the ionic perturbation of water remains elusive even after the categorization of ions as kosmotrope or chaotrope. Scattering-based techniques, such as neutron and X-ray diffraction, unveiled the geometrical distribution of water around an ion, enabling the determination of the number of water molecules in the first hydration shell, known as the coordination number (CN) [23–27]. The CN suggests the arrangement of water in the first hydration shell, but their H-bonding structure is not well understood by these measurements.

Vibrational spectroscopic techniques, such as IR, Raman and THz, are very sensitive to the H-bonded structure of water and have the potential to provide direct information by monitoring the intra- and inter-molecular vibrational modes of water. THz absorption spectroscopy, which measures the low-frequency intermolecular vibrational modes in the far-infrared region, revealed that the effect of a high-charge-density ion is extended beyond the first layer of water around the ion [13,28]. MD simulation in combination with THz spectroscopy further exposed interionic interactions that lead to the formation of a water–shared-ion pair (WSIP) and the dynamics of the shared water is cooperatively perturbed by the interacting ions [29]. The IR and Raman measurements of aqueous electrolyte solutions record the intramolecular modes (OH stretch (3400 cm^{-1}) and HOH bend (1640 cm^{-1})) and intermolecular (librational (470 and 670 cm^{-1}) and bend+librational combination (2110 cm^{-1})) and intermolecular H-bond stretch (185 cm^{-1})) vibration of water, irrespective of its position in bulk water or in the hydration shell of an ion. As a result, the experimental solution spectrum, especially for the diluted solution, contains a weak signal of the hydration water mixed with a huge signal of the bulk (unperturbed) water. This is why most early vibrational studies and even the recent THz studies were carried out with concentrated electrolyte solutions (a few-molar concentration) so that the relative contribution of the hydration water becomes appreciable in the experimental solution spectrum. However, at such a high electrolyte concentration, interionic interaction may lead to ion pairing, aggregation and the share/overlap of hydration shells, which invariably interfere with the native H-bonding of water in the hydration shell of a fully hydrated ion.

Analytical methods, such as spectral fitting, difference spectroscopy and factor analysis have been applied to extract the perturbed-water spectrum pertaining to the hydration shell of an ion [30–32]. In difference spectroscopy, the response of hydration water is obtained by subtracting the bulk water spectrum from the solution spectrum such that the subtracted spectrum provides the spectral signature of the water that are perturbed by the cation and anion. The improper (either excess or incomplete) subtraction of the bulk water spectrum may deform the band shape of the resulting salt-affected spectrum. While using salts of a fixed cation with varying anions (e.g., MgX_2 with X = F, Cl, Br, I) and vice-versa, the series of difference-spectra obtained by this method provide a trend of the spectral change in the hydration shell of the anion. The subsequent fitting of the difference-spectra provides the spectrum of water in the hydration shell of the cation and anion, respectively, though the interionic interactions on the respective ion hydration shell spectra cannot be ignored.

Raman spectroscopy, in combination with multivariate curve resolution analysis (Raman-MCR) [33], successfully extracted the minimum area vibrational spectrum of the ion-affected water [34–40]. The 'minimum area' signifies the response of the 'perturbed-water' only; in other words, the hydration shell water whose vibrational response is similar to the bulk water is not retained in the MCR-extracted hydration shell spectrum. An important advantage of Raman-MCR is that it does not require any spectral fitting. However, the MCR method used in the study of ion hydration shells is heavily biased towards the perturbed water [34]. As a result, the MCR-extracted hydration shell spectrum fails to appreciate the subtle spectral features of water in the hydration shell of an ion. This limitation can be overcome to a great extent by using spectral fitting-guided minimum-area difference spectroscopy, known as difference spectroscopy, with simultaneous curve fitting (DS-SCF) analysis [41]. The application of Raman-DS-SCF revealed new features of water in the hydration shell of ions [42], which were obscured in the Raman-MCR [43], as well as double difference infrared, spectroscopy (DDIR) [44].

This review article describes the perturbation of the H-bonding and the vibrational coupling of water in the hydration shell of different ions, using DS-SCF and MCR methods on the Raman spectra of the corresponding dilute electrolyte solutions. Herein, hydration water refers to water that are in the close vicinity of ions, collectively called hydration shell water, regardless of their positions in the first, second or third hydration spheres that may exist depending on the charge density of the ion. The subsequent discussions are structured as follows. Section 2 provides a comparative discussion of the Raman-DS-SCF method with other hydration shell spectroscopy techniques. Sections 3 and 4 describe the Raman spectrum of water in the OH stretch region and discuss the effect of isotopic dilution and electrolyte on the spectral line shape. Sections 5 and 6 discuss the perturbation of H-bonding and vibrational coupling in the hydration shell of the anion and cation, respectively, with their varying natures such as charge density, valency, atomic/molecular nature, etc. The restructuring of hydration shell water for the formation of WSIP is described in Section 7. Sections 8 and 9 discuss the hydration characteristics of a hydrophobic molecular ion ($(CH_3)_4N^+$;) and zwitterion ($(CH_3)_3N^+$-O^-), respectively. Finally, Section 10 provides a future perspective along with concluding remarks.

2. Raman-DS-SCF Analysis and Ion Hydration Shell

For the DS-SCF analysis, the Raman spectrum of a dilute electrolyte solution (~sub molar concentration) is considered as a linear combination of three spectral components: (1) vibrational response of the solute/ion (if any), (2) spectrum of the water that is perturbed by the ion/solute and (3) spectrum of the unperturbed water that is equivalent to the spectrum of the bulk water. Components 1 and 2 are associated with the solute and its hydration shell water, which is combinedly treated as the ion/solute-perturbed spectrum or the hydration shell spectrum. While subtracting the bulk-like water spectrum (reference spectrum; $R(\overline{\nu})$) from the solution spectrum ($S(\overline{\nu})$), the emerging perturbed-water spectrum, along with the inherent vibrational bands of the solute (if any), is fitted with a combination of Gaussian bands following Equation (1).

$$S(\overline{\nu}) - fR(\overline{\nu}) = \sum_{n=1}^{n}\left(A_n e^{-\frac{(\overline{\nu}-\overline{\nu}_n)^2}{2\Gamma_n^2}}\right) + A_0 \quad (1)$$

where f is the "appropriate fraction" of the reference spectrum ($R(\overline{\nu})$) subtracted from the dilute solution spectrum ($S(\overline{\nu})$). A_n, $\overline{\nu}_n$ and Γ_n are the amplitude, center frequency and *fwhm* of the nth band in components 1 and 2 (i.e., solute/ion's own vibrational bands and that of the perturbed water). A_0 is a constant that takes care of the difference in constant background (if any) between the solution and reference spectra.

Figure 1 depicts the $S(\overline{\nu})$ for 0.25 M aqueous $MgCl_2$, $R(\overline{\nu})$ corresponding to the 0.5 M aqueous NaCl and the Raman-DS-SCF-extracted hydration shell spectrum of Mg^{2+} ion. Instead of the neat water, a 0.5 M NaCl solution was chosen as the reference ($R(\overline{\nu})$), because the latter contains the same concentration of Cl^- (counter-ion) as in the 0.25 M aqueous $MgCl_2$, such that the effect of the counter ion (Cl^-) is nullified in the DS-SCF-extracted spectrum, representing the hydration shell water of Mg^{2+}. To be precise, the Mg^{2+} hydration shell spectrum shown in Figure 1 is the response of the Mg^{2+}-associated water relative to that of Na^+; however, due to the subtle influence of Na^+ (on water) [45], the perturbed water spectrum predominantly represents the hydration shell response of Mg^{2+}.

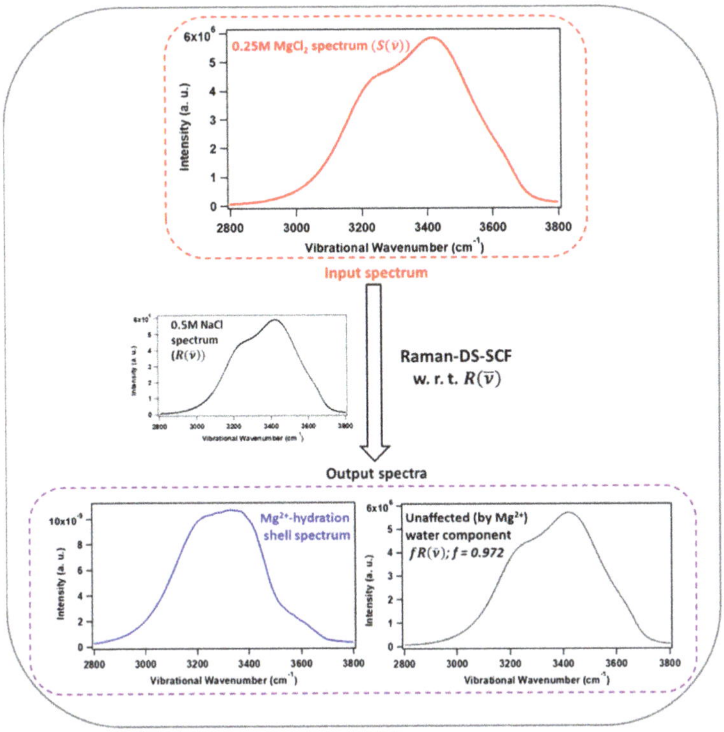

Figure 1. Input and output spectra corresponding to the extraction of Mg^{2+} hydration shell spectrum using Raman-DS-SCF analysis. Input: Raman spectra of 0.25 M $MgCl_2$ (red) as the solution spectrum and 0.5 M NaCl (black) as the reference spectrum. Output: Mg^{2+} hydration shell spectra (blue) and spectrum of water that are unperturbed by Mg^{2+} ($-$ f$R(\bar{\nu})$), which is basically the fraction of the reference spectrum removed from the solution spectrum. The coefficients of the Raman-DS-SCF fitting analysis are shown in Table 1 for reference.

Table 1. Fitting parameters of Mg^{2+}-hydration shell spectrum (OH stretch).

Coefficients	Peak 1	Peak 2	Peak 3	Peak 4
Amplitude, A_n, ($\times 10^{-4}$)	2.2	17.2	9.9	3.0
Peak center (cm^{-1}), $\bar{\nu}_n$	3048	3234	3399	3561
fwhm (cm^{-1}), Γ_n	155	113	65	92
A_0, ($\times 10^{-4}$)			−0.27	

As can be seen in the hydration shell spectrum of the Mg^{2+} ion (blue curve, Figure 1), a weak band in the high-frequency region (~3600 cm^{-1}) is distinctly visible in the Raman-DS-SCF extracted spectrum. Such subtle spectral features are not so obvious in the corresponding Raman-MCR-extracted Mg^{2+} hydration shell spectrum [43]. Moreover, unlike in the case of Raman-MCR, the Raman-DS-SCF extracted spectrum is not area-normalized, and hence retains quantitative information of the perturbed hydration shell water to a great extent [42]. The DDIR approach also applies difference spectroscopy and curve fitting, but the "fitting" analysis is carried out after the completion of the "subtraction" process [46–48], while in Raman-DS-SCF, the "subtraction" and "curve fitting" are carried out simultaneously with the gradual (small-step) increments of the 'f' factor while readjusting the Gaussian fitting coefficients. For the electrolytes with the dominant effect of the counter ion, unlike the DDIR, Raman-DS-SCF and Raman-MCR use an appropriate reference spectrum to automatically subtract the counter ion influence so that the subtracted spectrum is predominantly from the ion in question. Nevertheless, it is important to note that both Raman-MCR and Raman-DS-SCF are biased towards the perturbed water in the hydration shell, and the hydration water component having bulk-like vibrational response is not retained in the extracted spectrum.

3. OH Stretch Spectra of Neat Water

The Raman OH stretch (3000–3800 cm^{-1}) spectrum of water is affected in the presence of dissolved ions/solutes. These spectral changes are better understood by comparing the solution spectrum with the neat-water spectrum. Therefore, it important to briefly discuss the OH stretch spectrum of neat water. The Raman spectrum of neat water (Figure 2) shows a broad band in the OH stretch region, with a maximum at ~3420 cm^{-1} and shoulders at ~3250 and ~3620 cm^{-1}. The frequency of OH stretch vibration is strongly correlated with the H-bond strength: strengthening of the H-bond makes the parent O–H bond weaker, leading to a shift in the O–H stretch frequency to a lower wavenumber, termed the 'red-shift'. Similarly, the weakening of H-bonding causes a shift in the O–H stretch frequency to a higher wavenumber, termed the 'blue-shift'. However, the intramolecular vibrational coupling between the HOH-bend overtone and O–H stretch fundamental, called Fermi resonance (FR), as well as the intra- and intermolecular vibrational coupling of the OH-stretch, change the line shape of the OH stretch spectrum of water, barring its straightforward interpretation solely on the basis of the strength of H-bonding [37,49–52]. To correctly interpret the spectral change due to H-bonding, generally isotopically diluted water (HOD; a mixture of H_2O and D_2O) is used instead of isotopically pure H_2O or D_2O. An appropriate mixture of H_2O and D_2O (e.g., H_2O:D_2O = 1:9, *v/v*) is free from the intra- and intermolecular vibrational coupling due to the energetic mismatch of intramolecular vibrational modes (OD and OH stretches) and the bend overtones. As can be seen in Figure 2a, the peak-normalized OH stretch spectrum of HOD has substantially lower intensity around 3200 cm^{-1} compared to that of H_2O. This indicates that the 3200 cm^{-1} band of water is predominantly due to intra- and inter-molecular vibrational coupling and Fermi resonance. Polarized Raman measurement provides the symmetry information of the vibrational bands. The lower-frequency region (<3300 cm^{-1}) is dominated in the isotropic Raman spectrum, whereas the higher frequency region (>3300 cm^{-1}) is dominated in the anisotropic spectrum (Figure 2b). The isotropic Raman spectrum of neat water is dominated by the symmetric OH stretch that are coupled in-phase with the symmetric OH stretch modes of neighboring water molecules and that appear towards the lower-frequency region of the OH stretch band [49]. The band shape of the anisotropic Raman spectrum is similar to the unpolarized Raman spectrum of HOD and assignable to the antisymmetric OH stretch mode of water. Thus, the isotopic dilution and polarized Raman measurements show that the spectral response at ~3250 cm^{-1} region is dominated by the symmetric OH stretch that are intra- and intermolecularly coupled along with FR. The band at ~3420 cm^{-1} represents the average H-bond strength of bulk water (free from coupling effects). The extreme high-frequency shoulder around 3620 cm^{-1} mainly corresponds to the very weakly H-bonded water or the water–OH with broken H-bonds (free OH) that exists transiently during the switching of the H-bonded partner among water molecules [53,54]. Thus, at any instant, bulk water can be thought of a mixture of water species with continuously varying H-bond strengths.

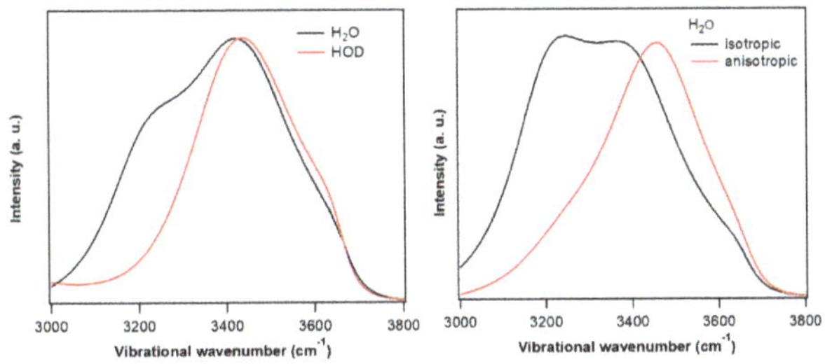

Figure 2. (**a**) Raman spectra of H_2O and HOD (H_2O:D_2O = 1:9 (v/v)) in the OH stretch region; (**b**) Isotropic (I_{iso})and anisotropic (I_{aniso}) Raman spectra of H_2O. All spectra are peak-normalized at 3420 cm^{-1}. The isotropic and anisotropic Raman spectra are calculated as $I_{iso} = I_{/\!/} - 4/3I_{\perp}$ and $I_{anirso} = 4/3I_{\perp}$. where $I_{/\!/}$ and $I_{\perp}$ are the parallel and perpendicular spectra recorded by placing laminated film polarizer (in the detection path) in the parallel and perpendicular orientation with respect to the polarization of the excitation beam.

4. Effect of Electrolytes on the OH Stretch Raman Cross-Section and Vibrational Coupling of Water

Coupling among the vibrational modes in liquid water has a close relation with its structure and properties. For instance, strong intramolecular vibrational coupling leads to ultrafast energy delocalization among different intramolecular vibrational modes of a water molecule; while strong intermolecular coupling does the same among several water molecules (collective vibration) [51,55]. These ultrafast elementary processes play pivotal roles in water-mediated energy dissipation processes. Figure 3a shows the effect of isotopic dilution on the OD stretch spectrum of D_2O. With an increasing percentage of H_2O in D_2O, the intensity around 2350 cm^{-1} (corresponding to the 3200 cm^{-1} for the OH stretch) decreases and that around 2500 cm^{-1} increases, making the overall spectrum narrower and blue-shifted compared to D_2O. These spectral changes suggest the reduction of Fermi resonance and vibrational decoupling in D_2O due to the addition of H_2O (isotopic dilution). Interestingly, the addition of electrolytes, such as NaBr, has a similar effect on the OD stretch of D_2O (Figure 3b). Except NaF, all other Na-halides lead to such a spectral change in the OD (or OH) stretch regions. Thus, the spectral change in Figure 3a,b indicates the reduced coupling (VC and FR) of water in the presence of NaX (X = Cl, Br, I) [37,45]. The reduction of coupling could be due to the change in OH stretch frequency (variation in H-bond strength) and the preferred orientation of water in the hydration shell of the ions. Quantitative estimation, on the basis of integrated Raman intensity, revealed that the Raman cross section increases in the presence of the halide ions, except F^-.

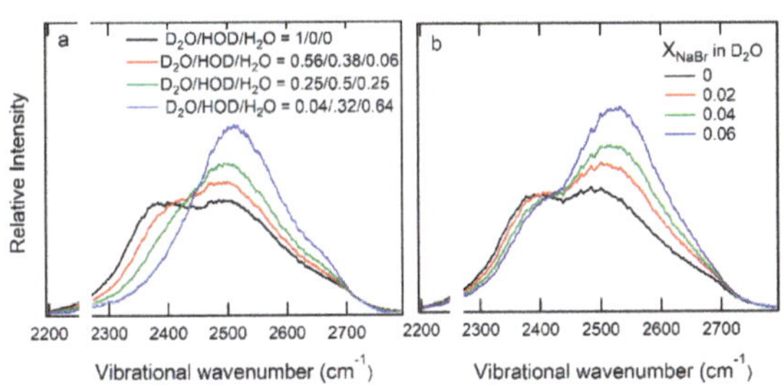

Figure 3. Raman spectra (OD stretch) of (**a**) D_2O-H_2O mixtures with different mole ratios and (**b**) D_2O with different mole fractions of NaBr. The spectra are normalized for the concentration of OD oscillator in NaBr solution and D_2O-H_2O mixture. (Reprinted with permission from ref. [37]. Copyright 2013 American Chemical Society).

5. Hydration Shell of Ions

5.1. Water in the Hydration Shell of a Low-Charge-Density Monovalent Anion

Water interacts with anions through its hydrogen atom by forming an H-bond with the latter. The E-field around the ion reorients the water in the hydration shell such that the water–Hs are pointed 'towards' the hydrated anion and 'away' from the hydrated cation, following the law of electrostatics. Figure 4a shows the Raman spectrum of neat water and the MCR-extracted hydration shell water spectra of different monovalent anions. It is important to note that, for the hydration shell of low-charge-density anions, the band shapes of the MCR-extracted spectra are very similar to the Raman-DS-SCF-extracted spectra (not shown here). It is obvious that the spectral responses of hydration water are markedly different from the Raman spectrum of the bulk water. To examine the H-bond strength of water in the vicinity of ions, it is reasonable to compare the OH stretch spectra of hydration water with that of HOD, in which the intermolecular coupling of OH stretch modes is largely suppressed. The maxima of the OH stretch Raman band of the hydration water of the monovalent halide anions (Br^-, I^-) lie in the higher-frequency region (blue-shift) compare to that of HOD (Figure 4a), indicating weaker average H-bond strength of water in the hydration shell of the halide anions [36,39,45]. A similar spectral change was also observed for Cl^- (spectrum not shown here) [39]. Moreover, the extent of the blue-shift of the hydration water spectra (OH stretch) follows the order of the increasing size of the anions, i.e., $Cl^- < Br^- < I^-$. A triatomic monovalent anion with lower charge density than I^- (e.g., I_3^-) shows a further decrease in the average H-bond strength of the hydration water than that of the I^- [56–58]. The monovalent NO_3^- anion also weakens the H-bond strength of its hydration shell water (Figure 4a) [36,44,59]. Further analysis of the integrated area of the 3250 cm^{-1} band with that of the H_2O and HOD (decoupled OH) suggested a significant reduction (~80%) of the combined coupling effects (intermolecular coupling and Fermi resonance) in the vicinity of halide ions relative to the bulk water [36].

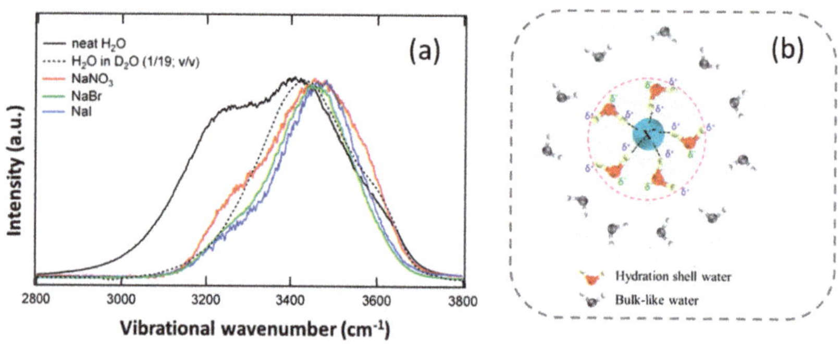

Figure 4. (**a**) Hydration shell water spectra (OH stretch) for I^-, Br^- and NO_3^- in the dilute aqueous solution (1.0 M) of the respective Na-salts. Raman spectra of neat H_2O and HOD ($H_2O/D_2O = 1/19$; v/v) are shown for comparison. (Reprinted with permission from ref. [36]. Copyright 2013 American Chemical Society). (**b**) Schematic representation of the first hydration shell of an anion (the number of water molecules shown in the hydration shells are just representative in nature and do not indicate the coordination number of the ions).

5.2. Hydration Shell of Monovalent Oxyhalide Anion

Although the halide ions weaken the H-bond strength of hydration water in the order $Cl^- < Br^- < I^-$, the oxyhalide ions affect the surrounding water differently. Figure 5a–c shows the hydration shell water spectra of ClO_3^-, BrO_3^- and IO_3^- anions in H_2O. The band maximum is gradually blue-shifted (shifted to the high-frequency side) as the central halogen atom changes from I to Br to Cl. The relative intensity of the 3200 cm^{-1} shoulder band is decreased due to reduced vibrational coupling and the Fermi resonance of water in the hydration shell of the oxyhalide anions, as has been observed with the halide ions [36,45].

The hydration shell spectra in HOD (Figure 5d–f) show that the OH stretch band maximum is blue-shifted by ~100 cm^{-1} and ~70 cm^{-1} for ClO_3^- and BrO_3^-, respectively, while in the case of IO_3^-, the relative intensity in the low-frequency region (below 3400 cm^{-1}) is enhanced. Hence, ClO_3^- and BrO_3^- weaken the H-bond strength of their hydration water following the order $ClO_3^- > BrO_3^-$, whereas IO_3^- is strengthened the same [60]. Using MD simulation and XAFS spectroscopy, Baer et al. showed that the central iodine atom of IO_3^- possesses a positive charge, while each of the three oxygen atoms bears a negative charge [61], which indicates two distinct hydration environments around the iodate ion corresponding to the cationic iodine and anionic oxygen atoms, respectively. MD simulation also revealed that the vibrational power spectra and distributions of different types of H-bonded water molecules hydrating the iodate-oxygens are very similar to that of bulk water [62]. Hence, the cationic iodine (of iodate) predominantly affects the structure of surrounding water. Thus, unlike the weakly interacting I^- anion, in spite of having an overall negative charge, IO_3^- effectively behaves as a hydrated cation in aqueous solution which makes the H-bonds of surrounding water stronger. The asymmetric hydration of IO_3^- in aqueous KIO_3 solution has also been suggested by Raman-MCR study [56]. The MD simulation of BrO_3^- ion in water suggested that anionic oxygens (of BrO_3^-) affect the hydrating water to a greater extent than the cationic bromine (of BrO_3^-) and that the average strength of BrO_3^-–water H-bonding is weaker than that of water–water in bulk [63].

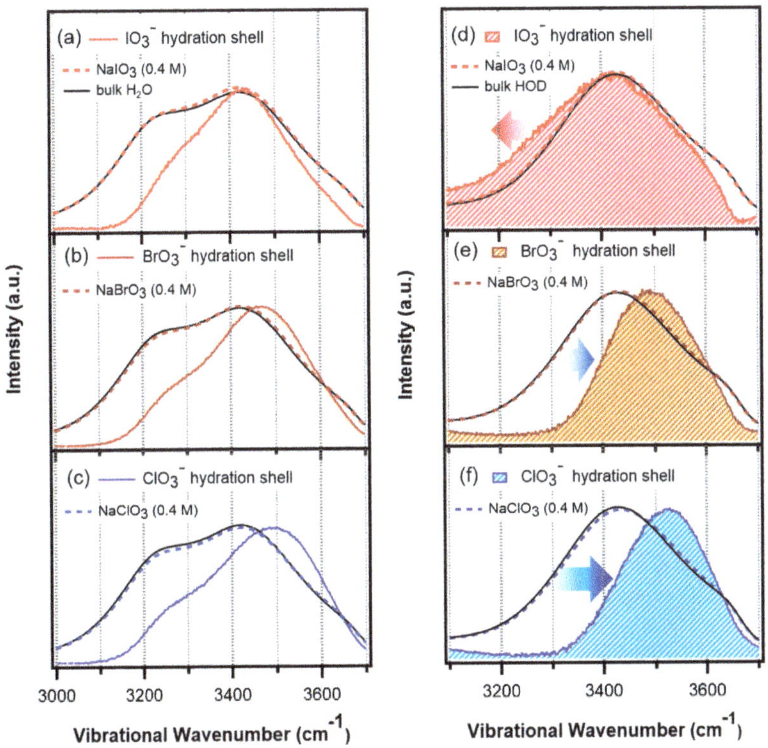

Figure 5. Raman-DS-SCF-extracted hydration shell spectra (peak intensity normalized) of oxyhalide anions in H_2O (**a–c**) and HOD (H_2O:D_2O = 1:6 (v/v); (**d–f**)); IO_3^- (red; (**a**,**d**)), BrO_3^- (brown; (**b**,**e**)), and ClO_3^- (blue; (**c**,**f**)). The experimental Raman spectra of the salt solutions (dashed curves) and that of the neat H_2O (HOD) (black curve) are shown in each panel for reference. (Reprinted with permission from ref. [60]. Copyright 2021 American Chemical Society).

5.3. Hydration Shell of High-Charge-Density Anions

For the high-charge-density anions, the Na-salt of F^- and $CO_3{}^{2-}$ anions has been analyzed with respect to the bulk water using Raman-DS-SCF analysis. The hydration water spectrum of F^- shows enhanced absorption (compared to neat water) in the low-frequency region of the OH stretch up to 2600 cm^{-1} (Figure 6) [64]. This indicates the strong H-bonding of water with the F^- ion. MD simulation also suggested that the reorientation dynamics of water in the first hydration shell of F^- is much slower than that of I^-, invoking stronger $F^- \cdots H_2O$ H-bonding [65]. Divalent $CO_3{}^{2-}$ anion perturbs the H-bond strength of hydration water quite similar to that of F^-, although the average number of water molecules perturbed per $CO_3{}^{2-}$ anion is greater than that of F^- (Figure 6) [64]. In the hydration shell Raman spectrum of $CO_3{}^{2-}$ and F^- ions in H_2O, a new band appears in the lower-frequency region (~3050 cm^{-1}) suggesting the presence of a very strong H-bond in hydration water [39,64]. Thus, unlike the low-charge-density monovalent ions (Cl^-, Br^-, I^-, $ClO_3{}^-$, $BrO_3{}^-$), higher-charge-density anions, such as F^- and $CO_3{}^{2-}$, increase the average H-bond strength of water in their hydration shell compared to that of bulk water. The dip feature around 3200 cm^{-1} in the hydration shell spectra in H_2O is due to the decoupling of the vibrational modes of water in the hydration shell of the ions; the dip feature is largely reduced in HOD. Hence, the structure of water in the hydration shell of $CO_3{}^{2-}$ and F^- is more inhomogeneous compared to bulk water. It is believed that two different types of water molecules exist in the hydration shell of the ions: water in the first hydration shell forms a very strong H-bond (~3050 cm^{-1}) with the carbonate oxygen (or F^-), and water in the second hydration shell has an H-bond strength weaker than the first hydration shell but stronger than bulk water. In agreement with this, the calculated (MD simulation) stretch frequency of the water–OD, directly H-bonded to the $CO_3{}^{2-}$ oxygen, is red-shifted (i.e., shifted towards low frequency region) from that of bulk-water–OD, whereas the stretch frequency of the second hydration shell water lies in between that of the bulk and first hydration shell water [66]. These observations certainly indicate that the $CO_3{}^{2-}$ anion has a long-range effect on neighboring water molecules that goes beyond the first hydration shell.

In the case of $SO_4{}^{2-}$, a distinct band around 3250 cm^{-1} is clearly visible is the hydration shell spectrum (in HOD; Figure 6e), though the band is not as intense as that with $CO_3{}^{2-}$. In agreement with the OH stretch, the analysis of the HOH bending band revealed stronger H-bonding in the hydration shell of $SO_4{}^{2-}$ [39]. The difference spectroscopy-based IR methods in the OD stretch region also suggested that water forms H-bonds with sulfate ions stronger than that between two water molecules in bulk water [44,59]. Kameda et al. showed that the intermolecular H-bonded O–O distance between sulfate and water (in aqueous sulfuric acid solution) is shorter than that of pure water, suggesting a strongly H-bonded intermolecular interaction [67].

As expected, the trivalent $PO_4{}^{3-}$ anion induces the strong H-bonding of water in the hydration shell [36]. The Raman spectrum of aqueous $PO_4{}^{3-}$ solution in the low-wavenumber region (hindered translation of the H-bonded water molecules or the H-bond stretch) also confirmed the formation of strong P–O ... H bonds in solution [68]. One interesting feature of the hydration shell water spectrum is that, unlike the case of other anions, the shoulder near 3600 cm^{-1}, which represents weakly interacting or under-coordinated water–OH, is not reduced for $PO_4{}^{3-}$ [36]. Furthermore, Raman measurements of aqueous K_3PO_4 solutions showed a polarized shoulder band at ~3620 cm^{-1}, which increased slightly in intensity with increasing concentration (range 0.7–5.3 M) of the salt [68]. Chandra and co-worker calculated the OD stretch frequencies in the hydration shells of $PO_4{}^{3-}$ [69]. They showed that the stretch frequency of OD in the first hydration shell that directly forms an H-bond with the anion is considerably red-shifted compared to the bulk water; whereas the other OD (of the same water forming an H-bond with the anion) stretch is blue-shifted and has significant overlap with the band of non-H-bonded OD in bulk water. On the basis of this observation, the authors speculated the existence of weakly

H-bonded (compared to water–water in bulk) water–OD in the first hydration shell of the PO_4^{3-} anion.

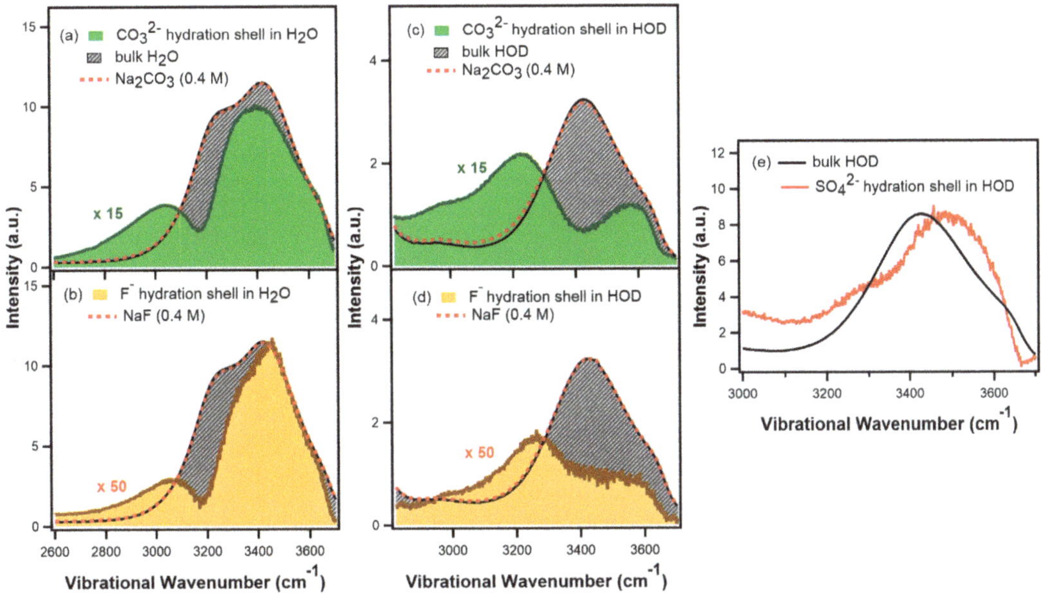

Figure 6. Raman-DS-SCF-extracted hydration shell spectra (OH stretch) of CO_3^{2-} and F^- in H_2O (**a**,**b**) and HOD (H_2O:D_2O = 1:6 (v/v); (**c**,**d**)); CO_3^{2-} (shaded green; (**a**,**c**)) and F^- (shaded orange; (**b**,**d**)). The experimental Raman spectra of the salt solutions (dashed curves) and that of the neat H_2O (HOD) (black shaded curve) are shown in each panel for reference. DS-SCF-extracted hydration shell spectra (OH stretch) of SO_4^{2-} (**e**) in HOD. (Figure 6a–d are reprinted with permission from ref. [64]. Copyright 2021 American Chemical Society).

5.4. HOH Bend and Bend+Libration Combination Band of Water in the Anion Hydration Shell

The restricted rotational motions of water in the H-bonded network are termed as librational motions that play an important role in the energy transfer mechanism in liquid water. It has been suggested that, unlike fast intermolecular stretch-to-stretch energy transfer in water, intermolecular bend-to-bend transfer is much slower [70]. Rather, librational motions randomize the orientation of the HOH bending mode transition dipole moment much faster, and thus, the vibrational energy of the HOH bending mode is primarily released to the librational modes.

In the Raman spectrum of neat water, the HOH bend vibration of water appears at ~1640 cm^{-1} [71]. As the H-bond strength of water increases, the HOH bending motion becomes more restricted. Therefore, the HOH bending band of water is expected to be blue-shifted (red-shifted) in the hydration shell of ions, where the water H-bond is strengthened (weakened). Upon isotopic dilution, the HOH bend maximum is blue-shifted by ~10 cm^{-1} and the spectral width is decreased due to vibrational decoupling (compare the black dotted and green dashed curves in Figure 7a,b). As can be seen in Figure 7a,b, the hydration shell spectra for Cl^-, I^- and SO_4^{2-} ions are blue-shifted and have lower spectral width than that of neat H_2O, indicating that the HOH bending mode of water in the hydration shell of these anions are decoupled from the neighboring water molecules. On the contrary, the hydration shell spectrum for CO_3^{2-} showed comparable spectral width to that of bulk H_2O (Figure 7b). It is likely that the larger spectral width (compared to that of other anions) for CO_3^{2-} is because of the greater distribution of H-bond strength in the hydration shell.

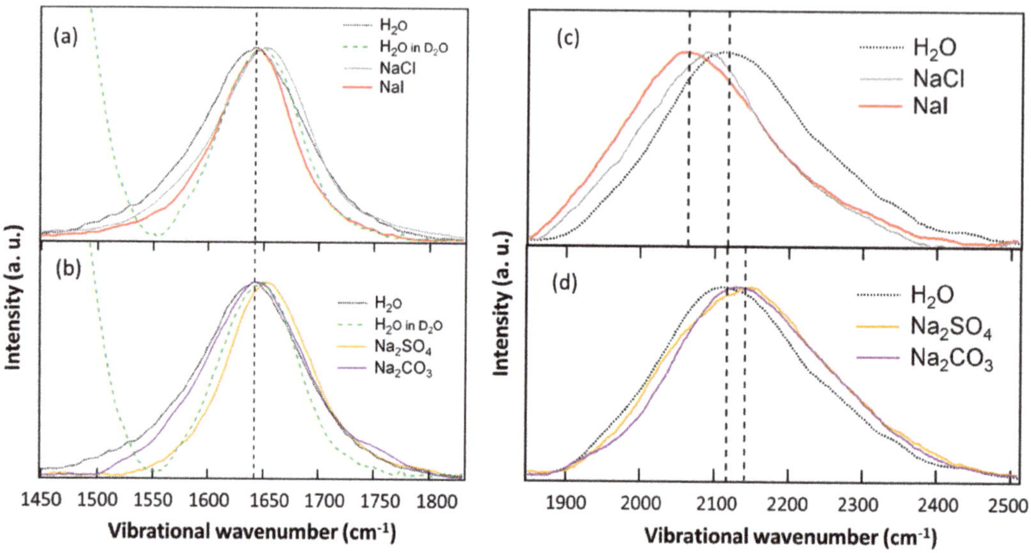

Figure 7. Hydration shell spectra (retrieved by Raman-MCR) of monovalent (Cl^- and I^-) and divalent (CO_3^{2-} and SO_4^{2-}) anions in the HOH bend (**a**,**b**) and bend+libration combination regions (**c**,**d**); Cl^- in NaCl (gray), I^- in NaI (red), SO_4^{2-} in Na_2SO_4 (orange) and CO_3^{2-} in Na_2CO_3 (purple) solutions (~1 M). The Raman spectrum of bulk H_2O (black dotted curve) is shown in each panel for reference. MCR-retrieved spectrum of H_2O in D_2O (green dashed curve) that bears the Raman response of decoupled H_2O (around 1650 cm^{-1}) are shown in panels (**a**,**b**) for reference. (Reprinted from ref. [39] with the permission of AIP Publishing).

Water librations represent a broad band over 300–1000 cm^{-1} in the Raman spectrum of neat water [8,72]. In fact, this broad librational response is composed of three overlapped bands (centered at ~450, ~580 and ~720 cm^{-1}) arising from water librations about the three axes of rotation [72–74]. The librational band of water is red-shifted with increasing temperature signifying increased librational freedom because of the weakened H-bonding of water at elevated temperatures [75]. Hence, it is expected that the ions that induce the weakening (strengthening) of the H-bond strength of neighboring water will increase (decrease) the librational freedom of the hydration water. The bend+libration combination band of hydration water is red-shifted for monovalent Cl^- and I^- compared to the neat water (~2130 cm^{-1}), while it is blue-shifted for divalent SO_4^{2-} and CO_3^{2-} (Figure 7c,d). The calculated librational band positions revealed that water molecules in the hydration shell of the monovalent Cl^- and I^- weakly interact with vicinal water and have higher librational freedom than bulk water [39]. Conversely, water in the immediate vicinity of divalent (strongly interacting) anions such as SO_4^{2-} and CO_3^{2-} have more restricted librational mobility compared to bulk water [39].

6. Water in the Hydration Shell of a Metal Cation

The local electric field around a cation orients hydration water dipole vectors to point radially away from the cation. The monovalent cations with low charge density (e.g., Na^+, Cs^+) exhibit a negligible effect on the H-bonded structure of the hydration water compared to the monovalent anions. Although monovalent Na^+ does not have significant influence on the neighboring water [76], the cations with higher charge density are expected to perturb the water in their hydration shell. In fact, the DDIR study of divalent transition metal ions (Mn^{2+}, Fe^{2+}, Co^{2+}) and trivalent lanthanide ions (La^{3+}, Nd^{3+}, Dy^{3+}, Yb^{3+}) with comparable ionic potential revealed quite a similar effect on the OD stretch band [77].

Rudolph and co-workers investigated a series of lanthanide ions (Ln^{3+}: La^{3+}, Ce^{3+}, Pr^{3+}, Nd^{3+}, Sm^{3+}, Ho^{3+}, Er^{3+}, Tm^{3+}, Yb^{3+} and Lu^{3+}) using Raman measurements that enabled determining the force constant of the Ln–O breathing modes of the aqua complexes [78,79]. It was observed that the strength of the force constant increases with decreasing Ln–O bond distances (bond distances: Lu–O < Yb–O < Tm–O < Er–O < Ho–O < Sm–O < Nd–O < Pr–O < Ce–O < La–O).

Recently, we have selectively extracted the OH stretch spectra of water in the hydration shell of a series of high-charge-density cations (Li^{+}, Mg^{2+}, Ca^{2+}, Sr^{2+}, Ba^{2+}, La^{3+}, Gd^{3+}, Dy^{3+} and Lu^{3+}, as well as H^{+}) using Raman-DS-SCF spectroscopy [42]. It was observed that these cations strongly interact with their hydration water and profoundly affect the local water structure. Interestingly, the strong cation–water interaction not only increases the average H-bond strength of hydration water but also produces a small fraction of extremely weakly interacting water–OH in the hydration shell, which increases the structural heterogeneity therein (see Figure 8). Figure 8a,b compare the hydration shell spectra of monovalent Cs^{+} (low charge density) to trivalent Lu^{3+} (high charge density), as extracted by the Raman-DS-SCF method. The hydration shell spectrum (OH stretch band-shape) of Cs^{+} is qualitatively similar to that of bulk water. However, the hydration shell spectrum of Lu^{3+} shows distinct changes compared to that of bulk water: the spectral intensity is enhanced in the low-frequency region (below 3400 cm^{-1}; strong H-bonding) and a small band appears at the high-frequency end near 3600 cm^{-1} (weakly interacting water–OH). The extremely weakly interacting water–OH is believed to have originated from the second hydration shell water–OH that donates an H-bond to the electron-deficient water–oxygen in the first hydration shell of the high-charge-density cations and distributed unevenly within the range of ionic influence (see Figure 9). In fact, this weakly interacting water–OH is different from the broken H-bonded OH transients (usually appearing at $\sim$3620 cm^{-1}) that is uniformly distributed in bulk water [54]. Hence, the high-charge-density metal ions induce the structural perturbation of water beyond their first hydration shell, which was also suggested by MD simulation studies [80,81]. The number of affected water molecules by the metal cations is correlated with the ionic potential of the ions: higher the ionic potential, the greater the number of affected water molecules by the cations. Interestingly, among the lanthanides, the number of affected water molecules marginally decreases with increasing ionic potential. On moving from La^{3+} to Lu^{3+}, the number of affected water molecules decreases following a similar trend as in the coordination numbers due to lanthanide contraction [82–84]. Nevertheless, the effects of cations on water cannot be described solely by the ionic potential. The ionic potential of Mg^{2+} is higher than that of La^{3+}; however, the latter affects a greater number of water molecules. In addition to the ionic potential, the absolute charge (i.e., oxidation state) and the size of an ion play roles in the ion's effect on water. Between two ions with the same ionic potential, the larger one affects a greater number of water molecules than the smaller one. Mg^{2+} and Dy^{3+} have the same ionic potential, but the latter ion affects a greater number of water molecules [42].

The perturbing influence of a proton (H^{+}) on the structure of water is very similar to that of high-charge-density metal ions [42,43]. Indeed, H^{+} and a high-charge-density metal ion are similar from the perspective of interaction with hydration water: both of them accept electrons from the oxygen of the water in the first hydration shell.

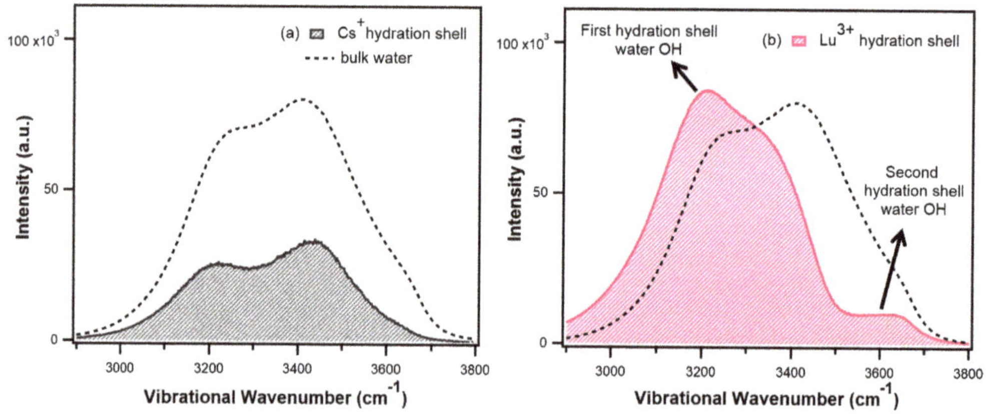

Figure 8. Raman-DS-SCF-extracted OH stretch spectra (shaded curves) of water in the hydration shell of (**a**) Cs^+ in CsCl (0.6 M) and (**b**) Lu^{3+} in $LuCl_3$ (0.2 M). The spectrum of the bulk water (black dashed curve; multiplied by a factor of 1/20) is shown in each panel for comparison. (Reprinted with permission from ref. [42]. Copyright 2020 American Chemical Society).

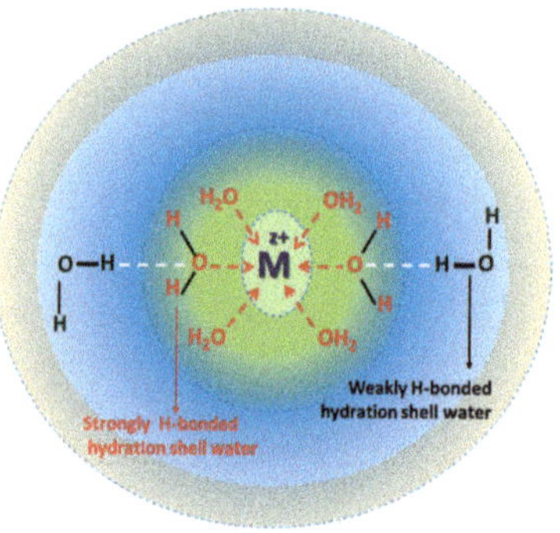

Figure 9. Schematic of the hydration shell of a high-charge-density cation: dashed white lines represent a very weak H-bond; the corresponding parent OH appears at ~3600 cm^{-1} in the hydration shell spectrum.

7. Water Shared Ion-Pair Formation and Restructuring of Hydration Shell Water

In a very dilute aqueous solution of an electrolyte, the cation and anion are hydrated separately and mostly do not influence the hydration shell structure of each other. In such a fully hydrated ion, the structural and dynamical properties of the hydrating water are perturbed by the physicochemical properties of the individual hydrated ion. However, with the increasing concentration of the electrolyte, the oppositely charged ions approach closer with their native hydration shells, which can mutually influence their properties [13,85,86]. In such cases, the native hydration shells of the oppositely charged ions overlap, and the intervening (shared) water becomes perturbed by both the interacting ions [87]. These ion-pairs are generally termed "solvent-shared ion-pairs (SSIP)". The structural perturbation of water in hydrated ion-pairs is important in understanding the ionic mobilities and thermodynamics of solutions.

Raman-DS-SCF analysis successfully enabled the identification of the formation of SSIPs and suitably extracted the vibrational response of the shared water. As can be seen in Figure 10a, the band shape of the OH stretch hydration shell water spectra (extracted by Raman-DS-SCF) of Mg^{2+} in its aqueous chloride solution undergoes gradual changes with the increasing concentration of $MgCl_2$ ($[MgCl_2] \geq 0.37$ M). The apparent band maxima of the Mg^{2+} hydration shell spectrum is shifted from ~3200 cm^{-1} to ~3400 cm^{-1} (as indicated by the gray dashed arrow in Figure 10a). This concentration-dependent spectral change indicates an interionic interaction between Mg^{2+} and Cl^-, forming SSIP. Further, a second-round DS-SCF analysis (Figure 10b) revealed that the shared water is affected by both Mg^{2+} and Cl^-; the average H-bonding of the shared water is weaker than the hydration shell water of Mg^{2+} but stronger than the hydration shell water of Cl^-. A similar observation was also observed for the case of an aqueous $LaCl_3$ solution [88]. The perturbing influences of both the cation and anion on the shared water have also been suggested in the aqueous solutions of nitrate and the chloride salt of Fe(III) [89].

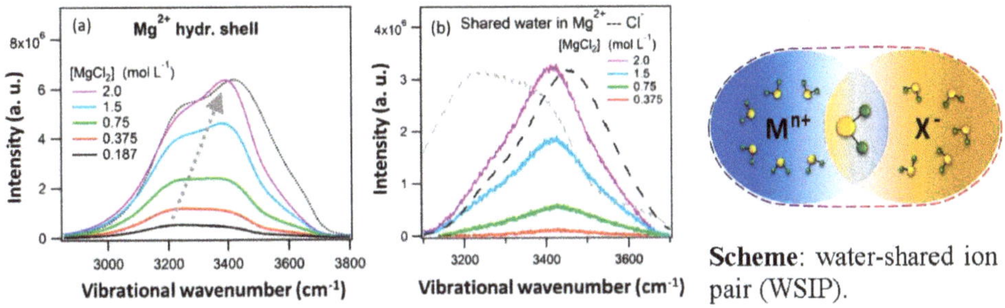

Figure 10. (**a**) Raman-DS-SCF-extracted hydration shell spectra (OH stretch) of Mg^{2+} in $MgCl_2$ solution corresponding to different salt concentrations. Raman spectrum of the bulk water (dotted curve) is shown for reference; (**b**) The second round Raman-DS-SCF-extracted OH stretch spectra (solid curves) of the shared water in SSIP formed between Mg^{2+} and Cl^-, corresponding to different salt concentrations. The hydration shell spectra of the fully hydrated ions corresponding to Mg^{2+} (gray dashed line) and Cl^- (black dashed line) are shown for reference. **Scheme:** Water-shared ion pair representing the hydration water along with the shared water in the hydration shell. (Reprinted with permission from ref. [88]. Copyright 2020 American Chemical Society).

The Raman-DS-SCF analysis also enabled the identification of the type of ion-pair formed in the solution from the band shape and the intensity of the hydration shell water spectra. In the formation of SSIP, a fraction of the hydration shell water molecules of an ion is shared by its counterion. Hence, the average number of water molecules affected by the ion is not expected to vary significantly. As a result, the integrated intensity of the hydration shell water spectra that is a quantitative measure of the ion-affected water varies linearly with concentration. On the contrary, a fraction of water molecules is excluded from the hydration shell of an ion during the formation of contact ion pairs (CIP) where the ion directly interacts with its counterion. Thus, the average number of affected water molecules is expected to decrease during CIP formation. The integrated intensity of the Cl^--affected water spectrum (OH stretch) linearly increases with concentrations up to 3 M NaCl while showing negative deviation above 3 M NaCl, indicating the formation of CIP between Na^+ and Cl^- (above 3M) [88]. On the contrary, in the cases of Mg^{2+}- and La^{3+}-salts, the predominant formation of SSIP was identified (data not shown here), in agreement with THz and dielectric relaxation spectroscopy studies [29,90]. X-ray absorption fine structure spectroscopy and neutron scattering investigations of divalent Ca^{2+} (chloride salt) solvation provided evidence for the predominant formation of Ca^{2+}-OH_2-Cl^- SSIPs compared to Ca^{2+}-Cl^- CIPs, even at high (~6 M) concentrations [91,92]. Presumably, the strong interaction (i.e., tight binding) of first hydration shell water with high-charge-density cations (e.g., Mg^{2+} and La^{3+}) promotes the preferential formation of SSIP, while weaker ion–water interaction favors the formation of CIP in NaCl solution.

Ion-pairing in aqueous solution has also been recently detected using Raman-MCR and IR-MCR analyses [93,94]. The analyses of the band shape (OH stretch) and intensity of the Raman-MCR extracted ion-perturbed water spectra of aqueous NaOH and LiOH solutions revealed that SSIPs are predominantly formed in these solutions, while CIPs only become significant at higher concentrations [93]. In fact, the strong H-bonding between OH^- and the first hydration shell water (HOH … OH^-) favors the predominant formation of SSIP. Moreover, the ion-pairing in aqueous NaOH and LiOH is enthalpically disfavored and entropically favored (i.e., positive ΔH and ΔS), implying that the thermodynamics of ion-pairing is dominated by ion–water interaction rather the electrostatic attraction between the oppositely charged ions.

8. Water in the Hydration Shell of Hydrophobic Molecular Cation ($(CH_3)_4N^+$)

Ions containing hydrophobic molecular groups exhibit distinct local effects on the neighboring water molecules. The hydration shell spectrum (OH stretch) of tetramethylammonium cation ($(CH_3)_4N^+$; TMA^+) closely resembles that of tert-butyl alcohol (TBA; an uncharged hydrophobe), showing "hydrophobic hydration" around the tetramethyl group (Figure 11). Water lacks favorable interactions with non-polar hydrophobes and is unable to form H-bonds. Accordingly, water molecules around a hydrophobe reorganize themselves to optimize the water–water interaction, resulting in increased average H-bond strength (i.e., the enhanced intensity ~3200 cm^{-1}; compare the green and black dashed curves of Figure 11a,b), along with some non-H-bonded "dangling OH" defects (small band ~3670 cm^{-1} in case of TBA hydration shell) [95,96]. Notably, the dangling OH defect of hydrophobic hydration in TMA^+ is largely suppressed due to the presence of a cationic charge. The ion–dipole interaction between the positively charged TMA^+ and hydration water preferentially orients the latter, pointing their hydrogens away from the TMA^+ surface, suppressing the propensity of dangling OH (3670 cm^{-1} band is absent in TMA^+ hydration shell).

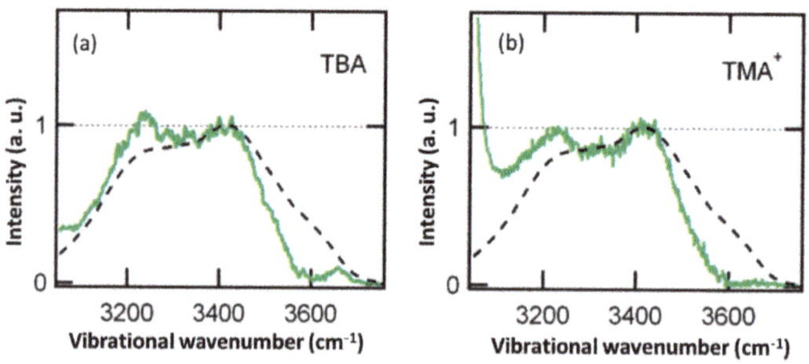

Figure 11. Hydration shell spectra (OH stretch; retrieved by Raman-MCR) of (**a**) TBA and (**b**) TMA^+ in aqueous solution (0.8 M). The spectrum of bulk water (black dashed curve) is shown in each panel for reference. (Reprinted with permission from ref. [97]. Copyright 2016 American Chemical Society).

9. Water in the Hydration Shell of Zwitterion, Trimethylamine N-Oxide ($(CH_3)_3N^+$-O^-)

In the case of zwitterions, such as trimethylamine N-oxide (TMAO), which contains a hydrophobic trimethyl and an zwitterionic N-oxide (N^+-O^-) group, shows different local hydration characteristics: hydrophobic hydration around the trimethyl group and hydrophilic (ionic) hydration of the N-oxide group (Figure 12) [97]. In addition to hydrophobic hydration, the TMAO–oxygen forms a stronger H-bond with the hydration water compared to that in bulk water (reflected by the markedly enhanced intensity ~3230 cm^{-1} in the TMAO hydration shell spectrum compared to both the TBA and the TMA^+ hydration shell; compare the green curves of Figures 11 and 12). In fact, a balance between these differential hydration characteristics (hydrophobic vs. ionic) of TMAO is believed to have implications in its interfacial properties and its interactions with a biological membrane/water interface [98,99]. Moreover, the hydration of such an amphoteric molecule is very sensitive to the local physicochemical environment. In aqueous solution, for example, TMAO experiences van der Waals attraction with TBA and hence, interacts through its trimethyl moiety. This is governed by their mutual hydrophobic interaction with water, which makes the trimethyl group of TMAO less exposed to water [100]. On the other hand, TMAO electrostatically interacts with cationic TMA^+ through its negatively charged oxygen, which orients the trimethyl group of TMAO away from TMA^+ and renders them exposed to water [100].

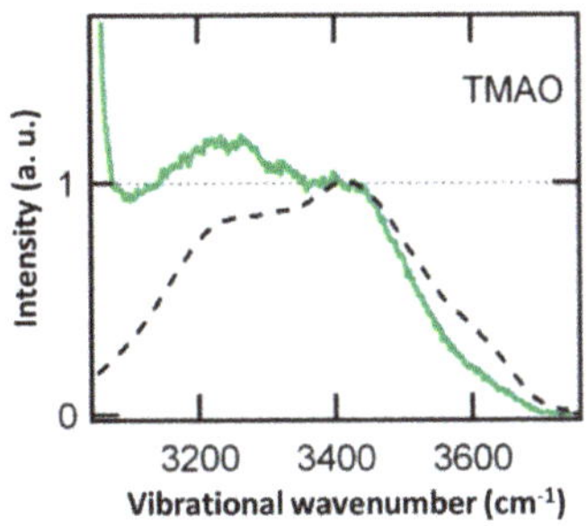

Figure 12. Hydration shell spectra (OH stretch; retrieved by Raman-MCR) of TMAO in 0.8 M aqueous solution. The spectrum of bulk water (black dashed curve) is shown for reference. (Reprinted with permission from ref. [97]. Copyright 2016 American Chemical Society).

10. Conclusions and Future Perspective

Anions and cations exert specific perturbation on the H-bonded structure of water. The strength of H-bonding and water libration are altered in the immediate vicinity of the ions, causing the decoupling of the intra- and intermolecular vibrational modes of water. The ion effect on water increases with increasing charge density for both anions as well as cations. Monovalent anions have relatively stronger influences on the hydration water compared to the monovalent cations with a similar charge density. Nevertheless, the multivalent cations strongly interact with water that not only increases the average H-bond strength of hydration water but also produces very weakly interacting water–OH beyond the first hydration shell. In the case of fully hydrated ions, the structure of hydration water becomes perturbed primarily by the individual hydrated ion, while at higher concentrations, interionic interaction between anions and cations starts to form ion-pairs (SSIP and CIP). In SSIP, the intervening shared water is perturbed by both ions, forming an ion-pair. In fact, the predominant formation of SSIP or CIP is driven by the specific nature of the individual ion–water interaction. Hydrophobic molecular ions exhibit distinct local hydration characteristics.

Ion-induced perturbation modifies coupling between vibrational modes of water. Anion–water interaction decouples the vibrational modes of the hydration water from its H-bonded partner. The vibrational coupling of water plays crucial roles in the biochemical reactions of enzymes [101,102]. Quantification of the vibrational coupling of perturbed water by metal ions with biochemical relevance is expected to provide a better understanding of such systems.

The majority of the IR and Raman spectroscopic investigations have used the OH stretch band of water as a marker of the ion effect on the water structure. Presumably, this is because of the high oscillator strength and the higher sensitivity of the OH stretch towards a structural perturbation of the water than that of its other vibrational modes. However, recent FTIR spectroscopic measurements indicated the potential of HOH bending and bend+libration combination band of water as powerful probes for solute-induced perturbations of a water H-bond network [103–105]. Moreover, probing low-frequency intermolecular H-bond stretch (~200 cm^{-1}) and librational modes can decipher the structural perturbation of water by ions (data not yet published). The low-frequency vibrational modes of water can be applied as complementary probes to the widely studied OH stretch mode in elucidating the structure of water.

One emerging trend is the combined investigation of ionic hydration in bulk as well as at aqueous interface. It is observed that the ion-induced perturbations of H-bonding structure in the bulk water and at the air/water interface are strongly correlated [56,60,64]. The local hydration characteristics of ions in bulk water dictate their surface affinity [56]. In fact, kosmotropic electrolytes perturb the structure of water at the air/water–electrolyte interface through their strongly solvated anion hydration shell [64]. DS-SCF analysis of the vibrational response from bulk water and the aqueous interface can provide a unified picture of a specific ion effect in bulk water and at the aqueous interfaces.

Funding: Bhabha Atomic Research Centre (BARC), Department of Atomic Energy, Government of India.

Acknowledgments: The authors gratefully acknowledge Awadhesh Kumar and A. K. Tyagi. N.G. thanks Alok Srivastava for his encouragement. S.R. and A.B. acknowledge DAE (HBNI, BARC) for Ph.D. fellowship. The financial support from BARC, Department of Atomic Energy, Government of India, is gratefully acknowledged.

Conflicts of Interest: The authors declare no conflict of interest.

References

1. Jacob, D. The role of water vapour in the atmosphere. A short overview from a climate modeller's point of view. *Phys. Chem. Earth Part A Solid Earth Geod.* **2001**, *26*, 523–527. [CrossRef]
2. Vaida, V. Perspective: Water cluster mediated atmospheric chemistry. *J. Chem. Phys.* **2011**, *135*, 020901. [CrossRef] [PubMed]
3. Wiggins, P.M. Role of water in some biological processes. *Microbiol. Rev.* **1990**, *54*, 432–449. [CrossRef]
4. Pal, S.K.; Zewail, A.H. Dynamics of Water in Biological Recognition. *Chem. Rev.* **2004**, *104*, 2099–2124. [CrossRef]
5. Walrafen, G.E. Raman Spectral Studies of the Effects of Solutes and Pressure on Water Structure. *J. Chem. Phys.* **1971**, *55*, 768–792. [CrossRef]
6. Marcus, Y. Effect of Ions on the Structure of Water: Structure Making and Breaking. *Chem. Rev.* **2009**, *109*, 1346–1370. [CrossRef]
7. Gaballa, G.A.; Neilson, G.W. The effect of pressure on the structure of light and heavy water. *Mol. Phys.* **1983**, *50*, 97–111. [CrossRef]
8. Walrafen, G.E. Raman Spectral Studies of the Effects of Temperature on Water and Electrolyte Solutions. *J. Chem. Phys.* **1966**, *44*, 1546–1558. [CrossRef]
9. Dudev, T.; Lim, C. Importance of Metal Hydration on the Selectivity of Mg^{2+} versus Ca^{2+} in Magnesium Ion Channels. *J. Am. Chem. Soc.* **2013**, *135*, 17200–17208. [CrossRef]
10. Cota, R.; Woutersen, S.; Bakker, H.J. Accelerated Vibrational Energy Relaxation of Water in Alkaline Environments. *J. Phys. Chem. B* **2021**, *125*, 11980–11986. [CrossRef]
11. Timmer, R.L.A.; Tielrooij, K.J.; Bakker, H.J. Vibrational Förster transfer to hydrated protons. *J. Chem. Phys.* **2010**, *132*, 194504. [CrossRef] [PubMed]
12. Wachter, W.; Kunz, W.; Buchner, R.; Hefter, G. Is There an Anionic Hofmeister Effect on Water Dynamics? Dielectric Spectroscopy of Aqueous Solutions of NaBr, NaI, $NaNO_3$, $NaClO_4$, and NaSCN. *J. Phys. Chem. A* **2005**, *109*, 8675–8683. [CrossRef] [PubMed]
13. Tielrooij, K.J.; Garcia-Araez, N.; Bonn, M.; Bakker, H.J. Cooperativity in Ion Hydration. *Science* **2010**, *328*, 1006–1009. [CrossRef] [PubMed]
14. Kim, J.S.; Wu, Z.; Morrow, A.R.; Yethiraj, A.; Yethiraj, A. Self-Diffusion and Viscosity in Electrolyte Solutions. *J. Phys. Chem. B* **2012**, *116*, 12007–12013. [CrossRef]
15. Cota, R.; van Dam, E.P.; Woutersen, S.; Bakker, H.J. Slowing Down of the Molecular Reorientation of Water in Concentrated Alkaline Solutions. *J. Phys. Chem. B* **2020**, *124*, 8309–8316. [CrossRef]
16. Collins, K.D.; Washabaugh, M.W. The Hofmeister effect and the behaviour of water at interfaces. *Q. Rev. Biophys.* **1985**, *18*, 323–422. [CrossRef]
17. Collins, K.D. Charge Density-Dependent Strength of Hydration and Biological Structure. *Biophys. J.* **1997**, *72*, 65–76. [CrossRef]
18. Collins, K.D. Why continuum electrostatics theories cannot explain biological structure, polyelectrolytes or ionic strength effects in ion–protein interactions. *Biophys. Chem.* **2012**, *167*, 43–59. [CrossRef]
19. Gurney, R.W. *Ionic Processes in Solution*; McGraw-Hill: New York, NY, USA, 1953.
20. Jones, G.; Dole, M. The viscosity of aqueous solutions of strong electrolytes with special reference to barium chloride. *J. Am. Chem. Soc.* **1929**, *51*, 2950–2964. [CrossRef]
21. Näslund, L.-Å.; Edwards, D.C.; Wernet, P.; Bergmann, U.; Ogasawara, H.; Pettersson, L.G.M.; Myneni, S.; Nilsson, A. X-ray Absorption Spectroscopy Study of the Hydrogen Bond Network in the Bulk Water of Aqueous Solutions. *J. Phys. Chem. A* **2005**, *109*, 5995–6002. [CrossRef]
22. Hofmesiter, F. About the science of the effects of salts: About the water withdrawing effect of the salts. *Arch. Exp. Pathol. Pharmacol.* **1888**, *24*, 247–260.

23. Mason, P.E.; Ansell, S.; Neilson, G.W.; Rempe, S.B. Neutron Scattering Studies of the Hydration Structure of Li^+. *J. Phys. Chem. B* **2015**, *119*, 2003–2009. [CrossRef] [PubMed]
24. Wang, Y.; Wang, G.; Bowron, D.T.; Zhu, F.; Hannon, A.C.; Zhou, Y.; Liu, X.; Shi, G. Unveiling the structure of aqueous magnesium nitrate solutions by combining X-ray diffraction and theoretical calculations. *Phys. Chem. Chem. Phys.* **2022**, *24*, 22939–22949. [CrossRef]
25. van der Vegt, N.F.A.; Haldrup, K.; Roke, S.; Zheng, J.; Lund, M.; Bakker, H.J. Water-Mediated Ion Pairing: Occurrence and Relevance. *Chem. Rev.* **2016**, *116*, 7626–7641. [CrossRef]
26. Ohtaki, H.; Radnai, T. Structure and dynamics of hydrated ions. *Chem. Rev.* **1993**, *93*, 1157–1204. [CrossRef]
27. Jalilehvand, F.; Spångberg, D.; Lindqvist-Reis, P.; Hermansson, K.; Persson, I.; Sandström, M. Hydration of the Calcium Ion. An EXAFS, Large-Angle X-ray Scattering, and Molecular Dynamics Simulation Study. *J. Am. Chem. Soc.* **2001**, *123*, 431–441. [CrossRef]
28. Schwaab, G.; Sebastiani, F.; Havenith, M. Ion Hydration and Ion Pairing as Probed by THz Spectroscopy. *Angew. Chem. Int. Ed.* **2019**, *58*, 3000–3013. [CrossRef]
29. Sebastiani, F.; Verde, A.V.; Heyden, M.; Schwaab, G.; Havenith, M. Cooperativity and ion pairing in magnesium sulfate aqueous solutions from the dilute regime to the solubility limit. *Phys. Chem. Chem. Phys.* **2020**, *22*, 12140–12153. [CrossRef]
30. Chen, Y.; Zhang, Y.-H.; Zhao, L.-J. ATR-FTIR spectroscopic studies on aqueous LiClO4, NaClO4, and $Mg(ClO_4)_2$ solutions. *Phys. Chem. Chem. Phys.* **2004**, *6*, 537–542. [CrossRef]
31. Max, J.-J.; Blois, S.d.; Veilleux, A.; Chapados, C. IR Spectroscopy of aqueous alkali halides. Factor analysis. *Can. J. Chem.* **2001**, *79*, 13–21. [CrossRef]
32. Yonehama, K.; Yoshimura, Y.; Takekiyo, T.; Kanno, H. Variation of the Uncoupled OD Stretching Frequency with Electrolyte Concentration in Aqueous Electrolyte Solutions. *Bull. Chem. Soc. Jpn.* **2009**, *82*, 563–569. [CrossRef]
33. Perera, P.; Wyche, M.; Loethen, Y.; Ben-Amotz, D. Solute-Induced Perturbations of Solvent-Shell Molecules Observed Using Multivariate Raman Curve Resolution. *J. Am. Chem. Soc.* **2008**, *130*, 4576–4577. [CrossRef] [PubMed]
34. Fega, K.R.; Wilcox, A.S.; Ben-Amotz, D. Application of Raman Multivariate Curve Resolution to Solvation-Shell Spectroscopy. *Appl. Spectrosc.* **2012**, *66*, 282–288. [CrossRef] [PubMed]
35. Rankin, B.M.; Ben-Amotz, D. Expulsion of Ions from Hydrophobic Hydration Shells. *J. Am. Chem. Soc.* **2013**, *135*, 8818–8821. [CrossRef] [PubMed]
36. Ahmed, M.; Namboodiri, V.; Singh, A.K.; Mondal, J.A.; Sarkar, S.K. How Ions Affect the Structure of Water: A Combined Raman Spectroscopy and Multivariate Curve Resolution Study. *J. Phys. Chem. B* **2013**, *117*, 16479–16485. [CrossRef]
37. Ahmed, M.; Singh, A.K.; Mondal, J.A.; Sarkar, S.K. Water in the Hydration Shell of Halide Ions Has Significantly Reduced Fermi Resonance and Moderately Enhanced Raman Cross Section in the OH Stretch Regions. *J. Phys. Chem. B* **2013**, *117*, 9728–9733. [CrossRef]
38. Ahmed, M.; Singh, A.K.; Mondal, J.A. Hydrogen-bonding and vibrational coupling of water in a hydrophobic hydration shell as observed by Raman-MCR and isotopic dilution spectroscopy. *Phys. Chem. Chem. Phys.* **2016**, *18*, 2767–2775. [CrossRef]
39. Ahmed, M.; Namboodiri, V.; Singh, A.K.; Mondal, J.A. On the Intermolecular Vibrational Coupling, Hydrogen Bonding, and Librational Freedom of Water in the Hydration Shell of Mono- and Bivalent Anions. *J. Chem. Phys.* **2014**, *141*, 164708. [CrossRef]
40. Ben-Amotz, D. Hydration-Shell Vibrational Spectroscopy. *J. Am. Chem. Soc.* **2019**, *141*, 10569–10580. [CrossRef]
41. Roy, S.; Biswas, B.; Ghosh, N.; Singh, P.C.; Mondal, J.A. Hydrophobic Hydration of Fluoroalkyl (C-F) is Distinctly Different from that of Its Hydrogenated Counterpart (C-H) as Observed by Raman Difference with Simultaneous Curve Fitting Analysis. *J. Phys. Chem. C* **2019**, *123*, 27012–27019. [CrossRef]
42. Patra, A.; Roy, S.; Saha, S.; Palit, D.K.; Mondal, J.A. Observation of Extremely Weakly Interacting OH (~3600 cm^{-1}) in the Vicinity of High Charge Density Metal Ions (M^{z+}; z = 1, 2, 3): A Structural Heterogeneity in the Extended Hydration Shell. *J. Phys. Chem. C* **2020**, *124*, 3028–3036. [CrossRef]
43. Daly, C.A.; Streacker, L.M.; Sun, Y.; Pattenaude, S.R.; Hassanali, A.A.; Petersen, P.B.; Corcelli, S.A.; Ben-Amotz, D. Decomposition of the Experimental Raman and Infrared Spectra of Acidic Water into Proton, Special Pair, and Counterion Contributions. *J. Phys. Chem. Lett.* **2017**, *8*, 5246–5252. [CrossRef] [PubMed]
44. Stangret, J.; Gampe, T. Ionic Hydration Behavior Derived from Infrared Spectra in HDO. *J. Phys. Chem. A* **2002**, *106*, 5393–5402. [CrossRef]
45. Perera, P.N.; Browder, B.; Ben-Amotz, D. Perturbations of Water by Alkali Halide Ions Measured using Multivariate Raman Curve Resolution. *J. Phys. Chem. B* **2009**, *113*, 1805–1809. [CrossRef]
46. Lindgren, J.; Hermansson, K.; Wojcik, M.J. Theoretical Simulation and Experimental-Determination of OH And OD Stretching Bands of Isotopically Diluted Hdo Molecules in Aqueous-Electrolyte Solutions. *J. Phys. Chem.* **1993**, *97*, 5254–5259. [CrossRef]
47. Kristiansson, O.; Lindgren, J.; De Villepin, J. A quantitative infrared spectroscopic method for the study of the hydration of ions in aqueous solutions. *J. Phys. Chem.* **1988**, *92*, 2680–2685. [CrossRef]
48. Śmiechowski, M.; Stangret, J. Vibrational spectroscopy of semiheavy water (HDO) as a probe of solute hydration. *Pure Appl. Chem.* **2010**, *82*, 1869–1887. [CrossRef]
49. Sokołowska, A.; Kęcki, Z. Inter- and intra-molecular coupling and Fermi resonance in the Raman spectra of liquid water. *J. Raman Spectrosc.* **1986**, *17*, 29–33. [CrossRef]

50. Auer, B.M.; Skinner, J.L. IR and Raman Spectra of Liquid Water: Theory and Interpretation. *J. Chem. Phys.* **2008**, *128*, 224511. [CrossRef]
51. Hare, D.E.; Sorensen, C.M. Interoscillator coupling effects on the OH stretching band of liquid water. *J. Chem. Phys.* **1992**, *96*, 13–22. [CrossRef]
52. Sovago, M.; Campen, R.K.; Wurpel, G.W.H.; Müller, M.; Bakker, H.J.; Bonn, M. Vibrational Response of Hydrogen-Bonded Interfacial Water is Dominated by Intramolecular Coupling. *Phys. Rev. Lett.* **2008**, *100*, 173901. [CrossRef] [PubMed]
53. Wang, Z.; Pakoulev, A.; Pang, Y.; Dlott, D.D. Vibrational Substructure in the OH Stretching Transition of Water and HOD. *J. Phys. Chem. A* **2004**, *108*, 9054–9063. [CrossRef]
54. Eaves, J.D.; Loparo, J.J.; Fecko, C.J.; Roberts, S.T.; Tokmakoff, A.; Geissler, P.L. Hydrogen bonds in liquid water are broken only fleetingly. *Proc. Natl. Acad. Sci. USA* **2005**, *102*, 13019–13022. [CrossRef] [PubMed]
55. Yang, M.; Skinner, J.L. Signatures of Coherent Vibrational Energy Transfer in IR and Raman Line Shapes for Liquid Water. *Phys. Chem. Chem. Phys.* **2010**, *12*, 982–991. [CrossRef]
56. Saha, S.; Roy, S.; Mathi, P.; Mondal, J.A. Polyatomic Iodine Species at the Air–Water Interface and Its Relevance to Atmospheric Iodine Chemistry: An HD-VSFG and Raman-MCR Study. *J. Phys. Chem. A* **2019**, *123*, 2924–2934. [CrossRef]
57. Jena, N.K.; Josefsson, I.; Eriksson, S.K.; Hagfeldt, A.; Siegbahn, H.; Björneholm, O.; Rensmo, H.; Odelius, M. Solvent-Dependent Structure of the I_3^- Ion Derived from Photoelectron Spectroscopy and Ab Initio Molecular Dynamics Simulations. *Chem.—Eur. J.* **2015**, *21*, 4049–4055. [CrossRef]
58. Josefsson, I.; Eriksson, S.K.; Ottosson, N.; Öhrwall, G.; Siegbahn, H.; Hagfeldt, A.; Rensmo, H.; Björneholm, O.; Odelius, M. Collective hydrogen-bond dynamics dictates the electronic structure of aqueous $I3^-$. *Phys. Chem. Chem. Phys.* **2013**, *15*, 20189–20196. [CrossRef]
59. Bergstroem, P.A.; Lindgren, J.; Kristiansson, O. An IR Study of the Hydration of Perchlorate, Nitrate, Iodide, Bromide, Chloride and Sulfate Anions in Aqueous Solution. *J. Phys. Chem.* **1991**, *95*, 8575–8580. [CrossRef]
60. Roy, S.; Mondal, J.A. "Breaking" and "Making" of Water Structure at the Air/Water–Electrolyte ($NaXO_3$; X = Cl, Br, I) Interface. *J. Phys. Chem. Lett.* **2021**, *12*, 1955–1960. [CrossRef]
61. Baer, M.D.; Pham, V.-T.; Fulton, J.L.; Schenter, G.K.; Balasubramanian, M.; Mundy, C.J. Is Iodate a Strongly Hydrated Cation? *J. Phys. Chem. Lett.* **2011**, *2*, 2650–2654. [CrossRef]
62. Sharma, B.; Chandra, A. Nature of hydration shells of a polyoxy-anion with a large cationic centre: The case of iodate ion in water. *J. Comput. Chem.* **2018**, *39*, 1226–1235. [CrossRef] [PubMed]
63. Sharma, B.; Chandra, A. Born–Oppenheimer Molecular Dynamics Simulations of a Bromate Ion in Water Reveal Its Dual Kosmotropic and Chaotropic Behavior. *J. Phys. Chem. B* **2018**, *122*, 2090–2101. [CrossRef] [PubMed]
64. Roy, S.; Mondal, J.A. Kosmotropic Electrolyte (Na_2CO_3, NaF) Perturbs the Air/Water Interface through Anion Hydration Shell without Forming a Well-Defined Electric Double Layer. *J. Phys. Chem. B* **2021**, *125*, 3977–3985. [CrossRef]
65. Boisson, J.; Stirnemann, G.; Laage, D.; Hynes, J.T. Water reorientation dynamics in the first hydration shells of F^- and I^-. *Phys. Chem. Chem. Phys.* **2011**, *13*, 19895–19901. [CrossRef] [PubMed]
66. Yadav, S.; Chandra, A. Structural and Dynamical Nature of Hydration Shells of the Carbonate Ion in Water: An Ab Initio Molecular Dynamics Study. *J. Phys. Chem. B* **2018**, *122*, 1495–1504. [CrossRef] [PubMed]
67. Kameda, Y.; Hosoya, K.; Sakamoto, S.; Suzuki, H.; Usuki, T.; Uemura, O. Hydrogen-bonded structure in aqueous sulfuric acid solutions. *J. Mol. Liq.* **1995**, *65–66*, 305–308. [CrossRef]
68. Rudolph, W.W.; Irmer, G. Raman and Infrared Spectroscopic Investigations on Aqueous Alkali Metal Phosphate Solutions and Density Functional Theory Calculations of Phosphate—Water Clusters. *Appl. Spectrosc.* **2007**, *61*, 1312–1327. [CrossRef]
69. Sharma, B.; Chandra, A. Ab Initio Molecular Dynamics Simulation of the Phosphate Ion in Water: Insights into Solvation Shell Structure, Dynamics, and Kosmotropic Activity. *J. Phys. Chem. B* **2017**, *121*, 10519–10529. [CrossRef]
70. Yu, C.-C.; Chiang, K.-Y.; Okuno, M.; Seki, T.; Ohto, T.; Yu, X.; Korepanov, V.; Hamaguchi, H.-O.; Bonn, M.; Hunger, J.; et al. Vibrational couplings and energy transfer pathways of water's bending mode. *Nat. Commun.* **2020**, *11*, 5977. [CrossRef]
71. Seki, T.; Chiang, K.-Y.; Yu, C.-C.; Yu, X.; Okuno, M.; Hunger, J.; Nagata, Y.; Bonn, M. The Bending Mode of Water: A Powerful Probe for Hydrogen Bond Structure of Aqueous Systems. *J. Phys. Chem. Lett.* **2020**, *11*, 8459–8469. [CrossRef]
72. Walrafen, G.E. Raman Spectral Studies of the Effects of Temperature on Water Structure. *J. Chem. Phys.* **1967**, *47*, 114–126. [CrossRef]
73. Cho, M.; Fleming, G.R.; Saito, S.; Ohmine, I.; Stratt, R.M. Instantaneous normal mode analysis of liquid water. *J. Chem. Phys.* **1994**, *100*, 6672–6683. [CrossRef]
74. Nash, C.P.; Donnelly, T.C.; Rock, P.A. Raman spectra in the libration region of concentrated aqueous solutions of lithium-6 and lithium-7 halides. Evidence for tetrahedral $Li(OH_2)^{4+}$. *J. Solut. Chem.* **1977**, *6*, 663–670. [CrossRef]
75. Carey, D.M.; Korenowski, G.M. Measurement of the Raman spectrum of liquid water. *J. Chem. Phys.* **1998**, *108*, 2669–2675. [CrossRef]
76. Cappa, C.D.; Smith, J.D.; Messer, B.M.; Cohen, R.C.; Saykally, R.J. Effects of Cations on the Hydrogen Bond Network of Liquid Water: New Results from X-ray Absorption Spectroscopy of Liquid Microjets. *J. Phys. Chem. B* **2006**, *110*, 5301–5309. [CrossRef] [PubMed]

77. Bergstroem, P.A.; Lindgren, J. Infrared study on the hydration of manganese (2+), iron (2+), cobalt (2+), lanthanum (3+), neodymium (3+), dysprosium (3+), and ytterbium (3+) ions in dilute aqueous solution. *Inorg. Chem.* **1992**, *31*, 1529–1533. [CrossRef]
78. Rudolph, W.W.; Irmer, G. Raman spectroscopic characterization of light rare earth ions: La^{3+}, Ce^{3+}, Pr^{3+}, Nd^{3+} and Sm^{3+}–hydration and ion pair formation. *Dalton Trans.* **2017**, *46*, 4235–4244. [CrossRef]
79. Rudolph, W.; Irmer, G. On the Hydration of Heavy Rare Earth Ions: Ho^{3+}, Er^{3+}, Tm^{3+}, Yb^{3+} and Lu^{3+}—A Raman Study. *Molecules* **2019**, *24*, 1953. [CrossRef]
80. Das, B.; Mondal, S.; Chandra, A. Two-Dimensional Infrared Spectroscopy of Aqueous Solutions of Metal Nitrates: Slowdown of Spectral Diffusion in the Presence of Divalent Cations. *J. Phys. Chem. B* **2020**, *124*, 7391–7404. [CrossRef]
81. Zeng, Y.; Jia, Y.; Yan, T.; Zhuang, W. Binary structure and dynamics of the hydrogen bonds in the hydration shells of ions. *Phys. Chem. Chem. Phys.* **2021**, *23*, 11400–11410. [CrossRef]
82. Habenschuss, A.; Spedding, F.H. The coordination (hydration) of rare earth ions in aqueous chloride solutions from x-ray diffraction. I.$TbCl_3$, $DyCl_3$, $ErCl_3$, $TmCl_3$, and $LuCl_3$. *J. Chem. Phys.* **1979**, *70*, 2797–2806. [CrossRef]
83. Habenschuss, A.; Spedding, F.H. The coordination (hydration) of rare earth ions in aqueous chloride solutions from x-ray diffraction. II. $LaCl_3$, $PrCl_3$, and $NdCl_3$a). *J. Chem. Phys.* **1979**, *70*, 3758–3763. [CrossRef]
84. Duvail, M.; Vitorge, P.; Spezia, R. Building a polarizable pair interaction potential for lanthanoids(III) in liquid water: A molecular dynamics study of structure and dynamics of the whole series. *J. Chem. Phys.* **2009**, *130*, 104501. [CrossRef] [PubMed]
85. Verde, A.V.; Santer, M.; Lipowsky, R. Solvent-shared pairs of densely charged ions induce intense but short-range supra-additive slowdown of water rotation. *Phys. Chem. Chem. Phys.* **2016**, *18*, 1918–1930. [CrossRef]
86. Yadav, S.; Chandra, A. Preferential solvation, ion pairing, and dynamics of concentrated aqueous solutions of divalent metal nitrate salts. *J. Chem. Phys.* **2017**, *147*, 244503. [CrossRef]
87. Marcus, Y.; Hefter, G. Ion Pairing. *Chem. Rev.* **2006**, *106*, 4585–4621. [CrossRef]
88. Roy, S.; Patra, A.; Saha, S.; Palit, D.K.; Mondal, J.A. Restructuring of Hydration Shell Water due to Solvent-Shared Ion Pairing (SSIP): A Case Study of Aqueous $MgCl_2$ and $LaCl_3$ Solutions. *J. Phys. Chem. B* **2020**, *124*, 8141–8148. [CrossRef]
89. Baumler, S.M.; Hartt, V.W.H.; Allen, H.C. Hydration of ferric chloride and nitrate in aqueous solutions: Water-mediated ion pairing revealed by Raman spectroscopy. *Phys. Chem. Chem. Phys.* **2019**, *21*, 19172–19180. [CrossRef]
90. Friesen, S.; Hefter, G.; Buchner, R. Cation Hydration and Ion Pairing in Aqueous Solutions of $MgCl_2$ and $CaCl_2$. *J. Phys. Chem. B* **2019**, *123*, 891–900. [CrossRef]
91. Fulton, J.L.; Heald, S.M.; Badyal, Y.S.; Simonson, J.M. Understanding the Effects of Concentration on the Solvation Structure of Ca^{2+} in Aqueous Solution. I: The Perspective on Local Structure from EXAFS and XANES. *J. Phys. Chem. A* **2003**, *107*, 4688–4696. [CrossRef]
92. Badyal, Y.S.; Barnes, A.C.; Cuello, G.J.; Simonson, J.M. Understanding the Effects of Concentration on the Solvation Structure of Ca^{2+} in Aqueous Solution. II: Insights into Longer Range Order from Neutron Diffraction Isotope Substitution. *J. Phys. Chem. A* **2004**, *108*, 11819–11827. [CrossRef]
93. de Oliveira, D.M.; Bredt, A.J.; Miller, T.C.; Corcelli, S.A.; Ben-Amotz, D. Spectroscopic and Structural Characterization of Water-Shared Ion-Pairs in Aqueous Sodium and Lithium Hydroxide. *J. Phys. Chem. B* **2021**, *125*, 1439–1446. [CrossRef] [PubMed]
94. Drexler, C.I.; Miller, T.C.; Rogers, B.A.; Li, Y.C.; Daly, C.A.; Yang, T.; Corcelli, S.A.; Cremer, P.S. Counter Cations Affect Transport in Aqueous Hydroxide Solutions with Ion Specificity. *J. Am. Chem. Soc.* **2019**, *141*, 6930–6936. [CrossRef] [PubMed]
95. Perera, P.N.; Fega, K.R.; Lawrence, C.; Sundstrom, E.J.; Tomlinson-Phillips, J.; Ben-Amotz, D. Observation of water dangling OH bonds around dissolved nonpolar groups. *Proc. Natl. Acad. Sci. USA* **2009**, *106*, 12230–12234. [CrossRef] [PubMed]
96. Davis, J.G.; Gierszal, K.P.; Wang, P.; Ben-Amotz, D. Water Structural Transformation at Molecular Hydrophobic Interfaces. *Nature* **2012**, *491*, 582–585. [CrossRef]
97. Ahmed, M.; Namboodiri, V.; Mathi, P.; Singh, A.K.; Mondal, J.A. How Osmolyte and Denaturant Affect Water at the Air–Water Interface and in Bulk: A Heterodyne-Detected Vibrational Sum Frequency Generation (HD-VSFG) and Hydration Shell Spectroscopic Study. *J. Phys. Chem. C* **2016**, *120*, 10252–10260. [CrossRef]
98. Mondal, J.A. Effect of Trimethylamine N-Oxide on Interfacial Electrostatics at Phospholipid Monolayer–Water Interfaces and Its Relevance to Cardiovascular Disease. *J. Phys. Chem. Lett.* **2016**, *7*, 1704–1708. [CrossRef]
99. Mondal, J.A. Metabolite-Affected Interfacial Electrostatics and Its Role in the Pathogenesis of Cardiovascular Disease: An Interface-Selective Vibrational Spectroscopic Study. *J. Phys. Chem. C* **2016**, *120*, 21642–21651. [CrossRef]
100. Roy, S.; Patra, A.; Palit, D.K.; Mondal, J.A. Interaction of Zwitterionic Osmolyte Trimethylamine N-oxide (TMAO) with Molecular Hydrophobes: An Interplay of Hydrophobic and Electrostatic Interactions. *J. Phys. Chem. B* **2021**, *125*, 10939–10946. [CrossRef]
101. Salazar-Salinas, K.; Baldera-Aguayo, P.A.; Encomendero-Risco, J.J.; Orihuela, M.; Sheen, P.; Seminario, J.M.; Zimic, M. Metal-Ion Effects on the Polarization of Metal-Bound Water and Infrared Vibrational Modes of the Coordinated Metal Center of Mycobacterium tuberculosis Pyrazinamidase via Quantum Mechanical Calculations. *J. Phys. Chem. B* **2014**, *118*, 10065–10075. [CrossRef]
102. Vergauwe, R.M.A.; Thomas, A.; Nagarajan, K.; Shalabney, A.; George, J.; Chervy, T.; Seidel, M.; Devaux, E.; Torbeev, V.; Ebbesen, T.W. Modification of Enzyme Activity by Vibrational Strong Coupling of Water. *Angew. Chem. Int. Ed.* **2019**, *58*, 15324–15328. [CrossRef] [PubMed]

103. Deng, G.-H.; Shen, Y.; Chen, H.; Chen, Y.; Jiang, B.; Wu, G.; Yang, X.; Yuan, K.; Zheng, J. Ordered-to-Disordered Transformation of Enhanced Water Structure on Hydrophobic Surfaces in Concentrated Alcohol–Water Solutions. *J. Phys. Chem. Lett.* **2019**, *10*, 7922–7928. [CrossRef] [PubMed]
104. Verma, P.K.; Kundu, A.; Puretz, M.S.; Dhoonmoon, C.; Chegwidden, O.S.; Londergan, C.H.; Cho, M. The Bend+Libration Combination Band Is an Intrinsic, Collective, and Strongly Solute-Dependent Reporter on the Hydrogen Bonding Network of Liquid Water. *J. Phys. Chem. B* **2018**, *122*, 2587–2599. [CrossRef] [PubMed]
105. Shattuck, J.; Shah, P.; Erramilli, S.; Ziegler, L.D. Structure Making and Breaking Effects of Cations in Aqueous Solution: Nitrous Oxide Pump–Probe Measurements. *J. Phys. Chem. B* **2016**, *120*, 10569–10580. [CrossRef]

liquids

MDPI

Article

Solute-Induced Perturbation of the Solvent Microstructure in Aqueous Electrolyte Solutions: Some Uses and Misuses of Structure Making/Breaking Criteria

Ariel A. Chialvo [1,*] and Oscar D. Crisalle [2]

[1] Independent Researcher, Knoxville, TN 37922-3108, USA
[2] Department of Chemical Engineering, University of Florida, Gainesville, FL 32611-6005, USA; crisalle@che.ufl.edu
* Correspondence: ovlaich@gmail.com

Abstract: In this article, we raise awareness about the misuses of frequently invoked criteria for structure making/breaking phenomena, resulting from the absence of any explicit cause–effect relationship between the proposed markers and the microstructural perturbation of the solvent induced by the solute. First, we support our assessment with rigorous molecular-based foundations to determine, directly and quantitatively, the solute-induced perturbation of the solvent structure leading to an unambiguous definition of a structure making/breaking event. Then, we highlight and discuss the sources of concealed ambiguities in two of the most frequently invoked structure making/breaking criteria, i.e., Hepler's thermal expansivity-based and Jones–Dole's *B* coefficient-based markers. Finally, we illustrate how the implementation of rigorous molecular-based arguments, in conjunction with the available experimental evidence on a variety of aqueous species at infinite dilution, rule out the validity of these two criteria as structure making/breaking markers and suggest their discontinuation to avoid the perpetuation of myths.

Keywords: aqueous electrolyte solutions; solute-induced microstructural perturbation; structure making/breaking events; solution non-idealities; solute–solvent intermolecular asymmetries; Krichevskii parameter; Hepler's criterion; Jones–Dole's *B* coefficient-based criterion

1. Introduction

The solvation of a solute in a solvent can be interpreted as the solute-induced (distortion) effect on the original microstructure of the solvent environment leading to the formation of the solvation structure. This distortion can be microstructurally described as a local perturbation of the solvent density originating in the molecular asymmetry between the strength of the original solvent–solvent and the ensuing solute–solvent interactions, thus leading macroscopically to diverse patterns of thermodynamic nonideality [1–3]. Unsurprisingly, researchers have invested significant effort toward the extraction of structural information of electrolyte and nonelectrolyte solutions from scattering experiments, such as X-ray spectroscopy (e.g., EXAFS and XANES) [4–7], neutron scattering with isotope substitution (NDIS) [8–12], hybrid methods involving empirical microstructural refinement of scattering data [13–19], and the interplay between NDIS and molecular simulation [20–26] as means to interpret the underlying links between the intermolecular interaction asymmetries and the resulting thermodynamic behavior.

Alternatively, researchers have used surrogate experimental techniques to study the solvent local environment around a solute in search for connections between the observed macroscopic (thermodynamic) behavior and the microscopic responses detected by different probes leading to the interpretation of the effect of the solute on the microstructure of the solution. However, these experimental tools either provide a rather limited (i.e., short-ranged and/or orientational) view of the solute effects, involve solvent-specific (e.g., hydrogen-bonding) approaches, or rely on other types of markers to infer a plausible (yet undefined) microscopic-to-macroscopic link. Two prominent examples of these alternative approaches are the focus of our critical analysis as we introduce them below, i.e., the markers associated with the behavior of the *B*–coefficient in Jones–Dole's equation [27] and the isobaric-thermal expansivity of the solute in solution [28].

On the one hand, soon after Jones and Dole [27] introduced the empirical expression to describe the concentration dependence of the relative viscosity (η/η_α^o) of dilute solutions in terms of the shear viscosity of the pure α–solvent η_α^o and the molar concentration c_β of the β–solute in solution, i.e.,

$$(\eta/\eta_\alpha^o) = 1 + Ac_\beta^{0.5} + Bc_\beta \tag{1}$$

followed by the theoretical interpretation of its positive definite *A*–coefficient [29,30], researchers were eager to find an explanation for the behavior of the *B*–coefficient, i.e., the origin of $B \lesseqgtr 0$ in particular, but not exclusively [31–35], for aqueous ionic solutes. From the early work of Cox and Wolfenden [36], who hypothesized that $B < 0$ could be traced back to "depolymerization" of water as a manifestation of the "rise of structural temperature", according to the theory of Bernal and Fowler available at that time [37], the *B*–coefficient was considered a measure of the solute–solvent interactions. In fact, for the case of aqueous ionic solutes, Wolfenden and collaborators [36,38] advanced the idea of additivity of the anion and cation contributions to the *B*–coefficient of the resulting aqueous salt and, despite the arbitrary nature of the anion–cation division of contributions, the concept was immediately adopted [39–44].

The above effort was likely the first attempt to link the sign of the *B*–coefficient of Jones–Dole's equation to some type of solvent structural motif as a means to explain the experimental observations. In fact, Bernal and Fowler conjectured that the ions affected their solvation water by either loosening or tightening its structure, e.g., a looser structure arising from a weakly solvated ions giving rise to a decrease of the relative viscosity while a tighter structure manifested as a stronger ion solvation translating into an increase of the relative viscosity of the aqueous solution. These ideas were further pursued by Frank and collaborators [45,46], who analyzed the phenomenon according to measures of free volume and "structural entropy" of the solution, and then interpreted through the introduction of the labels "structure-breaking" and "structure-promoting" species.

Furthermore, Tsangaris and Martin [47] introduced an alternative viscosity-based structure making/breaking criterion based on the combinations of signs for the *B*–coefficient and its temperature derivative $(\partial B/\partial T)_P$ by suggesting that "*the sign of* $(\partial B/\partial T)_P$ *appears to be a more straightforward indicator of structure-breaking or -making ability than sign or size of the B-coefficient*" and that "*a positive* $(\partial B/\partial T)_P$ *indicates a structure-breaking ion or molecule, and a negative sign, a structure making one*" without any explicit microstructural evidence to support it. Unfortunately, this criterion exacerbates the confusion as we have recently argued [48], given that the resulting four-sign combinations, comprising eight possible structure making/breaking outcomes, are devoid of any cause–effect relationship with the actual microstructural perturbation of the solvent structure around the solute.

On the other hand, early undertakings toward the understanding of the thermodynamic behavior of aqueous solutions were driven by the work of Frank and Evans [46], who proposed the formation of quasi-crystalline hydrogen-bonded structures (icebergs) around highly dilute aqueous nonpolar solutes in order to explain the observed endothermic negative entropy of solution of gases. In other words, this negative entropy of solution was interpreted as an increase of order in the aqueous environment around the solute, or equivalently, a promotor of water structure given that entropy was understood (at that time) as a measure of order. Following these ideas, and supplemented by the experimental evidence of the larger isobaric-thermal expansivity of heavy water over that of light water, Hepler [28] suggested that the thermal expansivity behavior of water was consistent with Frank and Evans' structural model of water, i.e., temperature and pressure increases break down "*structure and cause water to approach "normal" behavior*". Consequently, Hepler promoted its extension to dilute solutions as a means to interpret the structure-making/breaking behavior of solutes in solution.

In pursuit of this issue, Hepler invoked a Maxwell relation involving the partial molar volumes of the solute at infinite dilution, $\left(\partial \hat{c}^{\infty}_{P_\beta}/\partial P\right)_T = -T\left(\partial^2 \hat{v}^{\infty}_{\beta}/\partial T^2\right)_P$, to keep the analysis at the level of solute–solvent interactions, and assumed that $\left(\partial \hat{c}^{\infty}_{P_\beta}/\partial P\right)_T > 0$ and, consequently, $\left(\partial^2 \hat{v}^{\infty}_{\beta}/\partial T^2\right)_P < 0$, for a structure-breaking solute according to experimental evidence from some hand-picked aqueous systems. As surprising as it might be, and despite the evident lack of explicit microstructural support for the advocated structural marker compounded by the significant ambiguity of the supporting experimental evidence, the thermal expansivity criterion for structure-making/breaking species was adopted and has survived until today [49,50]. Unfortunately, the survival of this criterion is not the result of its accurate predictions, but perhaps for two other factors: its straightforward implementation based on the availability of volumetric data and, more importantly, the absence (until recently) of any rigorous (theoretical) development exposing its invalidity.

Immediately, we can identify a few commonalities between the attempts to link either the solute's B–coefficient or the isobaric-thermal expansivity to a structure-making/breaking behavior, including, (i) the chosen systems have been hand-picked to provide a self-consistent outcome by ignoring or overlooking systems not conforming to the preconceived connections, (ii) the structure-making/breaking process, i.e., the microstructural signature of the solvent perturbation around the solute, is vaguely described through a variety of narratives rather than by an explicit microscopic-to-macroscopic unambiguous connection, (iii) the lack of effort to make a direct connection between the actual microstructural behavior of the system and the physical property whose measurement will provide the information to decide the solute's structure-making/breaking ability, and consequently, (iv) the loss of interest in testing the adequacy and/or accuracy of those criteria by confronting their outcomes against precisely-defined model systems for which the answers are precisely known.

In light of this reality, the main goal of this article is to raise awareness of the large community of researchers in physical chemistry and solution chemistry to what can and cannot be construed from two popular and frequently invoked criteria for structure making/breaking phenomena, i.e., Hepler's thermal expansivity formula [28] and the behavior of Jones–Dole's B–coefficient [41,47]. More specifically, we would like to emphasize and demonstrate that (i) neither one of the above criteria provides any explicit link between the microstructural perturbation of the solvent caused by the presence of the solute; thus, they suffer from a lack of cause–effect connections between the actual microstructural perturbation and the proposed structure making/breaking marker; (ii) neither criterion can predict the correct solute-induced structural perturbation for the two simplest systems describing either the largest or the smallest solute–solvent intermolecular interaction asymmetry, i.e., those involving the ideal gas β–solute in a real α–solvent, and the real β–solute behaving as an α–solvent molecule, systems for which we know precisely the structure making/breaking behavior; and (iii) the macroscopic nature of the above criteria, compounded by the lack of any explicit relationship with the evolution of the solvent microstructure, preclude their reliable use as structure making/breaking markers.

Our presentation is organized as follows: in Section 2, we provide a rigorous molecular-based foundation for the quantitative determination of a solute-induced perturbation of the solvent structure leading to an unambiguous definition of a structure making/breaking event. Then, in Section 3, we highlight the sources of ambiguities of the criteria based on Hepler's isobaric-thermal expansivity and B–coefficient markers as descriptors of structure making/breaking events. In Section 4, we illustrate our arguments with experimental evidence for a variety of aqueous electrolyte species at infinite dilution. Finally, we provide some additional thoughts in Section 5.

2. Fundamentals Underlying the Description of the Microstructural Perturbation of the Solvent Environment upon Solute Solvation

Our first priority is to provide an unambiguous and precise description of what constitutes a microstructural perturbation of the solvent environment caused by the introduction of a solute species at the prevailing state conditions of the solution, and simultaneously, be able to link rigorously the observed microscopic behavior to the corresponding macroscopic manifestations which will ultimately be the object of the experimental measurements [49]. For that purpose, below, we lay out briefly the main ideas behind the statistical mechanics-based definition of the structure-making/breaking function and discuss the attributes of such a descriptor, including its universality.

2.1. What Does Really Mean That a Solute Strengthen/Weaken the Structure of the Solvent?

To address this issue, we invoke the Kirkwood–Buff (KB) fluctuation formalism of mixtures [51] as the statistical mechanical framework able to describe unambiguously the behavior of the system microstructure as volume integrals over pair correlation functions, and connect them to the system's macroscopic chemical and mechanical partial molar properties. In addition to being a rigorous formalism, KB imposes no restrictions to either the nature and type of the intermolecular interaction asymmetry, or the number of components in the system.

The key player here is the KB's total correlation function integral (TCFI), $G_{\beta\alpha}(T,P,x_\beta)$, defined as follows,

$$G_{\beta\alpha}(T,P,x_\beta) \equiv 4\pi \int_0^\infty h_{\beta\alpha}(r) r^2 dr \tag{2}$$

where x_β denotes the mole fraction of the β–species, while the radial correlation function $h_{\beta\alpha}(r) = g_{\beta\alpha}(r) - 1$ for the $\beta\alpha$–type interactions is defined in terms of the radial distribution functions $g_{\beta\alpha}(r)$ and the ideal gas (IG) uniform distribution $g_{\beta\alpha}^{IG}(r) = 1$ counterpart. Note that the α–species density ($\rho_\alpha = \rho x_\alpha$) weighted Equation (2), where $x_\alpha = 1 - x_\beta$, i.e.,

$$\begin{aligned} \mathcal{N}_{\beta\alpha}^{res}(T,P,x_\beta) &= 4\pi\rho x_\alpha \textstyle\int_0^\infty \left[g_{\beta\alpha}(r) - 1\right] r^2 dr \\ &= \mathcal{N}_{\beta\alpha}(T,P,x_\beta) - \mathcal{N}_\alpha(T,P,x_\beta) \end{aligned} \tag{3}$$

provides the first rigorous measure of the excess (or its negative counerpart also known as deficit) in the average number of α–molecules around any β–molecule relative to that when the α–molecule were uniformily distributed in the system, where $\mathcal{N}_\alpha(T,P,x_\beta) = \rho(1 - x_\beta)V$ and V denotes the volume of the system. Because $\mathcal{N}_{\beta\alpha}(T,P,x_\beta)$ describes the absolute average number of α–species around any β–species at the prevailing (T,P,x_β)–conditions, then Equation (3) defines the isobaric-isothermal residual $\mathcal{N}_{\beta\alpha}^{res}(T,P,x_\beta)$ counterpart of $\mathcal{N}_{\beta\alpha}(T,P,x_\beta)$, i.e., the effect of the intermolecular interactions on the average number of α–molecules around any β–molecule when the system goes from an ideal gas mixture to the actual mixture of interest.

According to the expressions (2) and (3), we can now focus on the following related quantity,

$$\begin{aligned} \mathcal{N}_{\beta\alpha}^{ex}(T,P,x_\beta) &\equiv \rho x_\alpha \left(G_{\beta\alpha} - G_{\alpha\alpha}\right)_{TPx} \\ &= \left(\mathcal{N}_{\beta\alpha}^{res} - \mathcal{N}_{\alpha\alpha}^{res}\right)_{TPx} \end{aligned} \tag{4}$$

which, according to the physical meaning of the involved TCFI, $\rho x_\alpha G_{\beta\alpha} = \mathcal{N}_{\beta\alpha}^{res}$, the quantity $\mathcal{N}_{\beta\alpha}^{ex}(T,P,x_\beta)$ represents the average number of α–molecules around any β–solute in excess/deficit to that around any α–molecule. Therefore, $\mathcal{N}_{\beta\alpha}^{ex}(T,P,x_\beta)$ becomes a versatile, unambiguous, and direct descriptor of the magnitude of the β–molecule induced-perturbation of the surrounding α–species environment, resulting from the intermolecular asymmetry between the $\alpha\alpha$– and the $\beta\alpha$–intermolecular interactions.

After identifying the α–species as the solvent and the β–species as the solute, immediately, Equation (4) indicates that, as the intermolecular asymmetry between the $\alpha\alpha$– and the

$\beta\alpha$–type interactions vanishes (i.e., the solute behavior becomes identical to that of the solvent), the excess quantity $\mathcal{N}^{ex}_{\beta\alpha}(T,P,x_\beta) \to 0$, as we might expect. However, when the intermolecular asymmetry between the $\alpha\alpha$– and the $\beta\alpha$–type interactions translates into stronger $\beta\alpha$– than $\alpha\alpha$–type interactions, then $G_{\beta\alpha}(T,P,x_\beta) > G_{\alpha\alpha}(T,P,x_\beta)$ and $\mathcal{N}^{ex}_{\beta\alpha}(T,P,x_\beta) > 0$ in Equation (4), which is the manifestation of a strengthening or enhancement of the α–solvent environment around the β–solute. Conversely, when $G_{\beta\alpha}(T,P,x_\beta) < G_{\alpha\alpha}(T,P,x_\beta)$ and $\mathcal{N}^{ex}_{\beta\alpha}(T,P,x_\beta) < 0$ in Equation (4), which is the manifestation of a weakening or depletion of the α–solvent environment around the β–solute. Moreover, we should highlight that an outstanding attribute of Equation (4) is its direct cause–effect connection between the microstructural changes of the mixture and its thermodynamic behavior, a feature that makes it powerful for the linking of the solute–solvent intermolecular interaction asymmetries of the mixture of interest, its microscopic manifestation, and the resulting macroscopic patterns of thermodynamic non-ideality as discussed below.

2.2. Need to Provide an Explicit Definition/Criterion for the Structure Making/Breaking Ability of a Solute Species

To decide whether a β–solute, forming a solution with an α–solvent, behaves as a structure making or breaking species, we must first set the molecular-based "meter" criterion that applies equally to any system, regardless of either the type or the magnitude of solute–solvent intermolecular interaction asymmetry [49,52]. In other words, it should apply equally to systems exhibiting either the smallest (i.e., neither making nor breaking such as a solute in a Lewis–Randall ideal solution) [2], the largest (i.e., an ideal gas solute in any real solvent) [3], or any magnitude of solute–solvent molecular interaction asymmetry in between regardless of how these interactions are microscopically described (e.g., hydrogen bonding, electrostatic, multipole, inductive, etc.) [53]. Moreover, the structure making/breaking definition must predict precisely the same answer regardless of the (experimental, theoretical, simulation) probe used in its implementation.

Given that the typical structure making/breaking experimental scenario involves solutes in solutions at infinite dilution, we proceed our analysis of binary systems comprising infinitely dilute solutes at isobaric-isothermal conditions, and define the following structural parameter $\mathcal{S}^{\infty}_{\beta\alpha}(T,P)$ as follows,

$$\begin{aligned} \mathcal{S}^{\infty}_{\beta\alpha}(T,P) &\equiv \lim_{x_\beta \to 0} \mathcal{N}^{ex}_{\beta\alpha}(T,P,x_\beta) \\ &= \rho^o_\alpha \left(G^{\infty}_{\beta\alpha} - G^o_{\alpha\alpha} \right)_{TP} \end{aligned} \tag{5}$$

so that the magnitude of $\mathcal{S}^{\infty}_{\beta\alpha}(T,P)$ quantifies rigorously and unambiguously how different the microstructure of the α–solvent around the β–solute becomes, relative to that around the α–solvent itself, while the sign of $\mathcal{S}^{\infty}_{\beta\alpha}(T,P)$ qualifies the behavior of the β–solute according to the three possible structural outcomes of the TCFI-difference $\left(G^{\infty}_{\beta\alpha} - G^o_{\alpha\alpha} \right)_{TP}$ as follows,

$$\left(G^{\infty}_{\beta\alpha} - G^o_{\alpha\alpha} \right)_{TP} \to \begin{cases} G^{\infty}_{\beta\alpha} > G^o_{\alpha\alpha} \to \mathcal{S}^{\infty}_{\beta\alpha} > 0 \to \textit{structure–maker} \\ G^{\infty}_{\beta\alpha} \cong G^o_{\alpha\alpha} \to \mathcal{S}^{\infty}_{\beta\alpha} \cong 0 \to \textit{neithermaker–norbreaker} \\ G^{\infty}_{\beta\alpha} < G^o_{\alpha\alpha} \to \mathcal{S}^{\infty}_{\beta\alpha} < 0 \to \textit{structure–breaker} \end{cases} \tag{6}$$

where the scheme embodied in Equations (5) and (6) provides a one-to-one correspondence between the actual solute-induced perturbation of the solvent microstructure and a meaningful, as well as experimentally accessible, structure-making/breaking function. In fact, by invoking the partial molar volumetric counterpart for the expressions for the TCFI-difference $\left(G^{\infty}_{\beta\alpha} - G^o_{\alpha\alpha} \right)_{TP}$, we immediately find the thermodynamic counterpart to Equation (5) that allows a straightforward experimental determination of $\mathcal{S}^{\infty}_{\beta\alpha}(T,P)$, i.e.,

$$\mathcal{S}^{\infty}_{\beta\alpha}(T,P) = 1 - \left(\bar{v}^{\infty}_{\beta} / \nu v^o_\alpha \right)_{TP} \tag{7}$$

where $\hat{v}_{\beta}^{\infty}(T,P)$ and $v_{\alpha}^{o}(T,P)$ describe the partial molar volumes of the β–solute at infinite dilution, $\hat{v}_{\beta}^{\infty}(T,P) = \nu\left(v_j^o + G_{jj}^o - G_{ij}^{\infty}\right)$ with the dissociative stoichiometric coefficient ν, and the corresponding to the pure α–solvent at the prevailing state conditions. Note that, because the ion–solvent TCFI's obey the identity $\left(G_{a\alpha}^{\infty} = G_{c\alpha}^{\infty} = G_{\beta\alpha}^{\infty}\right)_{TP}$ [49], the solute induced perturbation of the solvent microstructure is the same for either dissociated species, i.e., $\left(\mathcal{S}_{\beta\alpha}^{\infty} = \mathcal{S}_{a\alpha}^{\infty} = \mathcal{S}_{c\alpha}^{\infty}\right)_{TP}$ where subscripts a and c identify the anion and the cation species, respectively.

At this point, we can shed additional light onto the Kirkwood–Buff integrals involved in the definition of $\mathcal{S}_{\beta\alpha}^{\infty}(T,P)$, Equation (5), and the resulting microstructural scenarios described by Equation (6). We have already introduced the microscopic (statistical mechanical) definition of the Kirkwood–Buff integral for solvent–solvent interactions, $G_{\alpha\alpha}^{o}(T,P)$ [51], yet, after we invoke the solvation behavior of an ideal gas β–solute [54], we can show that $G_{\alpha\alpha}^{o}(T,P)$ embodies the following distinctive macroscopic meaning,

$$G_{\alpha\alpha}^{o} = \left(\nu^{-1}\hat{v}_{\beta}^{\infty,IG\_\beta} - v_{\alpha}^{o}\right)_{TP} \tag{8}$$

as the change of partial molar volume of a α–solvent molecule in the process of becoming an ideal gas β–solute at infinite dilution in an environment characterized by the prevailing (T,P) state conditions, where $\hat{v}_{\beta}^{IG\_\beta}(T,P) = \nu kT\kappa_{\alpha}^{o}$ with $\nu = 1$ for non-dissociative solutes [54]. Likewise, we can provide another distinctive macroscopic meaning to the Kirkwood–Buff integral of the solute–solvent interactions, $G_{\beta\alpha}^{\infty}(T,P)$, i.e.,

$$G_{\beta\alpha}^{\infty}(T,P) = \nu^{-1}\left(\hat{v}_{\beta}^{\infty,IG\_\beta} - \hat{v}_{\beta}^{\infty}\right)_{TP} \tag{9}$$

as the change of partial molar volume of the real β–solute at infinite dilution in a real α–solvent when the solute–solvent interactions vanish and the species becomes an ideal gas β–solute.

Alternatively, we could resort to the isothermal-isochoric rate of change of pressure, $(\partial P/\partial x_{\beta})_{T\rho}^{\infty}$, induced by the β–solute within the environment of pure α–solvent [55], a quantity that plays a crucial role in the understanding of solubility in highly-compressible solvent [2,56] whose finite critical value defines Krichevskii's parameter [57]. Therefore,

$$\mathcal{S}_{\beta\alpha}^{\infty}(T,P) = -\nu^{-1}\kappa_{\alpha}^{o}(\partial P/\partial x_{\beta})_{T\rho}^{\infty} \tag{10}$$

where κ_{α}^{o} denotes the isothermal compressibility of the pure α–solvent. In fact, the sign of $(\partial P/\partial x_{\beta})_{T\rho}^{\infty}$ has been pivotal in the characterization of solutes in near-critical solvents, so that according to Equation (10), a β–solute behaves as a *structure-maker* when $(\partial P/\partial x_{\beta})_{T\rho}^{\infty} < 0$, and the solute is depicted as non-volatile [58] or attractive [1]. Conversely, a β–solute behaves as a *structure-breaker* when $(\partial P/\partial x_{\beta})_{T\rho}^{\infty} > 0$, and the solute is described as volatile [58] or weakly-attractive and repulsive [1] in the jargon of supercritical fluid solutions [2,59].

Incidentally, Equation (10) predicts the divergence of $\mathcal{S}_{\beta\alpha}^{\infty}(T,P)$ as the state conditions of the pure solvent approach criticality, with the sign of $(\partial P/\partial x_{\beta})_{T\rho}^{\infty}$, resulting from the divergent behavior of the isothermal compressibility of the solvent $\kappa_{\alpha}^{o}(T,P)$. Moreover, while the typical structure making/breaking analysis involves state conditions where $\kappa_{\alpha}^{o}(T,P) \lesssim 10^{-5}/P$ (MPa), many novel chemical processes take place in highly compressible media [60–62]. Under these conditions, it becomes advantageous to avoid dealing with divergent quantities, while still capturing the structure making/breaking perturbing effect of the solute. In fact, from the fundamental expression given by Equation (10), we can split $\mathcal{S}_{\beta\alpha}^{\infty}(T,P)$ into its solvation (i.e., short-range local density perturbation, SR) contribution while isolating its diverging (i.e., long-range or compressibility driven, LR) contribution associated with the propagation of the density perturbation as follows [55],

$$\mathcal{S}^{\infty}_{\beta\alpha}(T,P) = \underbrace{-v^{-1}\kappa_{\alpha}^{o,IG}(\partial P/\partial x_{\beta})^{\infty}_{T\rho}}_{S_{\beta\alpha}(SR)} \underbrace{-v^{-1}\kappa_{\alpha}^{o,R}(\partial P/\partial x_{\beta})^{\infty}_{T\rho}}_{S_{\beta\alpha}(LR)} \quad (11)$$

In Equation (11), we identify $\kappa_{\alpha}^{o,IG} = (\rho_{\alpha}^{o}kT)^{-1}$ as the ideal gas compressibility at the prevailing state conditions, and $\kappa_{\alpha}^{o,R}(T,P) = \kappa_{\alpha}^{o} - \kappa_{\alpha}^{o,IG}$ as the corresponding isobaric-isothermal residual isothermal compressibility. Therefore, from Equations (10) and (11), we immediately find the desired explicit expression for the solvation finite contribution,

$$\mathcal{S}^{\infty}_{\beta\alpha}(SR) = \left(\kappa_{\alpha}^{o,IG}\Big/\kappa_{\alpha}^{o}\right)\mathcal{S}^{\infty}_{\beta\alpha} \quad (12)$$

whose divergent compressibility-driven contribution becomes

$$\mathcal{S}^{\infty}_{\beta\alpha}(LR) = \left(\kappa_{\alpha}^{o,R}\Big/\kappa_{\alpha}^{o}\right)\mathcal{S}^{\infty}_{\beta\alpha} \quad (13)$$

Note also that the isothermal-isochoric rate of change of pressure $(\partial P/\partial x_{\beta})^{\infty}_{T\rho}$ can also be written as $(\partial P/\partial x_{\beta})^{\infty}_{T\rho} = \nu\rho_{\alpha}^{o}\left(C^{o}_{\alpha\alpha} - C^{\infty}_{\beta\alpha}\right)\Big/\kappa_{\alpha}^{o,IG}$, where $C^{\oplus}_{ij}(T,P) \equiv 4\pi\int_0^{\infty} c^{\oplus}_{ij}(r)r^2dr$ defines the direct correlation function integral (DCFI) counterpart of Equation (2) for the ij–type of interactions at the prevailing (T,P) state conditions and composition, i.e., $\oplus = o$ for the pure component and $\oplus = \infty$ for the infinite dilution [2]. Thus, after invoking the following macroscopic interpretation for the TCFI [53],

$$C^{o}_{\alpha\alpha}(T,P) = v^{o}_{\alpha}\left[1 - \left(\nu v^{o}_{\alpha}\Big/\hat{v}_{\beta}^{\infty,IG\_\beta}\right)\right] \quad (14)$$

$$C^{\infty}_{\beta\alpha}(T,P) = v^{o}_{\alpha}\left[1 - \left(\hat{v}^{\infty}_{\beta}\Big/\hat{v}_{\beta}^{\infty,IG\_\beta}\right)\right] \quad (15)$$

which are the counterparts of the TCFI's given by Equations (8) and (9), we obtain

$$\begin{aligned}(\partial P/\partial x_{\beta})^{\infty}_{T\rho} &= \nu kT\rho^{o}_{\alpha}\left(\hat{v}^{\infty}_{\beta} - \nu v^{o}_{\alpha}\right)\Big/\hat{v}_{\beta}^{\infty,IG\_\beta} \\ &= \nu\left(\hat{v}^{\infty}_{\beta} - \nu v^{o}_{\alpha}\right)\Big/\left(\hat{v}_{\beta}^{\infty,IG\_\beta}\kappa_{\alpha}^{o,IG}\right)\end{aligned} \quad (16)$$

$$\begin{aligned}\mathcal{S}^{\infty}_{\beta}(SR) &= -\left(\hat{v}^{\infty}_{\beta} - \nu v^{o}_{\alpha}\right)\Big/\hat{v}_{\beta}^{\infty,IG\_\beta} \\ &= -\left(\hat{v}^{\infty}_{\beta} - \nu v^{o}_{\alpha}\right)\Big/(\nu kT\kappa^{o}_{\alpha})\end{aligned} \quad (17)$$

leading straightforwardly to

$$\begin{aligned}\mathcal{S}^{\infty}_{\beta\alpha}(LR) &= \left(\hat{v}^{\infty}_{\beta} - \nu v^{o}_{\alpha}\right)\Big/\hat{v}_{\beta}^{\infty,IG\_\beta} - \left(\hat{v}^{\infty}_{\beta} - \nu v^{o}_{\alpha}\right)\Big/\nu v^{o}_{\alpha} \\ &= \left(\hat{v}^{\infty}_{\beta} - \nu v^{o}_{\alpha}\right)\left(\left[1\Big/\hat{v}_{\beta}^{\infty,IG\_\beta}\right] - \left[1/\nu v^{o}_{\alpha}\right]\right)\end{aligned} \quad (18)$$

so that

$$\begin{aligned}\mathcal{S}^{\infty}_{\beta\alpha}(LR) &= (kT\rho^{o}_{\alpha}\kappa^{o}_{\alpha} - 1)\mathcal{S}^{\infty}_{\beta\alpha}(SR) \\ &= -\mathcal{S}_{\beta\alpha}^{\infty,IG\_\beta}\mathcal{S}^{\infty}_{\beta\alpha}(SR)\end{aligned} \quad (19)$$

Equation (19) tells us that the long-range contribution to the structure parameter of any real solute, $\mathcal{S}^{\infty}_{\beta\alpha}(LR)$, becomes proportional to its short-range counterpart $\mathcal{S}^{\infty}_{\beta\alpha}(SR)$ through the negative value of the structure parameter of the ideal gas β–solute at the prevailing state conditions, $\mathcal{S}_{\beta\alpha}^{\infty,IG\_\beta}(T,P)$. Consequently,

$$\begin{aligned} \mathcal{S}_{\beta\alpha}^{\infty}(T,P) &= \mathcal{S}_{\beta\alpha}^{\infty}(SR) + \mathcal{S}_{\beta\alpha}^{\infty}(LR) \\ &= \left(1 - \mathcal{S}_{\beta\alpha}^{\infty,IG\_\beta}\right)\mathcal{S}_{\beta\alpha}^{\infty}(SR) \\ &= kT\kappa_{\alpha}^{o}\rho_{\alpha}^{o}\mathcal{S}_{\beta\alpha}^{\infty}(SR) \end{aligned} \tag{20}$$

an outcome that confirms the contention that, even for a highly compressible solvent environment, the structure making/breaking behavior of a β–solute at infinite dilution is still defined by its short range (solvation) contribution. In other words, the isothermal compressibility of the α–solvent only magnifies its magnitude by the positive defined factor, $(kT\kappa_{\alpha}^{o}\rho_{\alpha}^{o})$, at the prevailing state conditions.

We can also identify a rigorous connection between $\mathcal{S}_{\beta\alpha}^{\infty}(SR)$ and the corresponding Krichevskii parameter given their common microstructural origin. In fact, from Equation (11) and the definition $\lim_{T,\rho_{\alpha}^{o}\to critical}(\partial P/\partial x_{\beta})_{T\rho}^{\infty} \equiv \mathcal{A}_{Kr}$, follows immediately that

$$\mathcal{A}_{Kr} = -\nu \lim_{T,\rho_{\alpha}^{o}\to critical}\left(\mathcal{S}_{\beta\alpha}^{\infty}\Big/\kappa_{\alpha}^{o}\right) \tag{21}$$

and consequently,

$$\mathcal{A}_{Kr} = -\nu\mathcal{A}_{Kr}^{IG\_\beta} \lim_{T,\rho_{\alpha}^{o}\to critical}\mathcal{S}_{\beta\alpha}^{\infty}(SR) \tag{22}$$

In summary, Equations (5)–(9) provide the sought-after rigorous microscopic-to-macroscopic connection that grants an unambiguous description of a β–solute' propensity to perturb, i.e., to *structure-make/break* the α–solvent environment around β–solute at any state conditions, and leads to the determination of its magnitude based on experimentally-available thermodynamic data, regardless of either the type of solute, solvent, or the nature of the intermolecular interactions. Moreover, it allows for the prediction of the structural response to changes in state conditions and composition of the system, based only on the knowledge of the partial molar volumetric behavior of the species and their T– or P–derivatives at the original (T,P,x_{β})–conditions [50].

In fact, we have shown that $\mathcal{S}_{\beta\alpha}^{\infty}(T,P)$ applies equaly to sub-, near-, and super-critical state conditions of the pure α–solvent, where the divergent $\mathcal{S}_{\beta\alpha}^{\infty}(T,P)$ becomes unambiguously and rigorous described (Equations (11)–(20)) by its short-ranged and finite (solvation) $\mathcal{S}_{\beta\alpha}^{\infty}(SR)$ counterpart. Most importantly, the $\mathcal{S}_{\beta\alpha}^{\infty}(T,P)$, or $\mathcal{S}_{\beta\alpha}(T,P,x_{\beta})$ for that matter, leads to the rigorous description of the thermodynamic non-ideality behavior of the mixture resulting from the solute–solvent intermolecular asymmetries [3,63]. For instance, we have derived the explicit connections between the structure parameter $\mathcal{S}_{\beta\alpha}^{\infty}(T,P)$ with the solute–solvent intermolecular asymmetry described by $\Delta_{\beta\alpha}^{\infty}(T,P) \equiv G_{\alpha\alpha}^{o} + G_{\beta\beta}^{\infty} - 2G_{\beta\alpha}^{\infty}$ and its associated limiting composition slope of the solute activity coefficient $\left(\partial \ln \gamma_{\beta}^{LR}\Big/\partial x_{\beta}\right)_{TP}^{\infty}$, the Krichevskii parameter $\mathcal{A}_{Kr} = \lim_{T,\rho_{\alpha}^{o}\to crit}\left(-\mathcal{S}_{\beta\alpha}^{\infty}\Big/\kappa_{\alpha}^{o}\right)$, the osmotic second virial coefficients associated with the composition perturbation expansion of the solute chemical potential along four distinctive thermodynamic paths, i.e., $B_{\beta}^{*}(T,\mu_{\alpha})$, $B_{\beta}'(T,P)$, $B_{\beta}''(T,\rho_{\alpha})$, and $B_{\beta}^{\sigma}(T_{\sigma})$, and the resulting patterns of thermodynamic non-ideality behavior [3,49,50,53,64]. Finally, note that for non-dissociative solutes, we simply need to set $\nu = 1$ in the corresponding expressions.

3. Critical Analysis of the Ambiguity of Two Widespread Structure Making/Breaking Markers

The major shortcoming behind the most frequently invoked structure-making/breaking markers is the absolute lack of precision and explicit characterization of what constitutes a structure-making/breaking event, and consequently, how this event might connect to the solute–solvent intermolecular interaction asymmetry of the solution, how it manifests macroscopically in terms of thermodynamic quantities, and how it could be probed/measured experimentally. The solute–solvent interaction asymmetry for a given β–solute in solution with an α–solvent could exhibit a significantly wide span, ranging from that of a non-interacting ideal gas species (i.e., the largest) to that of a species behaving identically to the α–solvent (i.e., the absolute smallest) [3], and finally, to either a weakly or a strongly interacting β–solute, regardless of how we describe the like- and unlike-pair interactions [2].

It becomes immediately obvious that any criterion for the analysis of the *structure making/breaking* process must be able the describe the transition between the structure-making and the structure-breaking perturbations through the crossing of the null-effect boundary, i.e., the condition described by an unperturbed solvent microstructure in the presence of a solute at the prevailing state conditions. To address these issues according to the theoretical developments of Section 2, we discuss below the inadequacy, and consequent failure, of the two most frequently invoked structure making/breaking markers. In particular, we identify and highlight their inability to predict the correct behavior of two precisely-defined model systems, involving the largest and the smallest solute–solvent intermolecular asymmetries, for which we know the exact answer.

3.1. Why Hepler's Isobaric-Thermal Expansivity Criterion Cannot Describe a Structure Making/Breaking Event?

As mentioned in the Introduction, the structure making/breaking ability of an aqueous species has been frequently interpreted in terms of the sign of the isothermal-pressure dependence of its partial molar heat capacity $\left(\partial \hat{c}^{\infty}_{P_\beta} / \partial P\right)_T$, or its more easily accessible Maxwell related expression $-\left(\partial^2 \hat{v}^{\infty}_{\beta} / \partial T^2\right)_P$, as suggested by Hepler [28], who proposed the criterion based on the conjecture that "*since increasing pressure would also break up the bulky aggregates, the same reasoning suggests that the heat capacity of pure water should decrease with increasing pressure*" this behavior could be extended to species in solution at infinite dilution. The fact that there is no cause–effect relation between either $\left(\partial \hat{c}^{\infty}_{P_\beta} / \partial P\right)_T$ or $-T\left(\partial^2 \hat{v}^{\infty}_{\beta} / \partial T^2\right)_P$ and the magnitude (and/or sign) of the solute-induced perturbation of the solvent microstructure [49] makes futile any attempt to interpret the structure making/breaking events from either volumetry or pressure perturbation calorimetry (PPC) [65].

As we have discussed previously [49,50], Hepler's criterion fails two fundamental requirements stemming from its inability to describe the structure making/breaking behavior of systems comprising either: (a) a β–solute species behaving identically to an α–solvent species resulting in an unperturbed solvent microstructure, or (b) the solvation of a non-interacting ideal gas β–solute in a real α–solvent, which represents the largest possible perturbation of the solvent microstructure. Consequently, Hepler's criterion cannot account for the structure making-to-structure breaking transition with changes in the state conditions of the system.

Beyond the questionable assumptions underlying this criterion [66], we can immediately: (i) demonstrate the failure of this thermal expansivity-based marker to predict the correct structure making/breaking behavior for the ideal gas β–solute, IG_β, in a real α–solvent, and that for the β–solute when behaving identically as the real α–solvent, as well as (ii) identify the isobaric-volumetric behavior of the pure α–solvent that would be required by the expansivity-based criterion to be obeyed.

On the one hand, for the largest solute–solvent intermolecular asymmetry, we have the IG_β solute, which according to Equation (7) and the identity $\hat{v}_{\beta}^{\infty,IG\_\beta}(T,P) = \nu kT\kappa_{\alpha}^{o}$ [54, 67], leads to the exact form for its structure parameter [49]

$$S_{\beta\alpha}^{\infty,IG\_\beta}(T,P) = (1 - kT\rho_{\alpha}^{o}\kappa_{\alpha}^{o}) \tag{23}$$

whose behavior leads to the following scheme,

$$S_{\beta\alpha}^{\infty,IG\_\beta}(T,P)\begin{cases} > 0 \rightarrow structure\ maker \rightarrow \kappa_{\beta}^{IG\_\beta} > \kappa_{\alpha}^{o} \\ = 0 \rightarrow unperturbed\ structure \rightarrow \kappa_{\beta}^{IG\_\beta} = \kappa_{\alpha}^{o} \\ < 0 \rightarrow structure\ breaker \rightarrow \kappa_{\beta}^{IG\_\beta} < \kappa_{\alpha}^{o} \end{cases} \tag{24}$$

The essential points from Equation (24) are three-fold: (i) the IG_β in a real α–solvent at ambient conditions will exhibit a structure-making behavior, (ii) the solvent microstructure stays unperturbed along the states where the solvent behaves as an ideal gas fluid, satisfying the $\kappa_{\alpha}^{o} = [kT\rho_{\alpha}^{o}]^{-1}$ condition, and (iii) after crossing this boundary, the ideal gas solute becomes a structure-breaker. Note also that, according to Equation (10), and the fact that $\lim_{P,T\rightarrow critical} (\partial P/\partial x_{\beta})_{T\rho}^{\infty,IG\_\beta} = \rho_{\alpha,c}^{o}kT_c$ [56], $S_{\beta\alpha}^{\infty,IG\_\beta}(T_c,P_c) \rightarrow -\infty$, indicating that the critical conditions of the α–solvent is located on the left side of the curve representing $S_{\beta\alpha}^{\infty,IG\_\beta}(T,P) = 0$ or its equivalent $\left(\kappa_{\alpha}^{o} = \kappa_{\beta}^{IG\_\beta}\right)_{T\rho_{\alpha}^{o}}$, as illustrated in Figures 8 and 9 in ref. [3] for the solvent water and the Lennard–Jones fluid, respectively.

The issue of interest here concerns the inability of the isobaric-thermal expansivity-based criterion to predict the transition across the curve $S_{\beta\alpha}^{\infty,IG\_\beta}(T,P) = 0$, as described by $-T\left(\partial^2\hat{v}_{\beta}^{\infty,IG\_\beta}/\partial T^2\right)_P = 0$ in Hepler's terms. In fact, we find that $\left(\partial^2\hat{v}_{\beta}^{\infty,IG\_\beta}/\partial T^2\right)_P^{\kappa_{\alpha}^{o}=\kappa_{\beta}^{IG\_\beta}} \neq 0$, in other words, Hepler's criterion cannot describe the structure making-to-breaking transition for the ideal gas solute IG_β in a real α–solvent along the $\left(\kappa_{\alpha}^{o} = \kappa_{\beta}^{IG\_\beta}\right)$–line given that

$$\left(\partial^2\hat{v}_{\beta}^{\infty,IG\_\beta}/\partial T^2\right)_P = k\left[2(\partial\kappa_{\alpha}^{o}/\partial T)_P + T\left(\partial^2\kappa_{\alpha}^{o}/\partial T^2\right)\right]^{\kappa_{\alpha}^{o}=\kappa_{\beta}^{IG\_\beta}} \neq 0 \tag{25}$$

after considering that $\left(\partial\kappa_{\alpha}^{o,IG}/\partial T\right)_P^{S_{\alpha\beta}^{\infty,IG\_\beta}=0} = \left(\partial^2\kappa_{\alpha}^{o,IG}/\partial T^2\right)_P^{S_{\alpha\beta}^{\infty,IG\_\beta}=0} = 0$ with $(\partial\kappa_{\alpha}^{o}/\partial T)_P^{S_{\alpha\beta}^{\infty,IG\_\beta}=0} \neq 0$ and $(\partial^2\kappa_{\alpha}^{o}/\partial T^2)_P^{S_{\alpha\beta}^{\infty,IG\_\beta}=0} \neq 0$.

On the other hand, for the smallest solute–solvent intermolecular asymmetry as characterized by the special case of a Lewis–Randall ideal solution, $LR - IS$, and regardless of the definition of structure making/breaking we choose, the outcome must be a null microstructural perturbation depicted here by $S_{\beta\alpha}^{\infty,LR-IS}(T,P) = 0$ in the schematic Equation (6). While the physical representation for the null solute–solvent intermolecular asymmetry is characterized by $\left(G_{\alpha\alpha}^{o} + G_{\beta\beta}^{\infty} - 2G_{\beta\alpha}^{\infty}\right)_{TP} = 0$ [68,69], its simplest and most advantageous case occurs when the Kirkwood–Buff integrals obey $\left(G_{\alpha\alpha}^{o} = G_{\beta\beta}^{\infty} = G_{\beta\alpha}^{\infty} \neq 0\right)_{TP}$ condition, not only resulting in the microstructural signature of a pure α–solvent [67], but also leading to the prototypical fingerprint for the microstructural transition between structure-making and structure-breaking processes driven by the solute–solvent intermolecular interaction asymmetry. This general and rigorous condition for the existence of a boundary between structure-making and structure breaking scenarios affords another opportunity to test the validity of the thermal expansivity-based criterion to predict this transition, which in Hepler's terms becomes described by $\left(\partial^2\hat{v}_{\beta}^{\infty}/\partial T^2\right)_P = (\partial^2 v_{\alpha}^{o}/\partial T^2)_P$.

Typically, pure fluids at normal (around ambient) conditions exhibit $T\left(\partial^2 v^o_\alpha/\partial T^2\right)_P > 0$, suggesting that for these fluids to describe the boundary $\left(\partial^2 v^o_\alpha/\partial T^2\right)_P = 0$ line, they would require meeting at least two rather unusual properties: an isobaric thermal expansivity $\beta^o_{T\alpha}$ independent of the state conditions, i.e., $\left(\partial \ln v^o_\alpha/\partial T\right)_P \neq \mathcal{F}(T,P)$, and a linear temperature-dependent isothermal compressibility, i.e., $\kappa^o_\alpha(T,P) \equiv \mathcal{G}(P) + \mathcal{K}(P)T$. We are not aware of any pure solvent following this highly unlikely type of behavior; therefore, we must conclude that Hepler's criterion cannot possibly describe (let alone predict) the structure-making to structure-breaking transition events involving common solvents (especially water). In other words, the conjectured Hepler's criterion cannot account for the solute-induced perturbation of the solvent microstructure of infinitely dilute solution comprising extreme solute–solvent intermolecular asymmetries, i.e., either the smallest (zero, for the $LR-IS$) or the largest (for the IG_β solute).

3.2. Why the Behavior of the Jones–Dole's B-Coefficient Cannot Be Taken as a Structure Making/Breaking Marker?

The Jones–Dole's equation for the relative (to pure solvent) viscosity of the dilute solution is represented by Equation (1); for the current analysis, it becomes advantageous to recast it as follows,

$$\begin{aligned} \eta_r(T,P,c_\beta) - 1 &= Ac_\beta^{0.5} + Bc_\beta \text{ as } c_\beta \to 0 \\ &\cong \ln \eta_r(T,P,c_\beta) \end{aligned} \tag{26}$$

where $\eta_r = \eta/\eta^o_j$ with $\eta^o_\alpha \equiv \eta(c_\beta = 0)$, c_β denotes the molar concentration of the β–solute, while the coefficients A and B account for the direct ion–ion and the (ion) solute–solvent interactions, respectively [27]. The Jones–Dole's equation has been empirically built, with an intentionally introduced [70] $c_\beta^{0.5}$–composition dependence to obey the Debye–Hückel limiting behavior when dealing with electrolyte solutes, provides a theoretical interpretation for the A–coefficient [30,71], and predicts $A > 0$ for all electrolytes [43,72], while $A = 0$ for non-electrolyte solutes.

If we accepted the Jones–Dole's equation to be an accurate representation of the shear viscosity of a dilute solution, and because the A–coefficient has been derived around the Debye–Hückel limiting behavior, the B–coefficient should also follow from the corresponding isothermal-isobaric composition limiting slope, i.e.,

$$\begin{aligned} B &= \lim_{c_\beta \to 0}\left[\left(\eta_r - 1 - Ac_\beta^{0.5}\right)/c_\beta\right] \\ &\cong \lim_{c_\beta \to 0}\left[\left(\ln \eta_r - Ac_\beta^{0.5}\right)/c_\beta\right] \end{aligned} \tag{27}$$

where the experimental evidence indicates that $B \gtrless 0$. Moreover, because the B–coefficient becomes the pre-factor behind the linear concentration dependence in the Jones–Dole's equation, this coefficient has been associated with the solvent-mediated solute–solute interactions, and consequently, with a conjectured structure marker, namely: $B > 0$ for a structure-maker species and $B < 0$ for a structure-breaker species [40–42,46], later supplemented with the alternative isobaric-temperature derivative $\left(\partial B/\partial T\right)_P$ [47].

The common feature between the two viscosity-based structure making/breaking markers mentioned above is the absolute lack of any explicit cause–effect link between either the B–coefficient, or its temperature derivative $\left(\partial B/\partial T\right)_P$, and the actual solute-induced perturbation of the solvent microstructure [48]. For that reason, we would like to provide a few observations and draw plausible connections between the experimental evidence on the behavior of the Jones–Dole's B–coefficient and the precisely-defined structure making/breaking parameter $S^\infty_{\beta\alpha}(T,P)$. Given the empirical nature of Jones–Dole's equation, we need to find whether or not the B–coefficient contains any embedded microstructural information, and also to connect this microstructural information to the alleged structure making/breaking makers based on the signs of both B–coefficient [40,42,46,73–82] and its isobaric-temperature derivative $(\partial B/\partial T)_P$ [47,83–100].

One plausible connection between the experimental evidence of the Jones–Dole's B–coefficient and a precisely-defined microstructural behavior of the β–solute at infinite dilution can be drawn by following Feakins et al.'s [101] transition state (TS)-based interpretation of the B–coefficient. In fact, Feakins et al. invoked the transition state theory [102] to derive an expression for the B–coefficient of a β–solute at infinite dilution in an α–solvent as follows (highlights of its derivation are given in Appendix A below),

$$B = \beta v_\alpha^o \left(\Delta\mu_\beta^{\natural,\infty} - \nu \Delta\mu_\alpha^{\natural,o} \right) - \left(\hat{v}_\beta^\infty - \nu v_\alpha^o \right) \tag{28}$$

with $\nu = \nu^+ + \nu^-$ for an electrolyte, i.e., $\nu = 1$ for non-dissociative solutes, where $\Delta\mu_\beta^{\natural,\oplus}$ and $\hat{v}_\beta^\oplus$ denote the molar Gibbs free energy of activation of viscous flow and the corresponding partial molar volume for the β–species, respectively, at the $\oplus$–composition condition, i.e., either an infinite dilution or a pure component.

Immediately, we find that the transition-state interpretation of the B–coefficient, as described by Equation (28), comprises two contributions, i.e., the $\left(\hat{v}_\beta^\infty - \nu v_\alpha^o \right)$ term as the solute-induced volumetric effect on the solvent structure, and its activation Gibbs free energy counterpart for the viscous flow, $\left(\Delta\mu_\beta^{\natural,\infty} - \nu \Delta\mu_\alpha^{\natural,o} \right)$. Obviously, the volumetric $\left(\hat{v}_\beta^\infty - \nu v_\alpha^o \right)$ term provides a direct link to the structure making/breaking parameter $\mathcal{S}_{\beta\alpha}^\infty(T,P)$ as discussed in Section 2.2, i.e.,

$$B = v_\alpha^o \left[\nu \mathcal{S}_{\beta\alpha}^\infty + \beta \left(\Delta\mu_\beta^{\natural,\infty} - \nu \Delta\mu_\alpha^{\natural,o} \right) \right] \tag{29}$$

which suggests that, under Feakins et al.'s framework, the B–coefficient comprises some information about the solute-induced effect on the solvent microstructure. However, even if we assumed the reliability of the Jones–Dole's representation for the composition dependent relative viscosity of the solution, we cannot in principle take the sign of either the B–coefficient or its temperature-derivative $(\partial B/\partial T)_P$ counterpart as a marker of the structure making/breaking nature of the dilute β–solute because the B–coefficient, and consequently $(\partial B/\partial T)_P$, involve the $\left(\Delta\mu_\beta^{\natural,\infty} - \nu \Delta\mu_\alpha^{\natural,o} \right)$ term that also contributes to the sought sign.

To support and illustrate this contention, we resort again to the system involving an ideal gas β–solute (IG_β) in a real α–solvent, $\Delta\mu_\beta^{\natural,\infty,IG\_\beta} = 0$ [102] and $\mathcal{S}_{\beta\alpha}^{\infty,IG\_\beta} = (1 - kT\rho_\alpha^o \kappa_\alpha^o)$ from Equation (23), so that Equation (29) becomes

$$\begin{aligned} B^{IG\_\beta} &= v_\alpha^o \left(\mathcal{S}_{\beta\alpha}^{\infty,IG\_\beta} - \beta\Delta\mu_\alpha^{\natural,o} \right) \\ &= v_\alpha^o \left[1 - \beta\Delta\mu_\alpha^{\natural,o} - \left(\kappa_\alpha^o / \kappa_\alpha^{IG} \right) \right] \end{aligned} \tag{30}$$

In this equation, we have $\beta\Delta\mu_\alpha^{\natural,o} = \ln(\eta_\alpha^o v_\alpha^o / \boldsymbol{h}\mathcal{N})$ [102], where η_α^o and v_α^o denote the shear viscosity and molar volume of the pure α–solvent, while $\boldsymbol{h}$ and $\mathcal{N}$ identify Planck and Avogadro constants, respectively. According to Equation (28), the B^{IG_β}–coefficient depends exclusively on the thermodynamic properties of the pure α–solvent at the prevailing state conditions. Therefore, we can explore the phase diagram of the α–solvent to find the $\rho_\alpha^o - T$ conditions where $\mathcal{S}_{\beta\alpha}^{\infty,IG\_\beta}(T,\rho_\alpha^o)$, and simultaneously $B^{IG\_\beta}(T,\rho_\alpha^o)$, to predict a structure-making or breaking behavior. For this task, we already know the location of the line $\mathcal{S}_{\beta\alpha}^{\infty,IG\_\beta}(T,\rho_\alpha^o) = 0$, where the isothermal compressibility of the solvent becomes equal to that of an ideal gas at the same state conditions, i.e., $\kappa_\alpha^o(T,\rho_\alpha^o) = (kT\rho_\alpha^o)^{-1}$ [3,50].

In what follows, we illustrate the lack of a one-to-one correspondence between the sign of $\mathcal{S}_{\beta\alpha}^{\infty,IG\_\beta}(T,\rho_{\alpha}^{o})$ and $B^{IG\_\beta}(T,\rho_{\alpha}^{o})$.

For this purpose, let us start by selecting the structure-making behavior, $\mathcal{S}_{\beta\alpha}^{\infty,IG\_\beta} = 1 - kT\rho_{\alpha}^{o}\kappa_{\alpha}^{o} > 0$ [49], that leads to either $\eta_{\alpha}^{o}v_{\alpha}^{o} < h\mathcal{N}$ (i.e., $\beta\Delta\mu_{\alpha}^{\natural,o} < 0$) or $\eta_{\alpha}^{o}v_{\alpha}^{o} > h\mathcal{N}$ (i.e., $\beta\Delta\mu_{\alpha}^{\natural,o} > 0$ with $0 < \beta\Delta\mu_{\alpha}^{\natural,o} < \mathcal{S}_{\beta\alpha}^{\infty,IG\_\beta}$). From a thermodynamic viewpoint, this condition means also that $\beta\Delta\mu_{\alpha}^{\natural,o} + kT\rho_{\alpha}^{o}\kappa_{\alpha}^{o} < 1$ and will result in a $B^{IG\_\beta}(T,\rho_{\alpha}^{o}) > 0$ in (sign) agreement with the starting condition $\mathcal{S}_{\beta\alpha}^{\infty,IG\_\beta}(T,\rho_{\alpha}^{o}) > 0$. However, when $\eta_{\alpha}^{o}v_{\alpha}^{o} > h\mathcal{N}$ but $\beta\Delta\mu_{\alpha}^{\natural,o} > \mathcal{S}_{\beta\alpha}^{\infty,IG\_\beta}$ then, $\beta\Delta\mu_{\alpha}^{\natural,o} + kT\rho_{\alpha}^{o}\kappa_{\alpha}^{o} > 1$ and $B^{IG\_\beta}(T,\rho_{\alpha}^{o}) < 0$, in contrast (opposite sign) to the starting condition $\mathcal{S}_{\beta\alpha}^{\infty,IG-\beta}(T,\rho_{\alpha}^{o}) > 0$.

Likewise, if we choose the structure-breaking behavior, $\mathcal{S}_{\beta\alpha}^{\infty,IG\_\beta} = 1 - kT\rho_{\alpha}^{o}\kappa_{\alpha}^{o} < 0$ [49], then, when $\eta_{\alpha}^{o}v_{\alpha}^{o} < h\mathcal{N}$ (i.e., $\beta\Delta\mu_{\alpha}^{\natural,o} < 0$ and therefore, $\beta\left|\Delta\mu_{\alpha}^{\natural,o}\right| > \left|\mathcal{S}_{\beta\alpha}^{\infty,IG\_\beta}\right|$), we obtain $B^{IG\_\beta}(T,\rho_{\alpha}^{o}) > 0$ whose thermodynamic meaning, $1 - kT\rho_{\alpha}^{o}\kappa_{\alpha}^{o} > \beta\Delta\mu_{\alpha}^{\natural,o} < 0$ with $kT\rho_{\alpha}^{o}\kappa_{\alpha}^{o} > 1$, leads to a $B^{IG\_\beta}(T,\rho_{\alpha}^{o}) > 0$ in contrast (opposite sign) to the starting condition $\mathcal{S}_{\beta\alpha}^{\infty,IG\_\beta}(T,\rho_{\alpha}^{o}) < 0$. However, when either $\eta_{\alpha}^{o}v_{\alpha}^{o} > h\mathcal{N}$ or $\eta_{\alpha}^{o}v_{\alpha}^{o} < h\mathcal{N}$ (i.e., $\beta\Delta\mu_{\alpha}^{\natural,o} < 0$ or $\beta\left|\Delta\mu_{\alpha}^{\natural,o}\right| < \left|\mathcal{S}_{\beta\alpha}^{\infty,IG\_\beta}\right|$), then we find that $\beta\Delta\mu_{\alpha}^{\natural,o} + kT\rho_{\alpha}^{o}\kappa_{\alpha}^{o} > 1$ and $B^{IG\_\beta}(T,\rho_{\alpha}^{o}) < 0$, in (sign) agreement with the starting $\mathcal{S}_{\beta\alpha}^{\infty,IG\_\beta}(T,\rho_{\alpha}^{o}) < 0$.

The preceding rigorous microscopic-to-macroscopic analysis associated with the behavior of the ideal gas β–solute in a real α–solvent illustrates the lack of uniqueness of the structure making/breaking marker based on the sign of the B–coefficient. In fact, this criterion would not be able to distinguish a structure-making, $\mathcal{S}_{\beta\alpha}^{\infty,IG\_\beta} > 0$, from a structure-breaking, $\mathcal{S}_{\beta\alpha}^{\infty,IG\_\beta} < 0$, behavior since either $B^{IG\_\beta} < 0$ or $B^{IG\_\beta} > 0$ could simultaneously describe a $\mathcal{S}_{\beta\alpha}^{\infty,IG\_\beta} > 0$ and a $\mathcal{S}_{\beta\alpha}^{\infty,IG\_\beta} < 0$ scenario depending on the state conditions of the pure α–solvent.

Obviously, we can follow the same argument to analyze the microstructural behavior of a real binary mixture whose B–coefficient is described by Equation (28). Indeed, if the system exhibited $\mathcal{S}_{\beta\alpha}^{\infty} > 0$, then $B > 0$ would represent a structure-making behavior whenever $\mathcal{S}_{\beta\alpha}^{\infty} > \left(\nu\Delta\mu_{\alpha}^{\natural,o} - \Delta\mu_{\beta}^{\natural,\infty}\right)/\nu kT$, otherwise, the structure-making condition $\mathcal{S}_{\beta\alpha}^{\infty} > 0$ would be described by $B < 0$ as long as $0 < \mathcal{S}_{\beta\alpha}^{\infty} < \left(\nu\Delta\mu_{\alpha}^{\natural,o} - \Delta\mu_{\beta}^{\natural,\infty}\right)/\nu kT$. However, if the system showed $\mathcal{S}_{\beta\alpha}^{\infty} < 0$, then $B < 0$ would describe a structure-breaking behavior whenever $\mathcal{S}_{\beta\alpha}^{\infty} < \left(\nu\Delta\mu_{\alpha}^{\natural,o} - \Delta\mu_{\beta}^{\natural,\infty}\right)/\nu kT$, otherwise, the structure breaking condition $\mathcal{S}_{\beta\alpha}^{\infty} < 0$ would require a $B > 0$. As before, due to the lack of a one-to-one correspondence between the sign of the structure making/breaking parameter $\mathcal{S}_{\beta\alpha}^{\infty}$ and that of the B–coefficient, the ad hoc assumption about the sign of the Jones–Dole's B–coefficient cannot be taken as a marker of structure making/breaking trends as has been traditionally done in the literature [40,42,46,73–82].

Finally, we provide a brief comment on the ability of the $(\partial B/\partial T)_P$ counterpart to discriminate between structure-making and structure-breaking solutes. We should note that, regardless of the theoretical description for the behavior of the B–coefficient, there are in principle four possible combinations of pair conditions $B \gtrless 0$ and $(\partial B/\partial T)_P \gtrless 0$ for a binary mixture, resulting in eight structure making/breaking scenarios whose thorough analysis has been presented elsewhere [48]. According to the arguments above, and those in SI-C of the Supporting Information in [48], we determine that neither $(\partial B/\partial T)_P$ alone nor the $[B, (\partial B/\partial T)_P]$ pair-combination can offer an unambiguous correspondence between their signs and the structure making/breaking nature of the β–solute as described here by the molecular-based parameter $\mathcal{S}_{\beta\alpha}^{\infty}$.

4. Experimental Evidence of Structure Making/Breaking Behavior in Aqueous Electrolytes and Comparison against Predictions from the Conjectured Markers

In what follows, we illustrate the theoretical developments of Section 2 and findings of Section 3 based on irrefutable experimental evidence from a variety of aqueous solutes at infinite dilution for which we have available volumetric and rheologic experimental data.

4.1. Illustration of the Behavior of the Short- and Long-Range Contributions to the Structure Parameter along the Liquid Branch of the Coexistence Phase Envelope of Water

For that purpose, we selected a few simple aqueous electrolyte β–solutes and invoked the accurate SOCW thermodynamic model [103,104] to describe the behavior of their partial molar volume at infinite dilution $\hat{v}_\beta^\infty\left(\rho_{H_2O}^o\right)_\sigma$, as described by Equation(A11) of Appendix B, using the parameterization from Table 4 of [103], along the liquid branch of the water phase envelope. These infinite dilute aqueous systems include the following β–solutes: *NaCl, NaBr, NaI*, $NaCH_3CO_2$ (*NaAc*), *LiCl, KCl, CsCl*, NH_4Cl, and *NaOH*.

In Figures 1 and 2, we plot the orthobaric (σ) density dependence of the short-range contribution to the structure making/breaking parameter, $\mathcal{S}_{\beta H_2O}^\infty(SR)$ (described by Equation (A13) in Appendix B), in comparison with its full value counterpart, $\mathcal{S}_{\beta H_2O}^\infty\left(\rho_{H_2O}^o\right)_\sigma$ (described by Equation (A12) in Appendix B), and identify the resulting finite $\mathcal{S}_{\beta H_2O}^{\infty,critical}(SR)$ quantity associated to the corresponding Krichevskii parameter via Equation (22). On the one hand, Figures 1 and 2 illustrate the divergent nature of $\mathcal{S}_{\beta H_2O}^\infty\left(\rho_{H_2O}^o\right)_\sigma$ as the orthobaric solvent density approaches its critical point given that $\hat{v}_\beta^\infty\left(\rho_{H_2O}^o \to \rho_{H_2O,crit}^o\right)_\sigma \sim \kappa_{H_2O}^o$, and consequently, $\mathcal{S}_{\beta H_2O}^\infty\left(\rho_{H_2O}^o \to \rho_{H_2O,crit}^o\right)_\sigma \to +\infty$ (right ordinate). On the other hand, the corresponding $\mathcal{S}_{\beta H_2O}^\infty(SR)$ exibits a finite critical limit (left ordinate), which defines the Krichevskii parameter $\mathcal{A}_{Kr}$ of the β–solute in water according to Equation (22). Note that, because $\mathcal{A}_{Kr}^{IG\_\beta} = k\left(T\rho_{H_2O}^o\right)_{crit}$, these figures also indicate that all these electrolyte solutes describe $\mathcal{A}_{Kr} < 0$, i.e., these electrolyte solutes behave as attractive [1] or non-volatile species [58].

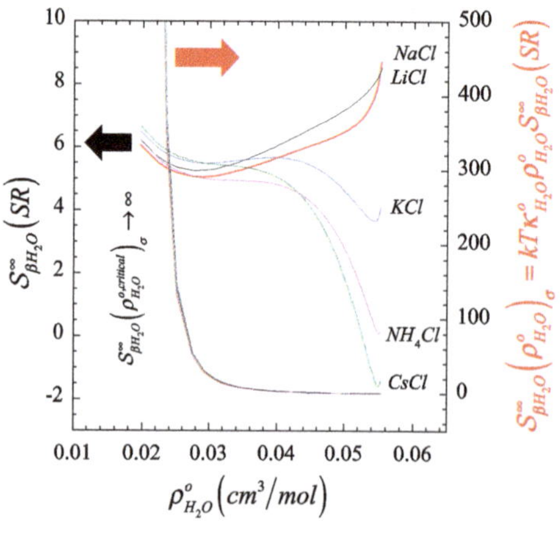

Figure 1. Color-coordinated comparison of the behavior of the liquid-phase orthobaric density dependence for the structure making/breaking parameter $\mathcal{S}_{\beta H_2O}^\infty\left(\rho_{H_2O}^o\right)_\sigma$, and its short-ranged counterpart $\mathcal{S}_{\beta H_2O}^\infty(SR)$, of a few aqueous univalent cation chlorides where we highlight (right) the divergent nature of the parameter at the critical point of the solvent, and (left) the corresponding finiteness of their short-range counterpart associated with the Krichevskii parameter.

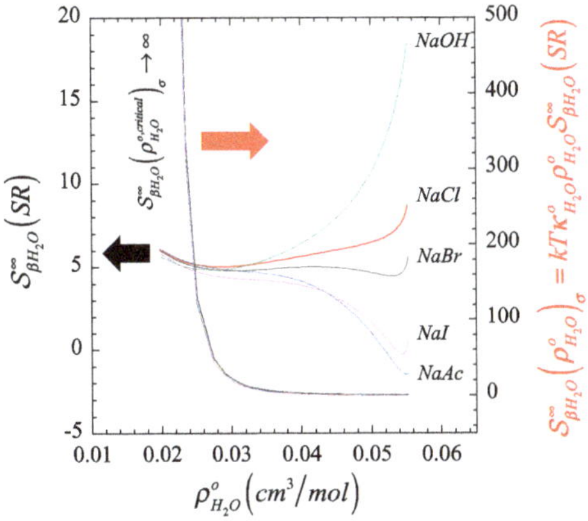

Figure 2. Color-coordinated comparison of the behavior of the liquid-phase orthobaric density dependence for the structure making/breaking parameter $\mathcal{S}^{\infty}_{\beta H_2O}\left(\rho^{o}_{H_2O}\right)_{\sigma}$, and its short-ranged counterpart $\mathcal{S}^{\infty}_{\beta H_2O}(SR)$, of a few aqueous univalent sodium salts where we highlight (right) the divergent nature of the parameter at the critical point of the solvent, and (left) the corresponding finiteness of their short-range counterpart associated with the Krichevskii parameter.

4.2. Comparison between the Predictions of the Two Common Structure Making/Breaking Markers and the Actual Microstructural Behavior

Here, we illustrate the incompatibility between the microstructural behavior and the volumetric, as well as rheological, data of a variety of dilute aqueous solutions and the predictions from the structure making/breaking discussed in Sections 2 and 3. These systems cover a wide range of solute–solvent intermolecular interaction asymmetries, and comprise organic as well as inorganic electrolyte solutes almost exclusively at ambient conditions. In Table 1, we display the reported volumetric experimental data in the form of the temperature-factored isobaric-temperature derivative $-T\left(\partial^2 v^{\infty}_{\beta}/\partial T^2\right)_P$ of the infinitely dilute solute in water as Hepler's structure making/breaking marker, and the structural parameter $\mathcal{S}^{\infty}_{\beta H_2O}(T,P)$ according to Equation (7) as the descriptor of the actual measure of the solute-induced microstructural perturbation. The rheologic data involve the B–coefficient of the Jones–Dole's equation and its isobaric temperature derivative $(\partial B/\partial T)_P$ as reported from the composition dependence of the shear viscosity of the solutions, via regression of the Jones–Dole's equation. For comparison purposes, we also included the predicted behavior from the systems involving the ideal gas solute, IG_β, and the β–solute behaving like a H_2O–molecule $(LR-IS)$ [3].

Table 1 illustrates the disparity of the results from the ad hoc criteria based on either the behavior of the isobaric-thermal expansivity or the B–coefficient and its isobaric-temperature derivative $(\partial B/\partial T)_P$, as compared against the actual microstructural responses of the systems accounted by $\mathcal{S}^{\infty}_{\beta H_2O}(T,P)$. In fact, the comparison between column 2 and columns 3–5 indicates that $-T\left(\partial^2 v^{\infty}_{\beta}/\partial T^2\right)_P$ as well as the B–coefficient and either its temperature derivative or their combination not only fail to predict unambiguously the structure making/breaking nature of the solutes, but also reveal, unsurprisingly, a broad inconsistency between them. This lack of consistency between the thermal expansivity- and the viscosity-based markers highlights the lack of one-to-one (i.e., uniqueness of the) connection between the solute-induced perturbation of the solvent microstructure and the markers.

Table 1. Experimental structure making/breaking parameter $\mathcal{S}^{\infty}_{\beta H_2O}(T,P)$ for aqueous infinitely dilute solutes at ambient conditions in comparison with predictions based on Hepler's isobaric-thermal expansivity marker as well as Jones–Dole's B– coefficient and $(\partial B/\partial T)_P$ derivative criteria.

β–*Solute*	$\mathcal{S}^{\infty}_{\beta H_2O}(T,P)$ [(a)]	$(\partial B/\partial T)_P$ [(b)]	B [(b)]	$-T(\partial^2 \bar{v}^{\infty}_{\beta}/\partial T^2)_P$	**Ref.**
Water (LR-IS)	0	0	0	maker	This work
Ideal gas	maker	<0	<0	breaker	This work
creatine	breaker	<0	>0	maker	[105]
creatinine	breaker	<0	>0	maker	[105]
nicotinic acid	breaker	<0	>0	maker	[106]
l-ascorbic acid	breaker	>0	>0	breaker	[50,94,106]
glycine	breaker	>0	>0	breaker	[107]
alanine	breaker	>0	>0	breaker	[107]
DTAB [(c)]	breaker	>0	<0	breaker	[107]
l-serine	breaker	<0	>0	maker	[96]
l-arginine	breaker	<0	>0	maker	[96]
choline-biotinate	breaker	<0	>0	maker	[108]
choline-nicotinate	breaker	<0	>0	maker	[108]
choline-ascorbate	breaker	>0	>0	maker	[108]
LiCy [(d)]	breaker	<0	>0	breaker	[109,110]
NaCy [(d)]	breaker	~0	>0	maker	[109,110]
KCy [(d)]	breaker	>0	>0	maker	[109,110]
$CaCl_2$	maker	>0	>0	breaker	[111,112]
$CdCl_2$	maker	<0	>0	breaker	[113,114]
$NiCl_2$	maker	>0	<0	breaker [(e)]	[114,115]
NH_4NO_3	breaker	>0	<0	breaker	[116,117]
$MgCl_2$	maker	<0	>0	breaker	[114,118]

[(a)] Defined to Equation (7); [(b)] See Equation (1); [(c)] Dodecyl-trimethyl-ammonium bromide; [(d)] Alkaline metal cyclohexyl sulfamate; [(e)] According to our 3rd-order polynomial regression of the $\bar{v}^{\infty}_{\beta}(T)$ data from Herrington et al. [113].

Indeed, the lack of uniqueness or complete ambiguity becomes clearly exposed as follows: either (a) from Table 1, we could choose four aqueous systems comprising structure-making solutes, $\mathcal{S}^{\infty}_{\beta H_2O}(T,P) > 0$, such as $\beta = (CaCl_2, NiCl_2, CdCl_2, IG\_\beta)$ and observe that their structure making/breaking ability are described by four different $[B, (\partial B/\partial T)_P]$–sign combinations, while Hepler's criterion describes all four solutes as structure-breakers; or (b) from Tables 1 and 2, we could choose four aqueous systems involving structure-breaking solutes, $\mathcal{S}^{\infty}_{\beta H_2O}(T,P) < 0$, such as $\beta = (glycine, DBTA, LiCy, IG\_\beta)$ and find that their structure making/breaking ability are described again by four different $[B, (\partial B/\partial T)_P]$–sign combinations, while Hepler's criterion describes the first three solutes as structure-breakers and the supercritical IG_β solute as a structure-maker.

Table 2. Representative systems illustrating the eight $[B, (\partial B/\partial T)_P]$ pair combinations and their resulting structure making/breaking parameter $\mathcal{S}^{\infty}_{\beta H_2O}(T,P)$ for infinitely dilute aqueous solutes at ambient conditions.

Aqueous Solute	$(\partial B/\partial T)_P$	B	$\mathcal{S}^{\infty}_{\beta H_2O}(T,P)$
$CaCl_2$	>0	>0	>0 (maker)
choline-ascorbate	>0	>0	<0 (breaker)
$CdCl_2$	>0	<0	>0 (maker)
NH_4NO_3	>0	<0	<0 (breaker)
$MgCl_2$	<0	>0	>0 (maker)
LiCy [(a)]	<0	>0	<0 (breaker)
IG_β	<0	<0	>0 (maker)
IG_β [(b)]	<0	<0	<0 (breaker)

[(a)] Alkaline metal cyclohexyl sulfamate; [(b)] Supercritical conditions where $\kappa^{o}_{H_2O} > \kappa^{o,IG}_{H_2O}$.

The lack of uniqueness in the structure making/breaking markers defined around the behavior of the B–coefficient becomes even more obvious in Table 2, where we identify four representative pairs of infinitely dilute β–solutes in which each pair displays precisely the same particular behavior for the B–coefficient, while the individual members of the pair exhibit opposite structure making/breaking parameter $\mathcal{S}^{\infty}_{\beta\alpha}(T,P)$. For example, $CaCl_2$ and choline-ascorbate share the same $\left[B, \left(\partial B/\partial T\right)_P\right] > 0$ behavior, although $\mathcal{S}^{\infty}_{CaCl_2\ H_2O} > 0$ while $\mathcal{S}^{\infty}_{Choline\ Ascorbate\ H_2O} < 0$.

The preceding assessment of the experimental evidence in conjunction with the rigorous definition of $\mathcal{S}^{\infty}_{\beta\alpha}(T,P)$, or $\mathcal{S}_{\beta\alpha}(T,P,x_\beta)$ for that matter, highlights the unreliability of the predictions from the two widespread structure making/breaking markers resulting from the lack of explicit microstructure-to-macroscopic relations in their definitions to confer an unambiguous description of the propensity of a β–solute to distort the microstructure of the α–solvent. It also emphasizes the fact that the sought unambiguous description according to Equation (7) only requires two pieces of volumetric information, namely the partial molar volumes of the pure α–solvent and of the β–solute at infinite dilution. For those readers eager to jump directly to the application of the parameter $\mathcal{S}^{\infty}_{\beta\alpha}(T,P)$ as a fundamentally-based tool for the description and measurement of the magnitude of a solute's ability to perturb the solvent structure, we provide in Appendix C the step-by-step procedure towards its straightforward calculation.

5. Final Remarks and Outlook

We must emphasize that our analysis does not judge the accuracy or the usefulness of the composition and temperature dependencies of either the viscosity coefficients or the volumetric behavior of dilute solutions, but rather assesses the validity of their microstructural interpretation based on either the resulting viscosity B–coefficient, and corresponding temperature derivative, or the isobaric thermal expansivity markers. Researchers have often been susceptible to adding a name to a phenomenon under investigation as if by inserting a label the phenomenon becomes intuitively understood. This practice has frequently led to misunderstanding and confusion, as we have repeatedly witnessed [2,119] during the early attempts to gain understanding of the solubility enhancement of sparingly soluble solutes in highly compressible solvents, a subject that bears striking similarities with the arguments of the present work. Indeed, the evolving, vague narrative intended to aid the microstructural interpretation of the alluded solvation phenomenon took a variety of names, including solvent clustering [120], densification/cavitation [121], density augmentation/depletion [122], and molecular charisma [123], which led to controversies resulting from the lack of precision in the meaning of "local or short-ranged effect, drastic or significant microstructural changes" [124,125]. The past events suggest that we should refrain from inserting a structure maker/breaker label (also known as kosmotropic/chaotropic) [52] to a solute species until we fully understand what that phenomenon means, by focusing on addressing the real issues, e.g., what we expect to learn about the thermodynamic behavior of the system by analyzing its microstructural behavior. In the words of Richard Feynman, [126] we must recognize "... *the difference between knowing the name of something and knowing something*".

In this work, we raised awareness of what can and cannot be inferred from Hepler's thermal expansivity and Jones–Dole's *B*–coefficient criteria as structure making/breaking markers, including: (i) neither one provides cause–effect connections between the actual microstructural perturbation and the proposed markers; (ii) neither criterion can predict the correct structure making/breaking answer for the two simplest systems describing either the largest or the smallest solute–solvent intermolecular interaction asymmetry, systems for which we know precisely the structure making/breaking behavior; and (iii) the macroscopic nature of the above criteria, compounded by the lack of any explicit link to the evolution of the solvent microstructure, preclude their reliable use as structure making/breaking markers and, as such, their use should be discontinued to avoid the perpetuation of confusion.

Author Contributions: Conceptualization, A.A.C.; methodology, A.A.C.; formal analysis, A.A.C.; resources, O.D.C.; writing—original draft preparation, A.A.C.; writing—review and editing, O.D.C. All authors have read and agreed to the published version of the manuscript.

Funding: This research received no external funding.

Institutional Review Board Statement: Not applicable.

Informed Consent Statement: Not applicable.

Data Availability Statement: Not applicable.

Acknowledgments: The authors express their gratitude to Josef Sedlbauer, Technical University Liberec (Czechia), for kindly commenting on the predictive ability of the SOCW equation used in this work to describe the orthobaric behavior of the partial molar volumes aqueous alkali halides at infinite dilution.

Conflicts of Interest: The authors declare no conflict of interest.

Appendix A. Transition State Interpretation of Jones–Dole's B-Coefficient

We must first highlight the approximations supporting Feakins et al. [101] transition-state interpretation of the *B*–coefficient. For that purpose, we first recast Jones–Dole's equation [27] as follows,

$$\ln(\eta/\eta_\alpha^o) \cong Ac_\beta^{0.5} + Bc_\beta \tag{A1}$$

where $\eta/\eta_\alpha^o - 1 \cong \ln(\eta/\eta_\alpha^o)$ and then invoke Glasstone–Laidler-Eyring's [102] transition state expressions for the viscosity of a pure α–solvent, η_α^o,

$$\eta_\alpha^o = (h\mathcal{N}/v_\alpha^o)\exp\left(\beta\Delta G_\alpha^{\natural,o}\right) \tag{A2}$$

and that of the dilute solution comprising a β–solute, η,

$$\eta = (h\mathcal{N}/v)\exp\left(\beta\Delta G_{\beta,\alpha}^{\natural}\right) \tag{A3}$$

where h and $\mathcal{N}$ denote Planck and Avogadro constants, while v_α^o and $\Delta G_\alpha^{\natural,o}$ describe the pure solvent molar volume and the corresponding Gibbs free energy of activation for the viscous flow process, respectively, while v and $\Delta G_{\beta,\alpha}^{\natural}$ denote the corresponding molar volume of the resulting solution and the average Gibbs free energy of activation for the viscous flow of the components in solution.

Equations (A1)–(A3) allow connecting the two Jones–Dole coefficients with the change of Gibbs free energy of activation for the viscous flow process and the relative (to the pure solvent) molar volume of the dilute solution, i.e.,

$$Ac_{\beta}^{0.5} + Bc_{\beta} \cong \ln(v_{\alpha}^{o}/v) + \beta\left(\Delta G_{\alpha,\beta}^{\natural} - \Delta G_{\alpha}^{\natural,o}\right) \tag{A4}$$

and

$$Ac_{\beta}^{0.5} + Bc_{\beta} \cong \ln(v_{\alpha}^{o}/v)_{c_{\beta}\to 0} + \beta\left(\Delta G_{\beta,\alpha}^{\natural} - \Delta G_{\alpha}^{\natural,o}\right)_{c_{\beta}\to 0} + \mathcal{F}\left(c_{\beta}^{0.5}\right) \tag{A5}$$

where $\mathcal{F}\left(c_{\beta}^{0.5}\right)$ will be identified below. Moreover, after recalling that $\nu = \nu^{+} + \nu^{-}$ for a general electrolyte solution, and $\nu = 1$ for a non-electrolyte solution, we have that

$$\begin{aligned} \Delta G_{\beta,\alpha}^{\natural} &= x_{\beta}\Delta\mu_{\beta}^{\natural} + x_{\alpha}\Delta\mu_{\alpha}^{\natural} \\ &= x_{\beta}\left(\Delta\mu_{\beta}^{\natural} - \nu\Delta\mu_{\alpha}^{\natural}\right) + \Delta\mu_{\alpha}^{\natural} \end{aligned} \tag{A6}$$

$$\begin{aligned} v &= x_{\beta}\hat{v}_{\beta} + x_{\alpha}\hat{v}_{\alpha} \\ &= x_{\beta}(\hat{v}_{\beta} - \nu\hat{v}_{\alpha}) + \hat{v}_{\alpha} \end{aligned} \tag{A7}$$

which lead to the following approximation of the relative molar volume,

$$\begin{aligned} \ln(v/v_{\alpha}^{o})_{x_{\beta}\to 0} &\cong -\ln\left[1 - x_{\beta}(\nu\hat{v}_{\alpha} - \hat{v}_{\beta})/v_{\alpha}^{o}\right]_{x_{\beta}\to 0} \\ &\cong c_{\beta}\left(\nu v_{\alpha}^{o} - \hat{v}_{\beta}^{\infty}\right) \end{aligned} \tag{A8}$$

where $c_{\beta} = x_{\beta}/v$ defines the molar concentration of the β–solute. Consequently, from Equations (A5)–(A8), we obtain,

$$\left(\Delta G_{\beta,\alpha}^{\natural} - \Delta G_{\alpha}^{\natural,o}\right)_{c_{\beta}\to 0} = c_{\beta}v_{\alpha}^{o}\left(\Delta\mu_{\beta}^{\natural,\infty} - \nu\Delta\mu_{\alpha}^{\natural,o}\right) \tag{A9}$$

which provides the identification of $\mathcal{F}\left(c_{\beta}^{0.5}\right) = Ac_{\beta}^{0.5}$ in A5 and the TS-interpretation of the B–coefficient in the Jones–Dole equation as follows,

$$B = \left(\nu v_{\alpha}^{o} - \hat{v}_{\beta}^{\infty}\right) + \beta v_{\alpha}^{o}\left(\Delta\mu_{\beta}^{\natural,\infty} - \nu\Delta\mu_{\alpha}^{\natural,o}\right) \tag{A10}$$

A direct comparison between A10 and Equation (18) in the original derivation of [101] highlights that these authors assumed $\nu = 1$ even when analyzing electrolyte solutions, a feature that has evaded the attention of many authors and might have contributed to errors in the calculation of transition state viscosity-related quantities and their interpretation in the literature. While the error introduced in the case of a dissociative solute comprising $\nu = 2$ is about 5–6% [96,109,127–129], depending on the relative ratio $\left(v_{\alpha}^{o}/\hat{v}_{\beta}^{\infty}\right)$, it becomes significantly larger, i.e., about 20–25% for dissociative solutes comprising $\nu = 4$.

Appendix B. Structure Making/Breaking Parameter from the SOCW Representation of the Partial Molar Volumes of Simple Electrolyte Solutes

We invoke the Sedlbauer–O'Connell–Wood (SOCW) [103,130] expression for the partial molar volume of the $(\nu^{+}{:}\nu^{-})$ ions conforming a β–electrolyte solute at infinite dilution in the α–solvent, i.e.,

$$\begin{aligned} \hat{v}_{i}^{\infty}(T,P) = &\{(1 - z_i) + a_i\rho_{\alpha}^{o} - d_i + b_i\rho_{\alpha}^{o}[\exp(\vartheta\rho_{\alpha}^{o}) - 1] + \\ &\delta\rho_{\alpha}^{o}[\exp(\lambda\rho_{\alpha}^{o}) - 1] + c_i\rho_{\alpha}^{o}\exp(\theta/T)\}kT\kappa_{\alpha}^{o} + d_i/\rho_{\alpha}^{o} \end{aligned} \tag{A11}$$

where the i–subindex identifies the individual ion bearing the electrostatic charge z_i, with the regressed parameters $[a_i, b_i, c_i, d_i]$ given in Table 4 in [103], the universal constants

$\theta = 1500$K, $\vartheta = 0.005$ m^3/kg, and $\lambda = -0.01$ m^3/kg, while $\delta = 0$ for cations and $\delta = -0.645$ m^3/kg for anions. While these parameters have been regressed from high-pressure liquid phases, they describe accurately the orthobaric behavior of electrolytes at infinite dilution [131]. Then, considering that the partial molar volume of the β–solute at infinite dilution equals to $\hat{v}_{\beta}^{\infty}(T,P) = \nu^{+}\hat{v}_{+}^{\infty} + \nu^{-}\hat{v}_{-}^{\infty}$, the structure making/breaking parameter from Equation (7) with $\nu^{+} = \nu^{-} = 1$ and $\nu = 2$ can be described by the following expression,

$$S_{\beta\alpha}^{\infty}(T,P) = 1 - 0.5d_{\pm} - 0.5\{\ 2 - d_{\pm} + \rho_{\alpha}^{o}a_{\pm} + \rho_{\alpha}^{o}b_{\pm}[\exp(\vartheta\rho_{\alpha}^{o}) - 1] + \delta\rho_{\alpha}^{o}[\exp(\lambda\rho_{\alpha}^{o}) - 1] + \rho_{\alpha}^{o}c_{\pm}\exp(\theta/T)\}kT\kappa_{\alpha}^{o}\rho_{\alpha}^{o} \quad \text{(A12)}$$

where $a_{\pm} = a_{+} + a_{-}$, $b_{\pm} = b_{+} + b_{-}$, $c_{\pm} = c_{+} + c_{-}$ and $d_{\pm} = d_{+} + d_{-}$. Consequently, from Equation (20) we find that

$$S_{\beta\alpha}^{\infty}(SR) = \kappa_{\alpha}^{o,IG}(1 - 0.5d_{\pm})/\kappa_{\alpha}^{o} - 0.5\{\ 2 - d_{\pm} + \rho_{\alpha}^{o}a_{\pm} + \rho_{\alpha}^{o}b_{\pm}[\exp(\vartheta\rho_{\alpha}^{o}) - 1] + \delta\rho_{\alpha}^{o}[\exp(\lambda\rho_{\alpha}^{o}) - 1] + \rho_{\alpha}^{o}c_{\pm}\exp(\theta/T)\} \quad \text{(A13)}$$

Appendix C. Practical Guide to the Calculation of the Fundamentally-Based Structure Making/Breaking Marker $S_{\beta\alpha}^{\infty}(T,P)$

For that purpose, we assume that we have available the isobaric-isothermal composition (molar concentration c_{β} or any alternative) dependence of either the molar volume of the dilute solution ($v(c_{\beta})$) or the apparent molar volume of the dilute β–solute in an α–solvent ($v_{\beta}^{\varphi}(c_{\beta})$) and proceed as follows:

1. Calculate the partial molar volume of the β–solute at infinite dilution $\hat{v}_{\beta}^{\infty}$ as the composition limiting behavior $\hat{v}_{\beta}^{\infty} = \lim_{c_{\beta}\to 0} v_{\beta}^{\varphi}(c_{\beta})$;
2. Calculate the partial molar volume of the pure α–solvent, i.e., $v_{\alpha}^{o} = \lim_{c_{\beta}\to 0} v(c_{\beta})$;
3. Calculate the structure making/breaking parameter $S_{\beta\alpha}^{\infty}(T,P) = 1 - \left(\hat{v}_{\beta}^{\infty}/\nu v_{\alpha}^{o}\right)_{TP}$, Equation (7), after considering the stoichiometric ν–parameter of the β–solute, either $\nu = \nu^{+} + \nu^{-}$ for an ionic solute or $\nu = 1$ for a non-dissociative solute;
4. Compare the ratio $\left(\hat{v}_{\beta}^{\infty}/v_{\alpha}^{o}\right)_{TP}$ with the stoichiometric ν–parameter:
   a. If $\left(\hat{v}_{\beta}^{\infty}/v_{\alpha}^{o}\right)_{TP} < \nu$, then $S_{\beta\alpha}^{\infty}(T,P) > 0$, i.e., the β–solute behaves as a structure making at the prevailing state conditions;
   b. If $\left(\hat{v}_{\beta}^{\infty}/v_{\alpha}^{o}\right)_{TP} > \nu$, then $S_{\beta\alpha}^{\infty}(T,P) < 0$, i.e., the β–solute behaves as a structure breaking at the prevailing state conditions;
   c. If $\left(\hat{v}_{\beta}^{\infty}/v_{\alpha}^{o}\right)_{TP} \cong \nu$, then $S_{\beta\alpha}^{\infty}(T,P) \cong 0$, i.e., the β–solute induces a negligible structure perturbation at the prevailing state conditions.

Note that these outcomes are completely independent on the nature of the solvent and the type of the solute–solvent intermolecular interactions, i.e., the $S_{\beta\alpha}^{\infty}(T,P)$ applies equally to aqueous and non-aqueous solvents, electrolyte and non-electrolyte solutions, and require no information whatsoever about any (solvent-specific or otherwise) interaction mechanism such as hydrogen bonding.

References

1. Debenedetti, P.G.; Mohamed, R.S. Attractive, Weakly Attractive and Repulsive Near-Critical Systems. *J. Chem. Phys.* **1989**, *90*, 4528–4536. [CrossRef]
2. Chialvo, A.A. Solvation Phenomena in Dilute Solutions: Formal, Experimental Evidence, and Modeling Implications. In *Fluctuation Theory of Solutions: Applications in Chemistry, Chemical Engineering and Biophysics*, 1st ed.; Matteoli, E., O'Connell, J.P., Smith, P.E., Eds.; CRC Press: Boca Raton, FL, USA, 2013; pp. 191–224.

3. Chialvo, A.A. On the Solvation Thermodynamics Involving Species with Large Intermolecular Asymmetries: A Rigorous Molecular-Based Approach to Simple Systems with Unconventionally Complex Behaviors. *J. Phys. Chem. B* **2020**, *124*, 7879–7896. [CrossRef] [PubMed]
4. Fulton, J.L.; Heald, S.M.; Badyal, Y.S.; Simonson, J.M. Understanding the Effects of Concentration on the Solvation Structure of Ca^{2+} in Aqueous Solution. I. The Perspective on Local Structure from EXAFS and XANES. *J. Phys. Chem. A* **2003**, *107*, 4688–4696. [CrossRef]
5. D'Angelo, P.; Roscioni, O.M.; Chillemi, G.; Della Longa, S.; Benfatto, M. Detection of second hydration shells in ionic solutions by XANES: Computed spectra for Ni^{2+} in water based on molecular dynamics. *J. Am. Chem. Soc.* **2006**, *128*, 1853–1858. [CrossRef]
6. Antalek, M.; Pace, E.; Hedman, B.; Hodgson, K.O.; Chillemi, G.; Benfatto, M.; Sarangi, R.; Frank, P. Solvation structure of the halides from x-ray absorption spectroscopy. *J. Chem. Phys.* **2016**, *145*, 044318. [CrossRef]
7. Zitolo, A.; Chillemi, G.; D'Angelo, P. X-ray Absorption Study of the Solvation Structure of Cu^{2+} in Methanol and Dimethyl Sulfoxide. *Inorg. Chem.* **2012**, *51*, 8827–8833. [CrossRef] [PubMed]
8. Enderby, J.E. Ion Solvation via Neutron Scattering. *Chem. Soc. Rev.* **1995**, *24*, 159–168. [CrossRef]
9. Badyal, Y.S.; Barnes, A.C.; Cuello, G.J.; Simonson, J.M. Understanding the effects of concentration on the solvation structure of Ca^{2+} in aqueous solutions. II: Insights into longer range order from neutron diffraction isotope substitution. *J. Phys. Chem. A* **2004**, *108*, 11819–11827. [CrossRef]
10. Ansell, S.; Barnes, A.C.; Mason, P.E.; Neilson, G.W.; Ramos, S. X-ray and neutron scattering studies of the hydration structure of alkali ions in concentrated aqueous solutions. *Biophys. Chem.* **2006**, *124*, 171–179. [CrossRef]
11. Mancinelli, R.; Botti, A.; Bruni, F.; Ricci, M.A.; Soper, A.K. Hydration of sodium, potassium, and chloride ions in solution and the concept of structure maker/breaker. *J. Phys. Chem. B* **2007**, *111*, 13570–13577. [CrossRef]
12. Bruni, F.; Imberti, S.; Mancinelli, R.; Ricci, M.A. Aqueous solutions of divalent chlorides: Ions hydration shell and water structure. *J. Chem. Phys.* **2012**, *136*, 064520. [CrossRef] [PubMed]
13. Pusztai, L.; McGreevy, R.L. MCGR: An inverse method for deriving the pair correlation function from the structure factor. *Physica B* **1997**, *234–236*, 357–358. [CrossRef]
14. Pusztai, L. Partial pair correlation functions of liquid water. *Phys. Rev. B* **1999**, *60*, 11851–11854. [CrossRef]
15. Soper, A.K. Tests of the empirical potential structure refinement method and a new method of application to neutron diffraction data on water. *Mol. Phys.* **2001**, *99*, 1503–1516. [CrossRef]
16. Soper, A.K. Joint structure refinement of x-ray and neutron diffraction data on disordered materials: Application to liquid water. *J. Phys. -Condens. Matter* **2007**, *19*, 335206. [CrossRef]
17. Harsanyi, I.; Pusztai, L. Hydration structure in concentrated aqueous lithium chloride solutions: A reverse Monte Carlo based combination of molecular dynamics simulations and diffraction data. *J. Chem. Phys.* **2012**, *137*, 204503–204509. [CrossRef]
18. Mile, V.; Gereben, O.; Kohara, S.; Pusztai, L. On the Structure of Aqueous Cesium Fluoride and Cesium Iodide Solutions: Diffraction Experiments, Molecular Dynamics Simulations, and Reverse Monte Carlo Modeling. *J. Phys. Chem. B* **2012**, *116*, 9758–9767. [CrossRef]
19. Pethes, I.; Pusztai, L. Reverse Monte Carlo modeling of liquid water with the explicit use of the SPC/E interatomic potential. *J. Chem. Phys.* **2017**, *146*, 064506. [CrossRef]
20. Chialvo, A.A.; Simonson, J.M. The Structure of Concentrated $NiCl_2$ Aqueous Solutions. What is Molecular Simulation Revealing about the Neutron Scattering Methodologies? *Mol. Phys.* **2002**, *100*, 2307–2315. [CrossRef]
21. Chialvo, A.A.; Simonson, J.M. The Structure of $CaCl_2$ Aqueous Solutions over a Wide Range of Concentrations. Interpretation of Difraction Experiments Via Molecular Simulation. *J. Chem. Phys.* **2003**, *119*, 8052–8061. [CrossRef]
22. Mason, P.E.; Neilson, G.W.; Dempsey, C.E.; Brady, J.W. Neutron diffraction and simulation studies of $CsNO_3$ and Cs_2CO_3 solutions. *J. Am. Chem. Soc.* **2006**, *128*, 15136–15144. [CrossRef] [PubMed]
23. Pluharova, E.; Fischer, H.E.; Mason, P.E.; Jungwirth, P. Hydration of the chloride ion in concentrated aqueous solutions using neutron scattering and molecular dynamics. *Mol. Phys.* **2014**, *112*, 1230–1240. [CrossRef]
24. Chialvo, A.A.; Vlcek, L. NO^{3-} Coordination in Aqueous Solutions by 15N/14N and 18O/natO Isotopic Substitution: What Can We Learn from Molecular Simulation? *J. Phys. Chem. B* **2015**, *119*, 519–531. [CrossRef] [PubMed]
25. Kohagen, M.; Pluhařová, E.; Mason, P.E.; Jungwirth, P. Exploring Ion–Ion Interactions in Aqueous Solutions by a Combination of Molecular Dynamics and Neutron Scattering. *J. Phys. Chem. Lett.* **2015**, *6*, 1563–1567. [CrossRef]
26. Chialvo, A.A.; Vlcek, L. "Thought experiments" as dry-runs for "tough experiments": Novel approaches to the hydration behavior of oxyanions. *Pure Appl. Chem.* **2016**, *88*, 163–176. [CrossRef]
27. Jones, G.; Dole, M. The viscosity of aqueous solutions of strong electrolytes with special reference to barium chloride. *J. Am. Chem. Soc.* **1929**, *51*, 2950–2964. [CrossRef]
28. Hepler, L.G. Thermal Expansion and Structure in Water and Aqueous Solutions. *Can. J. Chem.* **1969**, *47*, 4613–4617. [CrossRef]
29. Falkenhagen, H.; Dole, M. The root law of the internal friction of strong electrolytes. *Z. Fur Phys. Chem.-Abt. B-Chem. Der Elem. Aufbau Der Mater.* **1929**, *6*, 159–162.
30. Falkenhagen, H.; Dole, M. The internal friction of electrolytic solutions and their significance according to Debye's theory. *Phys. Z.* **1929**, *30*, 611–622.
31. Herskovits, T.T.; Kelly, T.M. Viscosity Studies of Aqueous Solutions of Alcohols, ureas, and Amides. *J. Phys. Chem.* **1973**, *77*, 381–388. [CrossRef]

32. Rupley, J.A. Effect of Urea+amides upon the Water Structure. *J. Phys. Chem.* **1964**, *68*, 2002–2003. [CrossRef]
33. Yoshida, K.; Tsuchihashi, N.; Ibuki, K.; Ueno, M. NMR and viscosity B coefficients for spherical nonelectrolytes in nonaqueous solvents. *J. Mol. Liq.* **2005**, *119*, 67–75. [CrossRef]
34. Verissimo, L.M.P.; Cabral, I.; Cabral, A.M.T.D.P.V.; Utzeri, G.; Veiga, F.J.B.; Valente, A.J.M.; Ribeiro, A.C.F. Transport properties of aqueous solutions of the oncologic drug 5-fluorouracil: A fundamental complement to therapeutics. *J. Chem. Thermodyn.* **2021**, *161*, 106533. [CrossRef]
35. Abbott, A.P.; Hope, E.G.; Palmer, D.J. Effect of solutes on the viscosity of supercritical solutions. *J. Phys. Chem. B* **2007**, *111*, 8114–8118. [CrossRef] [PubMed]
36. Cox, W.M.; Wolfenden, J.H. The viscosity of strong electrolytes measured by a differential method. *Proc. R. Soc. A.* **1934**, *145*, 488.
37. Bernal, J.D.; Fowler, R.H. A Theory of Water and Ionic Solution, with Particular Reference to Hydrogen and Hydroxyl Ions. *J. Chem. Phys.* **1933**, *1*, 515–548. [CrossRef]
38. Laurence, V.D.; Wolfenden, J.H. The viscosity of solutions of strong electrolytes. *J. Chem. Soc.* **1934**, 1144–1147. [CrossRef]
39. Asmus, E. Zur Frage des B-Koeffizienten Der Jones-Dole Gleichung. *Z. Fur Nat. Sect. A-A J. Phys. Sci.* **1949**, *4*, 589–594. [CrossRef]
40. Gurney, R.W. *Ionic Processes in Solution*; McGraw Hill: New York, NY, USA, 1953.
41. Kaminsky, M. Ion-Solvent Interaction and the Viscosity of Strong Electrolyte Solutions. *Discuss. Faraday Soc.* **1957**, *24*, 171–179. [CrossRef]
42. Nightingale, E.R. Phenomenological Theory of Ion Solvation: Effective Radii of Hydrated Ions. *J. Phys. Chem.* **1959**, *63*, 1381–1387. [CrossRef]
43. Stokes, R.H.; Mills, R. *Viscosity of Electrolytes and Related Properties*; Pergamon Press: Oxford, UK; New York, NY, USA, 1965.
44. Conway, B.E.; Verrall, R.E.; Desnoyer, J.E. Partial Molal Volumes of Tetraalkylamonium Halides and Assignment of Individual Ionic Contributions. *Trans. Faraday Soc.* **1966**, *62*, 2738–2749. [CrossRef]
45. Frank, H.S.; Robinson, A.L. The Entropy of Dilution of Strong Electrolytes in Aqueous Solutions. *J. Chem. Phys.* **1940**, *8*, 933–938. [CrossRef]
46. Frank, H.S.; Evans, M.W. Free Volume and Entropy in Condensed Systems. 3. Entropy in Binary Liquid Mixtures—Partial Molal Entropy in Dilute Solutions—Structure and Thermodynamics in Aqueous Electrolytes. *J. Chem. Phys.* **1945**, *13*, 507–532. [CrossRef]
47. Tsangaris, J.M.; Martin, R.B. Viscosities of Aqueous Solutions of Dipolar Ions. *Arch. Biochem. Biophys.* **1965**, *112*, 267–272. [CrossRef]
48. Chialvo, A.A.; Crisalle, O.D. Can Jones-Dole's B-coefficient Be a Consistent Structure Making/Breaking Marker?. Rigorous molecular-based analysis and critical assessment of its marker uniqueness. *J. Phys. Chem. B* **2021**, *125*, 12028–12041. [CrossRef]
49. Chialvo, A.A. On the Solute-Induced Structure-Making/Breaking Effect: Rigorous Links among Microscopic Behavior, Solvation Properties, and Solution Non-Ideality. *J. Phys. Chem. B* **2019**, *123*, 2930–2947. [CrossRef]
50. Chialvo, A.A.; Crisalle, O.D. Osmolyte-induced Effects on the Hydration Behavior and the Osmotic Second Virial Coefficients of Alkyl-Substituted Urea Derivatives. Critical assessment of their structure-making/breaking behavior. *J. Phys. Chem. B* **2021**, *125*, 6231–6243. [CrossRef]
51. Kirkwood, J.G.; Buff, F.P. The Statistical Mechanical Theory of Solution. I. *J. Chem. Phys.* **1951**, *19*, 774–777. [CrossRef]
52. Ball, P.; Hallsworth, J.E. Water structure and chaotropicity: Their uses, abuses and biological implications. *Phys. Chem. Chem. Phys.* **2015**, *17*, 8297–8305. [CrossRef]
53. Chialvo, A.A.; Crisalle, O.D. On density-based modeling of dilute non-electrolyte solutions involving wide ranges of state conditions and intermolecular asymmetries: Formal results, fundamental constraints, and the rationale for its molecular thermodynamic foundations. *Fluid Phase Equilib.* **2021**, *535*, 112969. [CrossRef]
54. Chialvo, A.A.; Cummings, P.T.; Kalyuzhnyi, Y.V. Solvation effect on kinetic rate constant of reactions in supercritical solvents. *AIChE J.* **1998**, *44*, 667–680. [CrossRef]
55. Chialvo, A.A.; Cummings, P.T. Comments on "Near Critical Phase Behavior of Dilute Mixtures". *Mol. Phys.* **1995**, *84*, 41–48. [CrossRef]
56. Chialvo, A.A.; Crisalle, O.D. On the Linear Orthobaric-density Representation of Near-critical Solvation Quantities: What Can We Conclude about the Accuracy of this Paradigm? *Fluid Phase Equilib.* **2020**, *514*, 112535. [CrossRef]
57. Chialvo, A.A. On the Krichevskii Parameter of Solutes in Dilute Solutions: Formal Links between its Magnitude, the Solute-solvent Intermolecular Asymmetry, and the Precise Description of Solution Thermodynamics. *Fluid Phase Equilib.* **2020**, *513*, 112546. [CrossRef]
58. Levelt Sengers, J.M.H. Solubility Near the Solvent's Critical Point. *J. Supercrit. Fluids* **1991**, *4*, 215–222. [CrossRef]
59. Chialvo, A.A. Solute-Solute and Solute-Solvent Correlations in Dilute Near-Critical Ternary Mixtures: Mixed Solute and Entrainer Effects. *J. Phys. Chem.* **1993**, *97*, 2740–2744. [CrossRef]
60. Brunner, G. (Ed.) *Supercritical Fluids as Solvents and Reaction Media*; Elsevier B.V.: Amsterdam, The Netherlands, 2004.
61. Anikeev, V.; Fan, M. *Supercritical Fluid Technology for Energy and Environmental Applications*; Elsevier Science: Amsterdam, The Netherlands, 2013.
62. Santos, D.T.; Santana, Á.L.; Meireles, M.A.A.; Gomes, M.T.M.S.; Torres, R.A.D.C.; Albarelli, J.Q.; Bakatselou, A.; Ensinas, A.V.; Maréchal, F. *Supercritical Fluid Biorefining: Fundamentals, Applications and Perspectives*; Springer: Berlin/Heidelberg, Germany, 2020.

63. Chialvo, A.A.; Crisalle, O.D. Solute-induced effects in solvation thermodynamics: Does urea behave as a structure-making or structure-breaking solute? *Mol. Phys.* **2019**, *117*, 3484–3492. [CrossRef]
64. Chialvo, A.A. Molecular-Based Description of the Osmotic Second Virial Coefficients of Electrolytes: Rigorous Formal Links to Solute–Solvent Interaction Asymmetry, Virial Expansion Paths, and Experimental Evidence. *J. Phys. Chem. B* **2022**, *126*, 4339–4353. [CrossRef]
65. Lin, L.N.; Brandts, J.F.; Brandts, J.M.; Plotnikov, V. Determination of the volumetric properties of proteins and other solutes using pressure perturbation calorimetry. *Anal. Biochem.* **2002**, *302*, 144–160. [CrossRef]
66. Holtzer, A.; Emerson, M.F. On Utility of Concept of Water Structure in Rationalization of Aqueous Solutions of Proteins and Small Molecules. *J. Phys. Chem.* **1969**, *73*, 26–33. [CrossRef]
67. Chialvo, A.A. Gas solubility in dilute solutions: A novel molecular thermodynamic perspective. *J. Chem. Phys.* **2018**, *148*, 174502. [CrossRef] [PubMed]
68. Mazo, R.M. Statistical Mechanical Theory of Solutions. *J. Chem. Phys.* **1958**, *29*, 1122–1128. [CrossRef]
69. Ben-Naim, A. *Molecular Theory of Solutions*; Oxford University Press: Oxford, UK, 2006.
70. Dole, M. Debye Contribution to the Theory of the Viscosity of Strong Electrolytes. *J. Phys. Chem.* **1984**, *88*, 6468–6469. [CrossRef]
71. Falkenhagen, H.; Vernon, E.L. LXII. The viscosity of strong electrolyte solutions according to electrostatic theory. *Lond. Edinb. Dublin Philos. Mag. J. Sci.* **1932**, *14*, 537–565. [CrossRef]
72. Falkenhagen, H.; Hertz, H.G.; Eberling, W. *Theorie der Elektrolyte*; Hirzel: Leipzig, Germany, 1971.
73. Robinson, R.A.; Stokes, R.H. *Electrolyte Solutions*, 2nd (revised) ed.; Butterworths: London, UK, 1970.
74. Conway, B.E. *Ionic Hydration in Chemistry and Biophysics*; Elsevier: Amsterdam, The Netherlands, 1981; Volume 12.
75. Franks, F. *Water: A Matrix of Life*; Royal Society of Chemistry: London, UK, 2000.
76. Waghorne, W.E. Viscosities of electrolyte solutions. *Philos. Trans. R. Soc. Lond. Ser. A-Math. Phys. Eng. Sci.* **2001**, *359*, 1529–1543. [CrossRef]
77. Hammadi, A. Electrical Conductance, Density, and Viscosity in Mixtures of Alkali-Metal Halides and Glycerol. *Int. J. Thermophys.* **2004**, *25*, 89–111. [CrossRef]
78. Marcus, Y. Effect of Ions on the Structure of Water: Structure Making and Breaking. *Chem. Rev.* **2009**, *109*, 1346–1370. [CrossRef]
79. Dhondge, S.S.; Zodape, S.P.; Parwate, D.V. Volumetric and viscometric studies of some drugs in aqueous solutions at different temperatures. *J. Chem. Thermodyn.* **2012**, *48*, 207–212. [CrossRef]
80. Jindal, R.; Singla, M.; Kumar, H. Transport behavior of aliphatic amino acids glycine/L-alanine/L-valine and hydroxyl amino acids L-serine/L-threonine in aqueous trilithium citrate solutions at different temperatures. *J. Mol. Liq.* **2015**, *206*, 343–349. [CrossRef]
81. Gadzuric, S.; Tot, A.; Armakovic, S.; Armakovic, S.; Panic, J.; Jovic, B.; Vranes, M. Uncommon structure making/breaking behaviour of cholinium taurate in water. *J. Chem. Thermodyn.* **2017**, *107*, 58–64. [CrossRef]
82. Zafarani-Moattar, M.T.; Shekaari, H.; Mostafavi, H.; Jafari, P. Thermodynamic and transport properties of aqueous solutions containing cholinium L-alaninate and polyethylene glycol dimethyl ether 250: Evaluation of solute-solvent interactions and phase separation. *J. Chem. Thermodyn.* **2019**, *132*, 9–22. [CrossRef]
83. Sarma, T.S.; Ahluwalia, J.C. Experimental Studies on Structure of Aqueous Solutions of Hydrophobic Solutes. *Chem. Soc. Rev.* **1973**, *2*, 203–232. [CrossRef]
84. Sanyal, S.K.; Mandal, S.K. Viscosity B-Coefficients of Alkyl Carboxylates. *Electrochim. Acta* **1983**, *28*, 1875–1876. [CrossRef]
85. Tamaki, K.; Ohara, Y.; Watanabe, S. Solution Properties of Sodium Perfluoralkanoates. Heats of Solution, Viscosity B-coefficients, and Surface Tensions. *Bull. Chem. Soc. Jpn.* **1989**, *62*, 2497–2501. [CrossRef]
86. Zhao, H. Viscosity B-coefficients and standard partial molar volumes of amino acids, and their roles in interpreting the protein (enzyme) stabilization. *Biophys. Chem.* **2006**, *122*, 157–183. [CrossRef]
87. Rajagopal, K.; Jayabalakrishnan, S.S. Volumetric, ultrasonic speed, and viscometric studies of salbutamol sulphate in aqueous methanol solution at different temperatures. *J. Chem. Thermodyn.* **2010**, *42*, 984–993. [CrossRef]
88. Kabiraz, D.C.; Biswas, T.K.; Huque, M.E. Physico-chemical properties of some electrolytes in water and aqueous sodiumdodecyl sulfate solutions at different temperatures. *J. Chem. Thermodyn.* **2011**, *43*, 1917–1923. [CrossRef]
89. Banipal, T.S.; Kaur, N.; Kaur, A.; Gupta, M.; Banipal, P.K. Effect of food preservatives on the hydration properties and taste behavior of amino acids: A volumetric and viscometric approach. *Food Chem.* **2015**, *181*, 339–346. [CrossRef]
90. Kumar, K.; Patial, B.S.; Chauhan, S. Interactions of Saccharides in Aqueous Glycine and Leucine Solutions at Different Temperatures of (293.15 to 313.15) K: A Viscometric Study. *J. Chem. Eng. Data* **2015**, *60*, 47–56. [CrossRef]
91. Sharma, S.K.; Singh, G.; Kumar, H.; Kataria, R. Densities, Sound Speed, and Viscosities of Some Amino Acids with Aqueous Tetra-Butyl Ammonium Iodide Solutions at Different Temperatures. *J. Chem. Eng. Data* **2015**, *60*, 2600–2611. [CrossRef]
92. Banipal, T.S.; Beni, A.; Kaur, N.; Banipal, P.K. Volumetric, Viscometric and Spectroscopic Approach to Study the Solvation Behavior of Xanthine Drugs in Aqueous Solutions of NaCl at T = 288.15–318.15 K and at p = 101.325 kPa. *J. Chem. Eng. Data* **2017**, *62*, 20–34. [CrossRef]
93. Zhu, C.Y.; Ren, X.F.; Ma, Y.G. Densities and Viscosities of Amino Acid plus Xylitol plus Water Solutions at 293.15 <= T/K <= 323.15. *J. Chem. Eng. Data* **2017**, *62*, 477–490. [CrossRef]
94. Sharma, R.; Thakur, R.C.; Kumar, H. Study of viscometric properties of L-ascorbic acid in binary aqueous mixtures of D-glucose and D-fructose. *Phys. Chem. Liq.* **2019**, *57*, 139–150. [CrossRef]

95. Sawhney, N.; Kumar, M.; Sandarve; Sharma, P.; Sharma, A.K.; Sharma, M. Structure-making behaviour of L-arginine in aqueous solution of drug ketorolac tromethamine: Volumetric, compressibility and viscometric studies. *Phys. Chem. Liq.* **2019**, *57*, 184–203. [CrossRef]
96. Rani, R.; Kumar, A.; Sharma, T.; Sharma, T.; Bamezai, R.K. Volumetric, acoustic and transport properties of ternary solutions of L-serine and L-arginine in aqueous solutions of thiamine hydrochloride at different temperatures. *J. Chem. Thermodyn.* **2019**, *135*, 260–277. [CrossRef]
97. Sarmad, S.; Zafarani-Moattar, M.T.; Nikjoo, D.; Mikkola, J.P. How Different Electrolytes Can Influence the Aqueous Solution Behavior of 1-Ethyl-3-Methylimidazolium Chloride: A Volumetric, Viscometric, and Infrared Spectroscopy Approach. *Front. Chem.* **2020**, *8*, 593786. [CrossRef]
98. Nain, A.K. Insight into solute-solute and solute-solvent interactions of l-proline in aqueous-D-xylose/L-arabinose solutions by using physicochemical methods at temperatures from 293.15 to 318.15 K. *J. Mol. Liq.* **2020**, *318*, 114190. [CrossRef]
99. Rajput, P.; Richu; Sharma, T.; Kumar, A. Temperature dependent physicochemical investigations of some nucleic acid bases (uracil, thymine and adenine) in aqueous inositol solutions. *J. Mol. Liq.* **2021**, *326*, 115210. [CrossRef]
100. Brinzei, M.; Ciocirlan, O. Volumetric and transport properties of ternary solutions of L-serine and L-valine in aqueous NaCl solutions in the temperature range (293.15–323.15) K. *J. Chem. Thermodyn.* **2021**, *154*, 106335. [CrossRef]
101. Feakins, D.; Freemantle, D.J.; Lawrence, K.G. Transition-State Treatment of Relative Viscosity of Electrolyte Solutions: Applications to Aqueous, Non-Aqueous and Metanol+Water Systems. *J. Chem. Soc.-Faraday Trans. I* **1974**, *70*, 795–806. [CrossRef]
102. Glasstone, S.; Laidler, K.J.; Eyring, H. *The Theory of Rate Processes: The Kinetics of Chemical Reactions, Viscosity, Diffusion and Electrochemical Phenomena*; McGraw-Hill Book Company, Incorporated: New York, NY, USA, 1941.
103. Sedlbauer, J.; O'Connell, J.P.; Wood, R.H. A New Equation of State for Correlation and Prediction of Standard Molal Thermodynamic Properties of Aqueous Species at High Temperature and Pressures. *Chem. Geol.* **2000**, *163*, 43–63. [CrossRef]
104. Sedlbauer, J.; Wood, R.H. Thermodynamic properties of dilute NaCl(aq) solutions near the critical point of water. *J. Phys. Chem. B* **2004**, *108*, 11838–11849. [CrossRef]
105. Panic, J.; Vranes, M.; Tot, A.; Ostojic, S.; Gadzuric, S. The organisation of water around creatine and creatinine molecules. *J. Chem. Thermodyn.* **2019**, *128*, 103–109. [CrossRef]
106. Richu; Kumar, A. A Comprehensive Study on Molecular Interactions of l-Ascorbic Acid/Nicotinic Acid in Aqueous [BMIm]Br at Varying Temperatures and Compositions: Spectroscopic and Thermodynamic Insights. *J. Chem. Eng. Data* **2021**, *66*, 3859–3880. [CrossRef]
107. Hossain, M.F.; Biswas, T.K.; Islam, M.N.; Huque, M.E. Volumetric and viscometric studies on dodecyltrimethylammonium bromide in aqueous and in aqueous amino acid solutions in premicellar region. *Monatsh. Chem.* **2010**, *141*, 1297–1308. [CrossRef]
108. Vranes, M.; Tot, A.; Jankovic, N.; Gadzuric, S. What is the taste of vitamin-based ionic liquids? *J. Mol. Liq.* **2019**, *276*, 902–909. [CrossRef]
109. Rudan-Tasic, D.; Klofutar, C.; Horvat, J. Viscosity of aqueous solutions of some alkali cyclohexylsulfamates at 25.0 degrees C. *Food Chem.* **2004**, *86*, 161–167. [CrossRef]
110. Horvat, J.; Bester-Rogac, M.; Klofutar, C.; Rudan-Tasic, D. Viscosity of aqueous solutions of lithium, sodium, potassium, rubidium and caesium cyclohexylsulfamates from 293.15 to 323.15 k. *J. Solut. Chem.* **2008**, *37*, 1329–1342. [CrossRef]
111. Dunn, L.A. Apparent Molar Volumes of Electrolytes. 3: Some 1-1 and 2-1 Electrolytes in Aqueous Solution at 0, 5. 15. 35, 45, 55, and 65 Degrees C. *Trans. Faraday Soc.* **1968**, *64*, 2951–2961. [CrossRef]
112. Abdulagatov, I.M.; Azizov, N.D. Viscosity of aqueous calcium chloride solutions at high temperatures and high pressures. *Fluid Phase Equilib.* **2006**, *240*, 204–219. [CrossRef]
113. Herrington, T.M.; Roffey, M.G.; Smith, D.P. Densities of Aqueous Electrolytes $MnCl_2$, $CoCl_2$, $NiCl_2$, $ZnCl_2$, and $CdCl_2$ from 25 degrees to 72 degrees at 1atm. *J. Chem. Eng. Data* **1986**, *31*, 221–225. [CrossRef]
114. Sonika; Thakur, R.C.; Chauhan, S.; Singh, K. Molecular Interactions of Transition Metal Chlorides in Water and Water–Ethanol Mixtures at 298–318 K on Viscometric Data. *Russ. J. Phys. Chem. A* **2018**, *92*, 2701–2709. [CrossRef]
115. Apelblat, A. A new two-parameter equation for correlation and prediction of densities as a function of concentration and temperature in binary aqueous solutions. *J. Mol. Liq.* **2016**, *219*, 313–331. [CrossRef]
116. Clegg, S.L.; Wexler, A.S. Densities and Apparent Molar Volumes of Atmospherically Important Electrolyte Solutions. 1. The Solutes H_2SO_4, HNO_3, HCl, Na_2SO_4, $NaNO_3$, NaCl, $(NH_4)_2SO_4$, NH4NO_3, and NH_4Cl from 0 to 50 degrees C, Including Extrapolations to Very Low Temperature and to the Pure Liquid State, and $NaHSO_4$, NaOH, and NH_3 at 25 degrees C. *J. Phys. Chem. A* **2011**, *115*, 3393–3460. [CrossRef] [PubMed]
117. Campbell, A.N.; Friesen, R.J. Conductance in the Range of Medium Concentration. *Can. J. Chem.-Rev. Can. De Chim.* **1959**, *37*, 1288–1293. [CrossRef]
118. Laliberte, M.; Cooper, W.E. Model for calculating the density of aqueous electrolyte solutions. *J. Chem. Eng. Data* **2004**, *49*, 1141–1151. [CrossRef]
119. Chialvo, A.A.; Cummings, P.T. Solute-induced Effects on the Structure and the Thermodynamics of Infinitely Dilute Mixtures. *AIChE J.* **1994**, *40*, 1558–1573. [CrossRef]
120. Kim, S.; Johnston, K.P. Clustering in Supercritical Fluid Mixtures. *AIChE J.* **1987**, *33*, 1603–1611. [CrossRef]
121. Wu, R.-S.; Lee, L.L.; Cochran, H.D. Solvent Structural Changes in Repulsive and Attractive Supercritical Mixtures. A Molecular Distribution Study. *J. Supercrit. Fluids* **1992**, *5*, 192–198. [CrossRef]

122. Brennecke, J.F.; Tomasko, D.L.; Peshkin, J.; Eckert, C.A. Fluorescence Spectroscopy Studies of Dilute Supercritical Solutions. *Ind. Eng. Chem. Res.* **1990**, *29*, 1682–1690. [CrossRef]
123. Eckert, C.A.; Knutson, B.L. Molecular Charisma in Supercritical Fluids. *Fluid Phase Equilib.* **1993**, *83*, 93–100. [CrossRef]
124. Brennecke, J.F.; Debenedetti, P.G.; Eckert, C.A.; Johnston, K.P. Letter to the Editor. *AIChE J.* **1990**, *36*, 1927–1928. [CrossRef]
125. Economou, I.G.; Donohue, M.D. Mean Field Calculation of Thermodynamic Properties of Supercritical Fluids. *AIChE J.* **1990**, *36*, 1920–1925. [CrossRef]
126. Feynman, R.P. *What Do You Care What Other People Think?: Further Adventures of a Curious Character*; Norton & Company, Incorporated: New York, NY, USA, 1988.
127. Vraneš, M.; Rackov, S.; Papović, S.; Pilić, B. The study of interactions in aqueous solutions of 1-alkyl-3-(3-butenyl)imidazolium bromide ionic liquids. *J. Chem. Thermodyn.* **2021**, *159*, 106479. [CrossRef]
128. Gupta, J.; Chand, D.; Nain, A.K. Study to reconnoiter solvation consequences of l-arginine/l-histidine and sodium salicylate in aqueous environment probed by physicochemical approach in the temperature range (293.15–318.15) K. *J. Mol. Liq.* **2020**, *305*, 112848. [CrossRef]
129. Ankita; Nain, A.K. Solute-solute and solute-solvent interactions of drug sodium salicylate in aqueous-glucose/sucrose solutions at temperatures from 293.15 to 318.15 K: A physicochemical study. *J. Mol. Liq.* **2020**, *298*, 112006. [CrossRef]
130. Majer, V.; Sedlbauer, J.; Wood, R.H. Chapter 4—Calculation of standard thermodynamic properties of aqueous electrolytes and nonelectrolytes. In *Aqueous Systems at Elevated Temperatures and Pressures*; Palmer, D.A., Fernández-Prini, R., Harvey, A.H., Eds.; Academic Press: London, UK, 2004; pp. 99–147. [CrossRef]
131. Sedlbauer, J. (Department of Chemistry, Technical University of Liberec, Liberec, Czechia). Personal communication, 2022.

liquids

MDPI

Article

Solvent Exchange around Aqueous Zn(II) from Ab Initio Molecular Dynamics Simulations

Adrian Malinowski and Maciej Śmiechowski *

Department of Physical Chemistry, Faculty of Chemistry, Gdańsk University of Technology, Narutowicza 11/12, 80-233 Gdańsk, Poland

* Correspondence: maciej.smiechowski@pg.edu.pl

Abstract: Hydrated zinc(II) cations, due to their importance in biological systems, are the subject of ongoing research concerning their hydration shell structure and dynamics. Here, ab initio molecular dynamics (AIMD) simulations are used to study solvent exchange events around aqueous Zn^{2+}, for which observation in detail is possible owing to the considerable length of the generated trajectory. While the hexacoordinated $Zn(H_2O)_6{}^{2+}$ is the dominant form of Zn(II) in an aqueous solution, there is a non-negligible contribution of the pentacoordinated $Zn(H_2O)_5{}^{2+}$ complex which presence is linked to the dissociative solvent exchange events around Zn^{2+}. The pentacoordinated Zn(II) has a much tighter hydration sphere and is characterized by a trigonal bipyramidal structure, in contrast to the usual octahedral symmetry of the hexacoordinated complex. In total, two full exchange events are registered in the analyzed trajectory. AIMD simulations on an adequate length scale thus provide a direct way of studying such solvent exchange events around ions in molecular detail.

Keywords: ab initio molecular dynamics; density functional theory; zinc(II) hydration; solvent exchange

1. Introduction

The coordination of zinc(II) cation is of paramount importance in biological systems, as zinc is the active center in hundreds of metalloenzymes, most often catalyzing hydrolysis reactions [1]. The importance of zinc goes beyond the role of cofactor in proteins, as the zinc cation is also actively utilized for intracellular signaling and subject to elaborate homeostasis mechanisms in the cell [2].

Due to the particular importance of water as the solvent in biological context, the structure and dynamics of hydrated ions is a topic of special significance in physical chemistry of solutions [3–5]. The influence of ions on water is often expressed in the terms of structural effects exerted by ions on the solvent, leading to the distinction between structure making and breaking ions [5].

The hydration of Zn(II) is also the subject of extensive research concerning its solvation shell structure and solvent exchange dynamics. To date, numerous experimental techniques have been applied to study Zn^{2+}(aq), including vibrational spectroscopy [6–12], extended X-ray absorption fine structure (EXAFS) [13–17], X-ray diffraction (XRD) [18–22], and neutron diffraction (ND) [23,24]. On the other hand, computational chemistry methods have also been used to obtain molecular-level data on zinc hydration. The static structure of aqueous complexes of Zn(II) has been comprehensively studied with quantum mechanical (QM) methods, using both wavefunction theory and density functional theory (DFT) [9,10,25–33]. Both structural and dynamical nature of the hydration phenomenona may be in turn investigated with molecular dynamics (MD) simulations with an increasing level of complexity. Zn(II) hydration has been studied with force field-based MD simulations [17,34–41], as well as quantum mechanics/molecular mechanics (QM/MM) MD simulations [42–46] and ab initio molecular dynamics (AIMD) simulations [47–50].

AIMD based on DFT is a particularly suited technique for studying aqueous ionic solutions [51,52]. Beginning with the pioneering study of Be^{2+}(aq) [53], many metal ions in an aqueous solution have been studied using AIMD simulations [54–58]. The current availability of fast and approximate DFT methods, such as density functional tight binding (DFTB) allows for the generation of extensive data sets on solvated ions in molecular solvents, including water [59]. The particular advantage of AIMD is the ability to study intermolecular polarization effects that ultimately influence many dynamic properties of the studied systems [54,55,60].

Computational methods are also indispensable for elucidating solvent exchange mechanisms around ions [61,62]. Experimental methods, rather than providing a mechanism of the exchange reaction, can only help in estimating the rate of this process [24]. The details of H_2O exchange around Zn(II) have been studied using MD simulations and static QM calculations [28–30,37]. Although AIMD simulations, combining the quantum description of electrons with the classical propagation of nuclei according to Newton's laws of motion, seem ideal for this purpose, the rather long timescale of water exchange in the first hydration shell of Zn(II) has thus far hampered direct observation of this process using AIMD.

In this work, in order to bridge this gap and obtain reliable data on the structure of Zn^{2+}(aq) and the dynamics of the solvent exchange process in its first hydration shell, we use extensive DFT-based AIMD simulations with a computational setup that is proven to be especially suitable for investigating liquid water. We hypothesize that the inability of hitherto equilibrium AIMD studies to register any solvent exchange around Zn(II) was a limitation of the short timescale of the simulations, rather than the flows in the method itself. We also aim to register the interplay of well-defined solvation complexes of Zn^{2+}(aq) that participate in the solvent exchange process.

2. Computational Methods

We studied aqueous Zn(II) in a system composed of a single Zn^{2+} cation and 100 H_2O molecules, which was contained in a cubic simulation supercell with applied periodic boundary conditions. The volume of the system was chosen to represent the experimental density of H_2O at 298 K (997 kg m^{-3} [63]) coupled with the absolute standard partial molar ionic volume of Zn^{2+}(aq) (-34.9 cm^3 mol^{-1} [64,65]), resulting in cell volume 14.33^3 Å^3. The starting configuration was prepared using GROMACS (v. 2018.5) tools [66] by solvating a centrally placed Zn^{2+} cation with a solvent configuration obtained from a well-equilibrated AMOEBA water simulation [67].

The system was initially subject to energy minimization and subsequently equilibrated using classical MD simulations in the canonical (NVT) ensemble at $T = 298.15$ K performed using Tinker-HP package (v. 1.2) [68]. The revised AMOEBA force field for liquid water [69] in combination with the AMOEBA parameterization of Zn(II) [38] were applied. The ultimate configuration from this NVT trajectory was then used to initialize the actual AIMD simulation.

AIMD simulations were performed with the cp2k computational suite (v. 6.0) [70–72], employing the DFT-based representation of the electronic structure as implemented in the Quickstep electronic structure module in cp2k [73]. The revPBE exchange-correlation functional [74,75] coded in the libXC library [76] was used, as it provides excellent reproduction of numerous static and dynamic properties of liquid water and aqueous ionic solutions [77–81]. A mixed Gaussian atomic orbitals with plane waves (GPW) representation of the electronic structure is applied in cp2k [82], and we used a molecularly optimized short-ranged double-zeta (DZVP-MOLOPT-SR-GTH) basis set for atomic orbitals [83], coupled with the auxiliary plane wave expansion of the electron density up to a 500 Ry cutoff. Though heavily contracted and featuring few diffuse primitive atomic orbitals, the chosen basis set performs favorably when compared to a more expensive triple-zeta polarized basis set for the description of liquid water [84]. Only valence electrons were treated explicitly, while core electrons were represented by norm-conserving GTH pseudopotentials [85] parameterized for the PBE functional. Additionally, the smoothing of the electron density and its derivative on the spatial integration grid was applied (via keywords `XC_SMOOTH_RHO NN50` and `XC_DERIV NN50_SMOOTH` in cp2k), as it was previously found to significantly improve the stability of the local energetics of liquid water [86]. Dispersion effects were included via two-body DFT-D3 empirical dispersion correction with zero damping term [87] with the cutoff set to 15 Å. The studied system has a non-zero total charge (+2) and the charge compensation is provided by an implicit neutralizing jellium background [88], as implemented in cp2k.

The system was first equilibrated for ∼35 ps in an AIMD simulation in the canonical (NVT) ensemble at $T = 298.15$ K using the canonical velocity rescaling (CSVR) thermostat [89] with the time constant set to 2000 cm^{-1} (≈16.67 fs). The simulation time step was set to 0.5 fs. After this equilibration period, the thermostat was removed and a production AIMD simulation in the microcanonical (NVE) ensemble was continued for 100 ps, with data collection every 2 fs. The centers of maximally localized Wannier functions (MLWFs) [90] were also computed and stored every 2 fs. Analyses were performed using in-house code and VMD (v. 1.9.4) [91], which also served as a visualization tool, along with gnuplot (v. 5.2) [92].

3. Results

The static structure of the hydration shells of aqueous Zn(II) is summarized by radial distribution functions (RDFs) for the Zn· · · O and Zn· · · H pairs, as depicted in Figure 1. The studied system is large enough to fully contain two hydration shells of Zn^{2+}, as clearly seen in the Zn· · · O RDF that features a prominent shallow minimum with $g(r) \approx 0$ separating the first hydration shell from the second one. At first glance, the running integration number of $g_{ZnO}(r)$ is very close to 6 along this minimum, implying a predominantly hexacoordinated Zn(II) ion. However, its exact value of 5.8 suggests that during as much as 20% of the trajectory Zn^{2+} might in fact be pentacoordinated. Detailed parameters of the analyzed RDFs are gathered in Table 1.

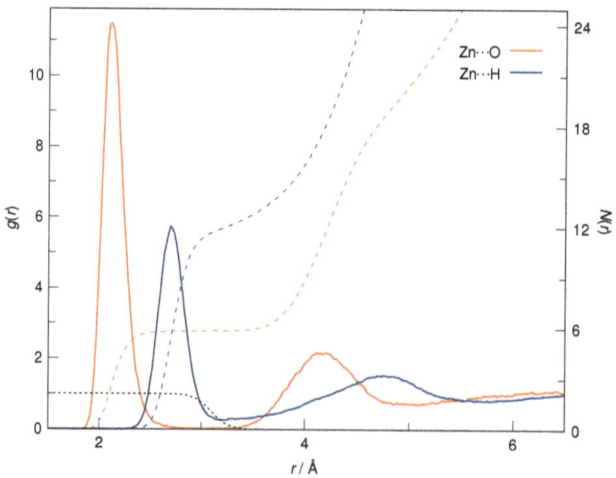

Figure 1. Radial distribution functions, $g(r)$, for Zn· · · O (orange) and Zn· · · H (blue) pairs (solid lines, left ordinate), as well as their running integrals, $N(r)$ (dashed lines, right ordinate). The black dotted line represents the smooth cutoff function used for water molecule coordination in Equation (1) (left ordinate).

Table 1. Most important parameters of the radial distribution functions depicted in Figure 1: location of the first and the second maxima and minima and the respective running integration numbers.

RDF	$r_{max,1}$ / Å	$r_{min,1}$ / Å	N_1 [1]	$r_{max,2}$ / Å	$r_{min,2}$ / Å	N_2 [2]
Zn···O	2.11	3.03	5.8	4.11	5.07	15.2
Zn···H	2.69	3.21	12.0	4.75	5.79	28.0

[1] $N_1 = N(r_{min,1})$; [2] $N_2 = N(r_{min,2}) - N(r_{min,1})$.

The first hydration shell of Zn^{2+} extends up to 3.03 Å and contains a non-integer number of water oxygen atoms. In order to determine the instantaneous coordination number (CN) of Zn(II), one could simply count O atoms found closer than the $r_{min,1}$ value, which would be tantamount to accepting the Heaviside step function as a selector function for H_2O molecules. However, this results in a discontinuous CN trajectory and adopting a smooth selector function instead can help visualize transitions between different coordination states of Zn^{2+}(aq), as previously proposed [50]. In this work, the coordinated solvent molecules are selected using a logistic cutoff function:

$$CN(t) = \sum_{i=1}^{N_{H_2O}} \frac{1}{1 + e^{\frac{r_{ZnO,i}(t) - r_0}{s}}}, \tag{1}$$

controlled by the cutoff radius r_0 and the sharpness parameter s. $r_{ZnO,i}(t)$ is the instantaneous Zn···O distance for the water molecule i. Adopting $r_0 = 3.1$ Å and $s = 0.08$ Å ensures a smooth transition between a fully coordinated H_2O molecule in the first hydration shell, where $CN_i(t) \to 1$, and a non-coordinated molecule in the second hydration shell, where $CN_i(t) \to 0$, cf. Figure 1.

The time evolution of thus obtained smooth coordination number of Zn^{2+} is shown in Figure 2. As clearly seen, even though a continuous $CN(t)$ definition is accepted, its value provides a clear-cut distinction between the hexa- and pentacoordinated states of aqueous Zn(II). The zinc cation spends most of the time in a $Zn(H_2O)_6{}^{2+}$ complex (about 81.2% of the trajectory). The pentacoordinated complex may be unanimously found in ~16.3% of the trajectory, while we ascribe the rest to transitional structures or short-time existence periods of one of the two dominating structures that are too short for inclusion in the analysis.

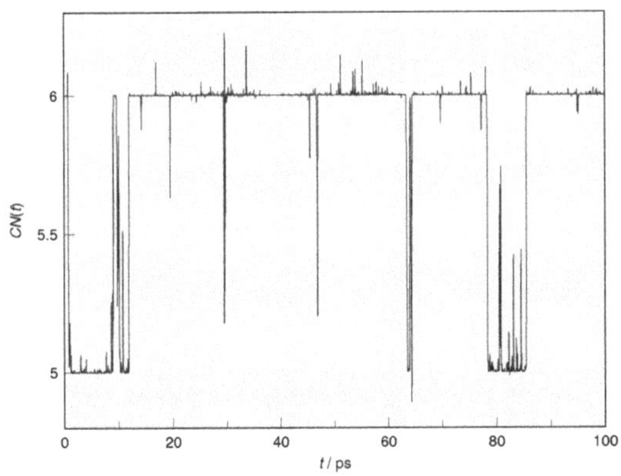

Figure 2. Time evolution of the smooth coordination number of Zn^{2+}, $CN(t)$, along the AIMD production trajectory. Light green and light purple areas indicate regions assigned to hexa- and pentacoordinated aqueous Zn(II) complexes, respectively, while white regions indicate transitional periods excluded from detailed analysis.

Armed with the knowledge about the existence of two distinct coordination states of Zn^{2+}(aq), we revisit Figure 1 in order to compare the Zn$\cdots$O RDFs in the regions of existence of the hexa- and pentacoordinated complexes. As seen in Figure 3, there is only a slight difference between the RDF based on the entire trajectory and the one based only on the fragments where $Zn(H_2O)_6{}^{2+}$ is found, which is unsurprising given the predominance of the hexacoordinated Zn(II). More importantly, the Zn$\cdots$O RDF restricted to the trajectory periods where $Zn(H_2O)_5{}^{2+}$ exists reveals a major compression of not only the first but also the second hydration shell of the zinc cation. The first maximum shifts to 2.07 Å, while the second one to 4.07 Å. Thus, the two distinct complexes are characterized with different hydration patterns extending up to the entire second hydration shell.

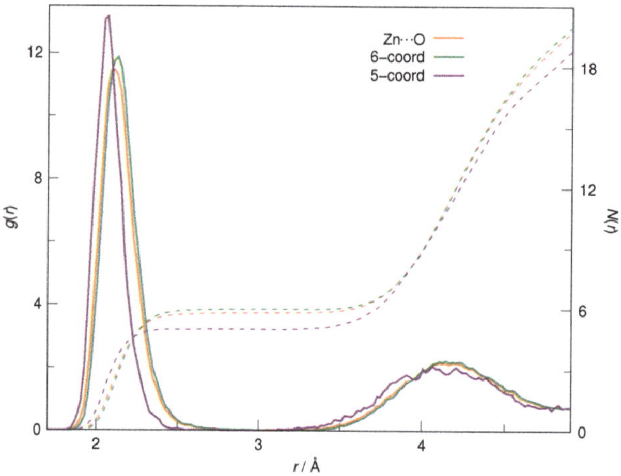

Figure 3. Radial distribution functions, $g(r)$, for Zn$\cdots$O pairs averaged over the entire AIMD trajectory (orange), as well as only over the regions of existence of hexacoordinated (green) and pentacoordinated (purple) aqueous Zn(II) complexes (solid lines, left ordinate), together with their running integrals, $N(r)$ (dashed lines, right ordinate).

Unlike static geometry optimizations, AIMD simulations can provide ensemble-averaged structures. In order to find the preferred geometry of hexa- and pentacoordinated Zn^{2+}(aq), we isolated the first shell complexes from the relevant trajectory fragments, rotated them to a common reference frame and averaged the geometry. The results can be seen in Figure 4. $Zn(H_2O)_6{}^{2+}$ forms a regular octahedral complex with the average Zn$\cdots$O distance equal 2.14 Å. On the contrary, $Zn(H_2O)_5{}^{2+}$ is characterized by a trigonal bipyramidal arrangement of water ligands. The Zn$\cdots$O distance for the three equatorial molecules is 2.04 Å, while the axial O atoms are located on average 2.14 Å from the metal cation. The geometries of the two complexes in Cartesian coordinates are available in the Supplementary Materials, see Figure S1.

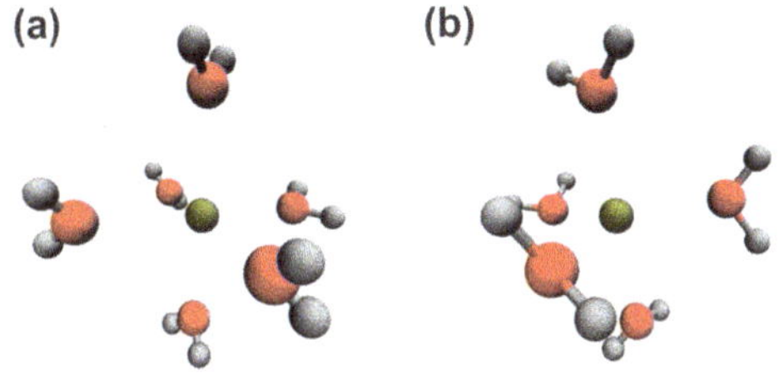

Figure 4. Time-averaged structure of (**a**) $Zn(H_2O)_6{}^{2+}$ and (**b**) $Zn(H_2O)_5{}^{2+}$ extracted from the respective regions of the AIMD trajectory.

While the positions of O atoms tend to be fixed in the Zn(II) solvation cage, the hydrogen atoms of the first shell water molecules are relatively free to bend from the coplanar arrangement with the Zn–O vector. We define the tilt angle, β, as the angle formed between the Zn–O vector and the dipole moment vector of the coordinated H_2O molecule. The two-dimensional distribution of water molecule configurations in the (r_{ZnO}, β) plane is shown in Figure 5. The distribution of angles is evidently bimodal, with a global maximum at 0°, indicating a coplanar arrangement of the H_2O molecular plane and the Zn–O vector, and a local maximum at ca. 30°, reflecting the presence of a substantial population of water molecules with tilted H atoms. We note here that the latter tend to be located slightly farther away from the central atom.

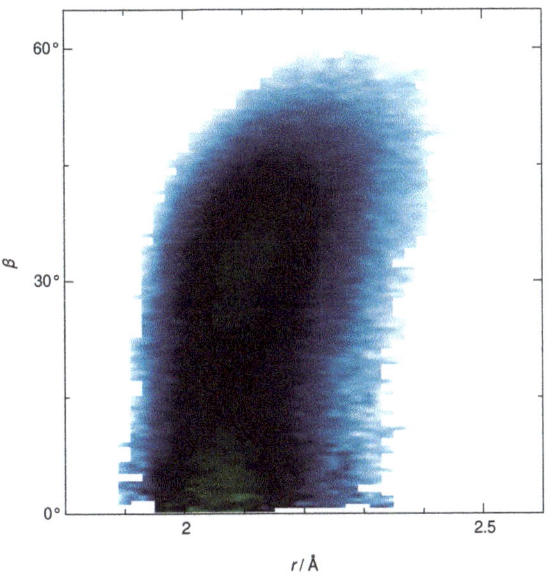

Figure 5. The two-dimensional distribution function, $g(r_{ZnO}, \beta)$, drawn on a logarithmic scale as $-\log g(r_{ZnO}, \beta)$ where β is the tilt angle of the coordinated H_2O molecule as defined in the text.

The presence of two complexes in aqueous solution of Zn(II) has far-reaching implications for the water molecules in its hydration shell. Ions are known to polarize water and the extent of this polarization is related to the changes in infrared spectrum of solution with respect to bulk water [54,55]. In Figure 6 we compare the probability distribution functions of the length of molecular dipoles of H_2O molecules in the entire system vs the ones in the first hydration shell of Zn^{2+}(aq). Unlike the dipole moment in bulk water, which has a symmetric Gaussian distribution [54,55,60], the curve for the studied system is right-skewed due to the substantial polarization of the neighboring H_2O molecules by the zinc cation. The distribution can be fitted to a split Pearson VII peak shape with a maximum located at 2.78 D. On the contrary, the distributions of dipole moments of water molecules found in the two complexes are symmetric and can similarly be fit to a Pearson VII analytical shape. It is found that H_2O molecules in $Zn(H_2O)_5{}^{2+}$ are polarized more significantly (on average 3.71 D) than the ones in $Zn(H_2O)_6{}^{2+}$ (mean 3.43 D).

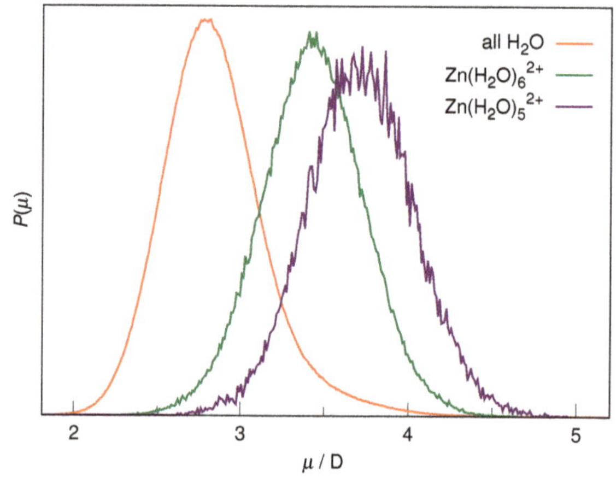

Figure 6. Normalized distribution functions of the molecular dipole moments of water molecules in the studied system, as well as water molecules in $Zn(H_2O)_6{}^{2+}$ (green) and $Zn(H_2O)_5{}^{2+}$ (purple).

We finally would like to estimate the time frame of a single exchange event based on the trajectories of the participating water molecules. As seen in Figure 2, we register two major exchange events connected to a change in CN of Zn(II), at around 0–12 ps and again at 78–86 ps. Additionally, a short-lived occurrence of $Zn(H_2O)_5{}^{2+}$ is found at ca. 64 ps. In order to shed light on the molecular details of these exchange events, we plotted the time evolution of the Zn$\cdots$O distance of involved H_2O molecules, as presented in Figure 7.

The first event begins as soon as the production trajectory starts. The water molecule W_{47} dissociates from the initially present $Zn(H_2O)_6{}^{2+}$ at 0.72–0.75 ps. Hereafter, the W_n notation indicates a H_2O molecule with an arbitrary index n in the trajectory. The solvent exchange process is obviously dissociative, as no new solvent molecule coordinates to the cation for an extended period of time. In this event, the released H_2O moves to the second hydration shell, as its separation from Zn^{2+} increases to ca. 4 Å (cf. Figure 1). This results in a transient $Zn(H_2O)_5{}^{2+}$ complex that persists up to ~8.8 ps, at which moment an interesting concerted mechanism is observed. Namely, W_{47} is kicked off to the weakly organized third hydration shell of the cation, at 6 Å and beyond, while another second shell H_2O, specifically W_{12}, coordinates to zinc(II), restoring the $Zn(H_2O)_6{}^{2+}$ complex. This, however, proves short-lived, as W_{12} dissociates already at ~10.1 ps and returns to the second hydration shell. Since a similar phenomenon is observed in further events, *vide infra*, such attempted exchange events that ultimately fail to stabilize the newly formed complex are most probably a common occurrence in Zn^{2+}(aq). Somewhat surprisingly, the entire exchange event is finalized by a third water molecule (W_{22}) that is captured by Zn^{2+} not from the second, but rather directly from the third hydration shell, initially >6 Å away from the cation. The hexacoordinated complex is restored at ~11.9 ps and remains stable for an extended period of time. Interestingly, the coordination of W_{22} is seemingly again assisted by W_{47}, since the time of arrival of the latter in the third hydration shell coincides with the start of movement of the former towards the inner hydration shells.

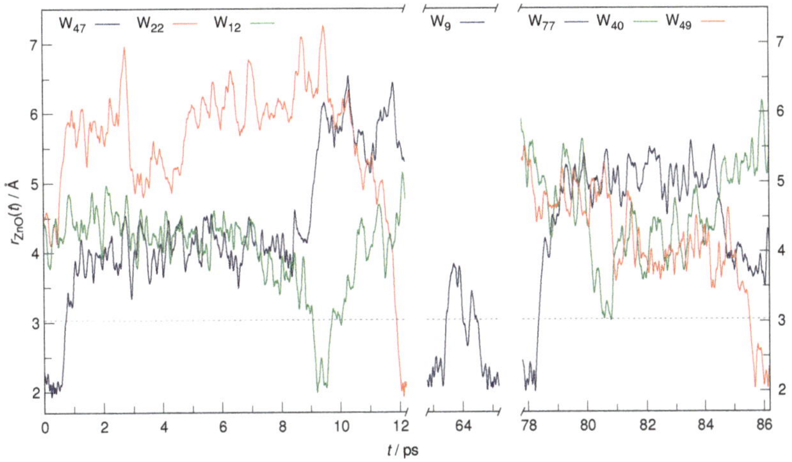

Figure 7. Major successful and attempted exchange events registered during the AIMD production trajectory as depicted by the Zn$\cdots$O distance of the participating water molecules. The W_n labels refer to the arbitrary index of the H_2O molecule in the trajectory. Line colors indicate the leaving molecule (blue), the incoming molecule (red), as well as the intervening molecule (green). Black dotted line shows the location of the first minimum in the Zn$\cdots$O radial distribution function (3.03 Å). Note the discontinuous abscissa axis.

The second exchange event is only an attempted escape of W_9 from $Zn(H_2O)_6^{2+}$ at ca. 64 ps. However, no other H_2O molecule tries to enter the first hydration shell of Zn(II) this time and ultimately, after transiently wandering off to the verge of the second shell, W_9 is recaptured by the complex and the stable octahedral arrangement of ligands persists.

The third exchange event (and the second successful one) resembles the first one to a great extent. It begins with the dissociation of W_{77} at ca. 78.3 ps. This H_2O molecule is released initially to the second hydration shell, but quickly travels further to ultimately stay at ca. 5 Å separation from zinc(II) for an extended period, which is the location of the shallow minimum between the second and the poorly defined third hydration shell of the cation (cf. Figure 1). Interestingly, this perturbation of the hydration shells leads to two concerted movements of water molecules towards the central complex, initially that of W_{40}, quickly followed by W_{49}. Similarly to the first exchange event, the intervening water molecule (here, W_{40}) ultimately does not manage to become properly coordinated to the cation. Instead, it remains at the very edge of the first hydration shell, only sligthly perturbing the newly formed $Zn(H_2O)_5^{2+}$ complex. After its release back into the second hydration shell at ca. 81 ps, the pentacoordinated complex remains stable for almost 4 ps. Then, an interesting cascade of concerted events can be noticed. W_{77} returns to the second hydration shell, apparently swapping its place with W_{40}, but at the same time W_{49} manages to get captured by zinc(II), reinstating the hexacoordinated complex. Again, concerted movement of H_2O molecules seems to be crucial for the mutually assisted exchanges between solvation shells leading to coordination number change of Zn^{2+}(aq).

4. Discussion

The coordination of Zn^{2+}(aq) has been hitherto the subject of numerous experimental and computational studies, as reviewed in the Introduction. There is an overwhelming consensus that Zn(II) forms a hexahydrated complex with the structure of a regular octahedron (point group O_h for the ZnO_6 core) [9–11,14–21,25,27,32,34–36,38,40–48,50,93]. Reports claiming predominantly pentacoordinated Zn^{2+}(aq) do surface [28,29,31,33,49,94], but appear to suffer from a size effect, i.e., the undercoordination of Zn(II) occurs most commonly in small aqueous clusters and the $Zn(H_2O)_6^{2+}$ complex is uniformly restored in the bulk limit [32,40]. Most notably for the present study, the coordination number of Zn(II) in an aqueous solution found from AIMD simulations seems to be dependent on the computational protocol, with dispersion corrections deemed particularly necessary to stabilize the $Zn(H_2O)_6{}^{2+}$ complex [50].

The Zn$\cdots$O distances found experimentally and in computational investigations have been compiled by several authors [3,9,15,23,42,46,48,95]. EXAFS measurements find r_{ZnO} = 2.05–2.13 Å [13–17], while a survey of reported XRD investigations leads to a possible range of r_{ZnO} = 2.08–2.17 Å [3,18–22]. The available AIMD simulations report also a considerable spread of r_{ZnO} = 2.02–2.18 Å [47–50] while a slightly narrower range (r_{ZnO} = 2.11–2.16 Å) is reported from QM/MM MD simulations [42–46]. In comparison, our global average of the position of the first peak in $g_{ZnO}(r)$ (2.11 Å, see Figure 1 and Table 1) is generally in the middle of the reported distances.

The particular advantage of our analysis is the rigorous separation of the trajectory fragments with hexa- vs. pentacoordinated Zn(II), thanks to the smooth instantaneous coordination number, see Figure 2. This allowed us to prove the tightening of the hydration shells of Zn^{2+}(aq) with the dissociation of one H_2O from the complex. Our results for the most probable Zn$\cdots$O distance in $Zn(H_2O)_6{}^{2+}$ vs $Zn(H_2O)_5{}^{2+}$ (2.13 Å and 2.07 Å, respectively, cf. Figure 3) are in good agreement with earlier AIMD simulations, a similar separation of trajectory fragments was proposed (2.11 Å and 2.05 Å, respectively [50]). Static geometry optimizations also lead to a general conclusion that the water molecules are located closer to Zn(II) in $Zn(H_2O)_5{}^{2+}$ than in $Zn(H_2O)_6{}^{2+}$ [26,27,29,31,32]. The hexacoordinated complex always maintains the octahedral arrangement of water oxygens (although the inclusion of hydrogens lowers the point group to T_h [9]), which is also found in our time-averaged structure, cf. Figure 4. More controversial is the preferred geometry of $Zn(H_2O)_5{}^{2+}$. Based on geometry optimizations, both trigonal bipyramid (point group D_{3h}) [26,31,32] and square pyramid (point group C_{2v}) [27,29,94] have been confirmed as minima of the potential energy surface. It appears though that the presence of second shell H_2O molecules stabilizes the former structure [29,40,94]. The presence of external hydration shells is thus crucial to the observation of the time-averaged trigonal bipyramidal structure that we report here, see Figure 4. The discrepancy between the Zn$\cdots$O distance to equatorial and axial molecules (2.04 Å and 2.14 Å, respectively) is similar to that found previously in an isolated cluster (2.05 Å and 2.13 Å, respectively [26]).

Notably for AIMD simulations, our considerable system size with 100 H_2O molecules allows us to record the RDF for the entire second hydration shell, as well as for the weakly formed third shell, see Figure 1. Due to the structure-making nature of Zn^{2+}(aq), the second shell is also accessible experimentally. Previous EXAFS results indicate a second shell of 12.2 O atoms at an average distance of 4.40 Å [16] or 11.2 O atoms at 4.10 Å [14]. The distance to second shell from XRD measurements equals 4.02–4.26 Å [20–22]. Results from AIMD and QM/MM MD simulations generally support these experimental findings, with second hydration shell composed of 13–15 molecules at an average distance 4.3–4.5 Å [42,43,48,50]. In comparison, our global average of the position of the second peak in $g_{ZnO}(r)$ (4.11 Å, see Figure 1 and Table 1) is generally in the lower limit of the reported distances, while the number of O atoms (15.2) indicates a fairly crowded second hydration shell of Zn(II). Most importantly, the lowering of CN of the cation results in a major compression of this shell, with the RDF peak shifted to 4.07 Å (see Figure 3) and at least two water molecules released to the bulk solution, as the number of O atoms in the second shell decreases to 13.

Earlier studies on monovalent ions found that they are strongly polarized by water, but in turn do not polarize the solvent significantly (in the sense of only slight alteration of H_2O dipole moment) [54,55,60]. In strike contrast to these findings, the water molecules in the first hydration shell of Zn^{2+}(aq) are heavily polarized by the electric field of the cation, see Figure 6. For all H_2O molecules, we find an average dipole moment of 2.78 D, which is slightly lower than the value usually found in AIMD simulations of liquid water ($\sim$3 D) [54,60]. However, mean dipole moment of liquid H_2O for revPBE/revPBE0 functionals is indeed slightly lower and in agreement with the present findings [96,97]. While the increase of dipole moment of H_2O near the cation was previously found in MD simulations with the polarizable Amoeba force field, as well as AIMD simulations [38,48], the clear advantage of our analysis is the ability to separate the two forms of hydrated Zn(II). In the hexacoordinated complex, the dipole moment of the first shell water molecules increases by 0.65 D with respect to the bulk. Dissociation of this complex leads to a further increase of the average dipole moment and it is almost 1 D greater than for non-coordinated H_2O. Thus, the water molecules in $Zn(H_2O)_5{}^{2+}$ are evidently more heavily polarized than in $Zn(H_2O)_6{}^{2+}$. The 0.6 D increase of the mean dipole moment of first hydration shell H_2O has been noted previously in AIMD simulations [48]. Simulations with the Amoeba force field indicate that the water molecules closest to the cation have their dipole moments raised by more than 1 D [38].

Computational methods are excellent tools for studying solvent exchange mechanism around solvated ions [62]. The experimental estimate of water residence time on Zn(II) from quasi-elastic neutron scattering is a rather broad range of 0.1–5 ns [24]. Assuming the actual residence time is near the lower limit, it is in principle accessible to longer MD simulations. Previously, only force field description of the potential energy surface could provide the possibilities to reach the required time scale. However, the exact formulation of Zn(II)–H_2O interactions is crucial for enabling solvent exchange events and they were frequently not observed even in prolonged MD simulations [16,34–36]. On the other hand, a water residence time of 2.2 ns was found in an MD simulation with the Amoeba force field [38] and a rate constant for water exchange equal 4.1×10^8 s^{-1} was also obtained [37]. Most AIMD or QM/MM MD simulations to date suffered from length limitations and no exchange events were observed [42–48]. In stark contrast to these earlier results, we managed to register two full solvent exchange events in the first hydration shell of Zn(II). While it is impossible to quantitatively estimate the water residence time from such a small sample, we note that the starts of successful exchange events are separated by ca. 78 ps, so that our results are not in disagreement with the experimental estimates provided the true residence time is close to the lower limit proposed previously [24]. We note here that the solvent exchange may be studied using accelerated sampling, e.g., metadynamics [37,50,93]; however, while such an approach provides the details of the potential energy surface of the hydrated cation, only equilibrium simulations can capture the actual timescale of the observed events.

Unfortunately, the question of how the two complexes differ on the free energy scale cannot be answered directly from our equilibrium simulation. However, we can roughly estimate this free energy difference from the fractions of the trajectory attributed to the penta- and hexacoordinated structures as $\Delta G = -RT\ln(0.163/0.812) \approx 4$ kJ mol^{-1}. On the one hand, this is high enough that the pentacoordinated structure might prove too elusive to be registered experimentally, e.g., with EXAFS, noting also that the average coordination number of Zn(II) (5.8) is within the typical uncertainty of structure determination methods from the value of 6.0, indicative of perfect hexacoordination. On the other hand, this is low enough (only 1.6RT) to allow relatively free interconversion between the two structures. We note here that the free energy difference between $[UO_2(H_2O)_5]^{2+}$(aq) and $[UO_2(H_2O)_4]^{2+}$(aq) from ab initio metadynamics is $\sim$3 kJ mol^{-1}, with similarly inconclusive experimental results as in the case of Zn(II) [98].

Our results indicate that the solvent exchange around Zn(II) follows a dissociative mechanism, cf. Figure 7. The release of a single water molecule from the first hydration shell leads to a transient $Zn(H_2O)_5{}^{2+}$ complex, the lifetime of which may be tentatively estimated as 8–10 ps, based on the limited sample of events we registered. A similar timescale of dissociative exchange events was noted before by Fatmi et al. [43]. A similar mechanism of solvent exchange around hexacoordinated Zn(II) was also inferred from static QM calculations [29,30,62]. Interestingly, the dissociation of a water molecule does not lead to increased CN in the second shell. On the contrary, it undergoes compression and major structural changes, as evident from Figure 3. Consequently, the registered exchange events require concerted movement of assisting solvent molecules, cf. Figure 7.

This cooperative movement of H_2O molecules enables the capture of a newly coordinated water molecule from as far as the third solvation shell of Zn(II).

5. Conclusions

In this paper, the structure of the hydrated zinc(II) cation has been studied using extensive AIMD simulations. In contrast to numerous earlier MD investigations at different levels of theory, which frequently found only hexacoordinated zinc with fixed first hydration shell waters, we clearly identify the presence of both hexa- and pentacoordinated complexes of Zn^{2+}. While the octahedral $Zn(H_2O)_6{}^{2+}$ remains the dominant form of Zn(II) in an aqueous solution, $Zn(H_2O)_5{}^{2+}$ is present in about 16% of the trajectory. The average structure of the latter is a trigonal bipyramid and its presence results in major structural changes reaching up to the second hydration shell and beyond. Radial distribution functions indicate major compression of the second hydration shell of the pentacoordinated complex and molecular dipole distributions show profound polarization of first shell H_2O molecules that is also heavily dependent on CN.

Owing to the considerable length of the generated trajectory, we were able to register two full solvent exchange events around aqueous Zn^{2+}. They undoubtedly follow a dissociative mechanism and are driven by concerted solvent movement that may feature molecules up to the third hydration shell of Zn(II). While, due to the limited sampling, we cannot quantitatively estimate the water residence time on the zinc cation, the separation of successful events by ca. 80 ps seems to be consistent with the lower limit of experimental residence times (about 0.1 ns). The single exchange event, accompanied by the presence of $Zn(H_2O)_5{}^{2+}$, seems to take about 8–10 ps.

We conclude that equilibrium AIMD simulations on an adequate length scale provide a direct way of studying solvent exchange events around aqueous ions in molecular detail. The smooth selector function for coordinated solvent is a powerful tool that allows the identification of solvent exchange events and their preferred (associative or dissociative) mechanism. The analysis framework presented here may be tremendously useful for other solvated ions and more complicated systems containing Zn(II), such as concentrated zinc halide solutions.

Supplementary Materials: The following supporting information can be downloaded at: https://www.mdpi.com/article/10.3390/liquids2030015/s1, Figure S1: Cartesian coordinates of the structures from Figure 4.

Author Contributions: Conceptualization, methodology, supervision, writing—original draft preparation, M.Ś.; investigation, formal analysis, visualization, writing—review and editing, A.M. and M.Ś. All authors have read and agreed to the published version of the manuscript.

Funding: This research received no external funding.

Data Availability Statement: The data presented in this study are available on request from the corresponding author.

Acknowledgments: Calculations were performed at the Academic Computer Center in Gdańsk (TASK).

Conflicts of Interest: The authors declare no conflict of interest.

Abbreviations

The following abbreviations are used in this manuscript:

AIMD	ab initio molecular dynamics
CN	coordination number
DFT	density functional theory
DFTB	density functional tight binding
EXAFS	extended X-ray absorption fine structure

GTH	Goedecker–Teter–Hutter (pseudopotentials)
MD	molecular dynamics
MLWF	maximally localized Wannier function
ND	neutron diffraction
QM	quantum mechanics
QM/MM	quantum mechanics/molecular mechanics
revPBE	revised Perdew–Burke–Ernzerhof (functional)
XRD	X-ray diffraction

References

1. Coleman, J.E. Zinc enzymes. *Curr. Opin. Chem. Biol.* **1998**, *2*, 222–234. [CrossRef]
2. Maret, W. Zinc Biochemistry: From a Single Zinc Enzyme to a Key Element of Life. *Adv. Nutr.* **2013**, *4*, 82–91. [CrossRef]
3. Ohtaki, H.; Radnai, T. Structure and Dynamics of Hydrated Ions. *Chem. Rev.* **1993**, *93*, 1157–1204. [CrossRef]
4. Bakker, H.J. Structural Dynamics of Aqueous Salt Solutions. *Chem. Rev.* **2008**, *108*, 1456–1473. [CrossRef]
5. Marcus, Y. Effect of Ions on the Structure of Water: Structure Making and Breaking. *Chem. Rev.* **2009**, *109*, 1346–1370. [CrossRef]
6. Kristiansson, O.; Eriksson, A.; Lindgren, J. Hydration of Ions in Aqueous Solutions Studied by Infrared Spectroscopy. II. Application. *Acta Chem. Scand. A* **1984**, *38*, 613–618. [CrossRef]
7. Stangret, J.; Libuś, Z. Influence of Zn(II), Mn(II), and Mg(II) Cations on the Vibrational Spectra of Water in Aqueous Perchlorate Solutions. *Spectrosc. Lett.* **1988**, *21*, 397–412. [CrossRef]
8. Å. Bergström, P.; Lindgren, J.; Sandström, M.; Zhou, Y. Infrared Spectroscopic Study on the Hydration of Mercury(II), Cadmium(II), and Zinc(II) in Aqueous Solution and in the Hexahydrated Perchlorate Salts. *Inorg. Chem.* **1992**, *31*, 150–152. [CrossRef]
9. Rudolph, W.W.; Pye, C.C. Zinc(II) hydration in aqueous solution. A Raman spectroscopic investigation and an ab-initio molecular orbital study. *Phys. Chem. Chem. Phys.* **1999**, *1*, 4583–4593. [CrossRef]
10. Rudolph, W.W.; Pye, C.C. Zinc(II) Hydration in Aqueous Solution: A Raman Spectroscopic Investigation and An ab initio Molecular Orbital Study of Zinc(II) Water Clusters. *J. Solut. Chem.* **1999**, *28*, 1045–1070. [CrossRef]
11. Mink, J.; Németh, C.; Hajba, L.; Sandström, M.; Goggin, P.L. Infrared and Raman spectroscopic and theoretical studies of hexaaqua metal ions in aqueous solution. *J. Mol. Struct.* **2003**, *661–662*, 141–151. [CrossRef]
12. Wei, Z.F.; Zhang, Y.H.; Zhao, L.J.; Liu, J.H.; Li, X.H. Observation of the First Hydration Layer of Isolated Cations and Anions through the FTIR-ATR Difference Spectra. *J. Phys. Chem. A* **2005**, *109*, 1337–1342. [CrossRef]
13. Miyanaga, T.; Watanabe, I.; Ikeda, S. Amplitude in EXAFS and Ligand Exchange Reaction of Hydrated 3d Transition Metal Complexes. *Chem. Lett.* **1988**, *17*, 1073–1076. [CrossRef]
14. noz Páez, A.M.; Pappalardo, R.R.; Marcos, E.S. Determination of the Second Hydration Shell of Cr^{3+} and Zn^{2+} in Aqueous Solutions by Extended X-ray Absorption Fine Structure. *J. Am. Chem. Soc.* **1995**, *117*, 11710–11720. [CrossRef]
15. Kuzmin, A.; Obst, S.; Purans, J. X-ray absorption spectroscopy and molecular dynamics studies of Zn^{2+} hydration in aqueous solutions. *J. Phys. Condens. Matter* **1997**, *9*, 10065–10078. [CrossRef]
16. D'Angelo, P.; Barone, V.; Chillemi, G.; Sanna, N.; Meyer-Klaucke, W.; Pavel, N.V. Hydrogen and Higher Shell Contributions in Zn^{2+}, Ni^{2+}, and Co^{2+} Aqueous Solutions: An X-ray Absorption Fine Structure and Molecular Dynamics Study. *J. Am. Chem. Soc.* **2002**, *124*, 1959–1967. [CrossRef]
17. Migliorati, V.; Mancini, G.; Tatoli, S.; Zitolo, A.; Filipponi, A.; De Panfilis, S.; Di Cicco, A.; D'Angelo, P. Hydration Properties of the Zn^{2+} Ion in Water at High Pressure. *Inorg. Chem.* **2013**, *52*, 1141–1150. [CrossRef]
18. Bol, W.; Gerrits, G.J.A.; Van Panthaleon Van Eck, C.L. The Hydration of Divalent Cations in Aqueous Solution. An X-ray Investigation with Isomorphous Replacement. *J. Appl. Cryst.* **1970**, *3*, 486–492. [CrossRef]
19. Ohtaki, H.; Yamaguchi, T.; Maeda, M. X-Ray Diffraction Studies of the Structures of Hydrated Divalent Transition-Metal Ions in Aqueous Solution. *Bull. Chem. Soc. Jpn.* **1976**, *49*, 701–708. [CrossRef]
20. Licheri, G.; Paschina, G.; Piccaluga, G.; Pinna, G. X-ray Diffraction Study of Aqueous Solutions of $ZnSO_4$. *Z. Naturforsch.* **1982**, *37*, 1205–1210. [CrossRef]
21. Radnai, T.; Palinkas, G.; Caminiti, R. X-Ray Diffraction Study on Hydration and Ion-Pairing in Aqueous $ZnSO_4$ Solution. *Z. Naturforsch.* **1982**, *37*, 1247–1252. [CrossRef]
22. Dagnall, S.P.; Hague, D.N.; Towl, A.D.C. X-ray Diffraction Study of Aqueous Zinc(II) Nitrate. *J. Chem. Soc. Faraday Trans. 2 Mol. Chem. Phys.* **1982**, *78*, 2161–2167. [CrossRef]
23. Powell, D.H.; Gullidge, P.M.N.; Neilson, G.W.; Bellissent-Funel, M.C. Zn^{2+} hydration and complexation in aqueous electrolyte solutions. *Mol. Phys.* **1990**, *71*, 1107–1116. [CrossRef]
24. Salmon, P.S.; Bellissent-Funel, M.C.; Herdman, G.J. The dynamics of aqueous Zn^{2+} solutions: A study using incoherent quasi-elastic neutron scattering. *J. Phys. Condens. Matter* **1990**, *2*, 4297–4309. [CrossRef]
25. Mhin, B.J.; Lee, S.; Cho, S.J.; Lee, K.; Kim, K.S. $Zn(H_2O)_6^{2+}$ is very stable among aqua-Zn(II) ions. *Chem. Phys. Lett.* **1992**, *197*, 77–80. [CrossRef]
26. Bock, C.W.; Katz, A.K.; Glusker, J.P. Hydration of Zinc Ions: A Comparison with Magnesium and Beryllium Ions. *J. Am. Chem. Soc.* **1995**, *117*, 3754–3765. [CrossRef]

27. Lee, S.; Kim, J.; Park, J.K.; Kim, K.S. Ab Initio Study of the Structures, Energetics, and Spectra of Aquazinc(II). *J. Phys. Chem.* **1996**, *100*, 14329–14338. [CrossRef]
28. Hartmann, M.; Clark, T.; van Eldik, R. Theoretical Study of the Water Exchange Reaction on Divalent Zinc Ion using Density Functional Theory. *J. Mol. Model.* **1996**, *2*, 354–357. [CrossRef]
29. Hartmann, M.; Clark, T.; van Eldik, R. Hydration and Water Exchange of Zinc(II) Ions. Application of Density Functional Theory. *J. Am. Chem. Soc.* **1997**, *119*, 7843–7850. [CrossRef]
30. Rotzinger, F.P. Mechanism of Water Exchange for the Di- and Trivalent Metal Hexaaqua Ions of the First Transition Series. *J. Am. Chem. Soc.* **1997**, *119*, 5230–5238. [CrossRef]
31. Pavlov, M.; Siegbahn, P.E.M.; Sandström, M. Hydration of Beryllium, Magnesium, Calcium, and Zinc Ions Using Density Functional Theory. *J. Phys. Chem. A* **1998**, *102*, 219–228. [CrossRef]
32. De, S.; Ali, S.M.; Ali, A.; Gaikar, V.G. Micro-solvation of the Zn^{2+} ion—A case study. *Phys. Chem. Chem. Phys.* **2009**, *11*, 8285–8294. [CrossRef] [PubMed]
33. Cooper, T.E.; O'Brien, J.T.; Williams, E.R.; Armentrout, P.B. Zn^{2+} Has a Primary Hydration Sphere of Five: IR Action Spectroscopy and Theoretical Studies of Hydrated Zn^{2+} Complexes in the Gas Phase. *J. Phys. Chem. A* **2010**, *114*, 12646–12655. [CrossRef]
34. Obst, S.; Bradaczek, H. Molecular Dynamics Simulations of Zinc Ions in Water Using CHARMM. *J. Mol. Model.* **1997**, *3*, 224–232. [CrossRef]
35. Chillemi, G.; D'Angelo, P.; Pavel, N.V.; Sanna, N.; Barone, V. Development and Validation of an Integrated Computational Approach for the Study of Ionic Species in Solution by Means of Effective Two-Body Potentials. The Case of Zn^{2+}, Ni^{2+}, and Co^{2+} in Aqueous Solutions. *J. Am. Chem. Soc.* **2002**, *124*, 1968–1976. [CrossRef]
36. Arab, M.; Bougeard, D.; Smirnov, K.S. Molecular dynamics study of the structure and dynamics of Zn^{2+} ion in water. *Chem. Phys. Lett.* **2003**, *379*, 268–276. [CrossRef]
37. Inada, Y.; Mohammed, A.M.; Loeffler, H.H.; Funahashi, S. Water-Exchange Mechanism for Zinc(II), Cadmium(II), and Mercury(II) Ions in Water as Studied by Umbrella-Sampling Molecular-Dynamics Simulations. *Helv. Chim. Acta* **2005**, *88*, 461–469. [CrossRef]
38. Wu, J.C.; Piquemal, J.P.; Chaudret, R.; Reinhardt, P.; Ren, P. Polarizable Molecular Dynamics Simulation of Zn(II) in Water Using the AMOEBA Force Field. *J. Chem. Theory Comput.* **2010**, *6*, 2059–2070. [CrossRef]
39. Yu, H.; Whitfield, T.W.; Harder, E.; Lamoureux, G.; Vorobyov, I.; Anisimov, V.M.; MacKerell, A.D.; Roux, B. Simulating Monovalent and Divalent Ions in Aqueous Solution Using a Drude Polarizable Force Field. *J. Chem. Theory Comput.* **2010**, *6*, 774–786. [CrossRef]
40. Jana, C.; Ohanessian, G.; Clavaguéra, C. Theoretical insight into the coordination number of hydrated Zn^{2+} from gas phase to solution. *Theor. Chem. Acc.* **2016**, *135*, 141. [CrossRef]
41. Xu, M.; Zhu, T.; Zhang, J.Z.H. Molecular Dynamics Simulation of Zinc Ion in Water with an ab Initio Based Neural Network Potential. *J. Phys. Chem. A* **2019**, *123*, 6587–6595. [CrossRef] [PubMed]
42. Mohammed, A.M.; Loeffler, H.H.; Inada, Y.; Tanada, K.; Funahashi, S. Quantum mechanical/molecular mechanical molecular dynamic simulation of zinc(II) ion in water. *J. Mol. Liquids* **2005**, *119*, 55–62. [CrossRef]
43. Fatmi, M.Q.; Hofer, T.S.; Randolf, B.R.; Rode, B.M. An extended *ab initio* QM/MM MD approach to structure and dynamics of Zn(II) in aqueous solution. *J. Chem. Phys.* **2005**, *123*, 054514. [CrossRef] [PubMed]
44. Brancato, G.; Rega, N.; Barone, V. Microsolvation of the Zn(II) ion in aqueous solution: A hybrid QM/MM MD approach using non-periodic boundary conditions. *Chem. Phys. Lett.* **2008**, *451*, 53–57. [CrossRef]
45. Rega, N.; Brancato, G.; Petrone, A.; Caruso, P.; Barone, V. Vibrational analysis of x-ray absorption fine structure thermal factors by ab initio molecular dynamics: The Zn(II) ion in aqueous solution as a case study. *J. Chem. Phys.* **2011**, *134*, 074504. [CrossRef]
46. Riahi, S.; Roux, B.; Rowley, C.N. QM/MM Molecular Dynamics Simulations of the Hydration of Mg(II) and Zn(II) Ions. *J. Chem. Phys.* **2013**, *99*, 1–9. [CrossRef]
47. Fujiwara, T.; Mochizuki, Y.; Komeiji, Y.; Okiyama, Y.; Mori, H.; Nakano, T.; Miyoshi, E. Fragment molecular orbital-based molecular dynamics (FMO-MD) simulations on hydrated Zn(II) ion. *Chem. Phys. Lett.* **2010**, *490*, 41–45. [CrossRef]
48. Cauët, E.; Bogatko, S.; Weare, J.H.; Fulton, J.L.; Schenter, G.K.; Bylaska, E.J. Structure and dynamics of the hydration shells of the Zn^{2+} ion from ab initio molecular dynamics and combined ab initio and classical molecular dynamics simulations. *J. Chem. Phys.* **2010**, *132*, 194502. [CrossRef]
49. Liu, X.; Lu, X.; Wang, R.; Meijer, E.J. Understanding hydration of Zn^{2+} in hydrothermal fluids with ab initio molecular dynamics. *Phys. Chem. Chem. Phys.* **2011**, *13*, 13305–13309. [CrossRef]
50. Ducher, M.; Pietrucci, F.; Balan, E.; Ferlat, G.; Paulatto, L.; Blanchard, M. van der Waals Contribution to the Relative Stability of Aqueous Zn(2+) Coordination States. *J. Chem. Theory Comput.* **2017**, *13*, 3340–3347. [CrossRef]
51. Marx, D.; Hutter, J. *Ab Initio Molecular Dynamics;* Cambridge University Press: Cambridge, UK, 2009.
52. Hassanali, A.A.; Cuny, J.; Verdolino, V.; Parrinello, M. Aqueous solutions: State of the art in ab initio molecular dynamics. *Phil. Trans. R. Soc. A* **2014**, *372*, 20120482. [CrossRef] [PubMed]
53. Marx, D.; Sprik, M.; Parrinello, M. Ab initio molecular dynamics of ion solvation. The case of Be^{2+} in water. *Chem. Phys. Lett.* **1997**, *273*, 360–366. [CrossRef]
54. Śmiechowski, M.; Forbert, H.; Marx, D. Spatial decomposition and assignment of infrared spectra of simple ions in water from mid-infrared to THz frequencies: Li^{+}(aq) and F^{-}(aq). *J. Chem. Phys.* **2013**, *139*, 014506. [CrossRef] [PubMed]

55. Śmiechowski, M.; Sun, J.; Forbert, H.; Marx, D. Solvation shell resolved THz spectra of simple aqua ions – distinct distance- and frequency-dependent contributions of solvation shells. *Phys. Chem. Chem. Phys.* **2015**, *17*, 8323–8329. [CrossRef]
56. Chaudhari, M.I.; Soniat, M.; Rempe, S.B. Octa-Coordination and the Aqueous Ba^{2+} Ion. *J. Phys. Chem. B* **2015**, *119*, 8746–8753. [CrossRef]
57. Schienbein, P.; Schwaab, G.; Forbert, H.; Havenith, M.; Marx, D. Correlations in the Solute–Solvent Dynamics Reach Beyond the First Hydration Shell of Ions. *J. Phys. Chem. Lett.* **2017**, *8*, 2373–2380. [CrossRef]
58. Giacobello, F.; Mollica-Nardo, V.; Foti, C.; Ponterio, R.C.; Saija, F.; Trusso, S.; Sponer, J.; Cassone, G.; Giuffrè, O. Hydrolysis of Al^{3+} in Aqueous Solutions: Experiments and Ab Initio Simulations. *Liquids* **2022**, *2*, 26–38. [CrossRef]
59. Gregory, K.P.; Elliott, G.R.; Wanless, E.J.; Webber, G.B.; Page, A.J. A quantum chemical molecular dynamics repository of solvated ions. *Sci. Data* **2022**, *9*, 430. [CrossRef]
60. Śmiechowski, M. Unusual Influence of Fluorinated Anions on the Stretching Vibrations of Liquid Water. *J. Phys. Chem. B* **2018**, *122*, 3141–3152. [CrossRef]
61. Helm, L.; Merbach, A.E. Water exchange on metal ions: Experiments and simulations. *Coord. Chem. Rev.* **1999**, *187*, 151–181. [CrossRef]
62. Rotzinger, F.P. Treatment of Substitution and Rearrangement Mechanisms of Transition Metal Complexes with Quantum Chemical Methods. *Chem. Rev.* **2005**, *105*, 2003–2037. [CrossRef] [PubMed]
63. Kell, G.S. Density, Thermal Expansivity, and Compressibility of Liquid Water from 0 °C to 150 °C: Correlations and Tables for Atmospheric Pressure and Saturation Reviewed and Expressed on 1968 Temperature Scale. *J. Chem. Eng. Data* **1975**, *20*, 97–105. [CrossRef]
64. Pan, P.; Tremaine, P.R. Thermodynamics of aqueous zinc: Standard partial molar heat capacities and volumes of Zn^{2+}(aq) from 10 to 55 °C. *Geochim. Cosmochim. Acta* **1994**, *58*, 4867–4874. [CrossRef]
65. Marcus, Y. The Standard Partial Molar Volumes of Ions in Solution. Part 4. Ionic Volumes in Water at 0–100 °C. *J. Phys. Chem. B* **2009**, *113*, 10285–10291. [CrossRef] [PubMed]
66. Abraham, M.J.; Murtola, T.; Schulz, R.; Páll, S.; Smith, J.C.; Hess, B.; Lindahl, E. GROMACS: High performance molecular simulations through multi-level parallelism from laptops to supercomputers. *SoftwareX* **2015**, *1–22*, 19–25. [CrossRef]
67. Śmiechowski, M. Anion–water interactions of weakly hydrated anions: Molecular dynamics simulations of aqueous $NaBF_4$ and $NaPF_6$. *Mol. Phys.* **2016**, *114*, 1831–1846. [CrossRef]
68. Lagardère, L.; Jolly, L.H.; Lipparini, F.; Aviat, F.; Stamm, B.; Jing, Z.F.; Harger, M.; Torabifard, H.; Cisneros, G.A.; Schnieders, M.J.; et al. Tinker-HP: A massively parallel molecular dynamics package for multiscale simulations of large complex systems with advanced point dipole polarizable force fields. *Chem. Sci.* **2018**, *9*, 956–972. [CrossRef]
69. Laury, M.L.; Wang, L.P.; Pande, V.S.; Head-Gordon, T.; Ponder, J.W. Revised Parameters for the AMOEBA Polarizable Atomic Multipole Water Model. *J. Phys. Chem. B* **2015**, *119*, 9423–9437. [CrossRef]
70. The cp2k Developers Group. cp2k v. 6.0, 2001–2018. Available online: http://www.cp2k.org/ (accessed on 13 September 2022).
71. Hutter, J.; Iannuzzi, M.; Schiffmann, F.; VandeVondele, J. cp2k: Atomistic simulations of condensed matter systems. *WIREs Comput. Mol. Sci.* **2014**, *4*, 15–25. [CrossRef]
72. Kühne, T.D.; Iannuzzi, M.; Ben, M.D.; Rybkin, V.V.; Seewald, P.; Stein, F.; Laino, T.; Khaliullin, R.Z.; Schütt, O.; Schiffmann, F.; et al. CP2K: An electronic structure and molecular dynamics software package—Quickstep: Efficient and accurate electronic structure calculations. *J. Chem. Phys.* **2020**, *152*, 194103. [CrossRef]
73. VandeVondele, J.; Krack, M.; Mohamed, F.; Parrinello, M.; Chassaing, T.; Hutter, J. Quickstep: Fast and accurate density functional calculations using a mixed Gaussian and plane waves approach. *Comp. Phys. Commun.* **2005**, *167*, 103–128. [CrossRef]
74. Zhang, Y.; Yang, W. Comment on "Generalized Gradient Approximation Made Simple". *Phys. Rev. Lett.* **1998**, *80*, 890. [CrossRef]
75. Perdew, J.P.; Burke, K.; Ernzerhof, M. Generalized Gradient Approximation Made Simple. *Phys. Rev. Lett.* **1996**, *77*, 3865–3868. [CrossRef] [PubMed]
76. Marques, M.A.L.; Oliveira, M.J.T.; Burnus, T. Libxc: A library of exchange and correlation functionals for density functional theory. *Comp. Phys. Commun.* **2012**, *183*, 2272–2281. [CrossRef]
77. Marsalek, O.; Markland, T.E. Quantum Dynamics and Spectroscopy of Ab Initio Liquid Water: The Interplay of Nuclear and Electronic Quantum Effects. *J. Phys. Chem. Lett.* **2017**, *8*, 1545–1551. [CrossRef] [PubMed]
78. Duignan, T.T.; Baer, M.D.; Schenter, G.K.; Mundy, C.J. Real single ion solvation free energies with quantum mechanical simulation. *Chem. Sci.* **2017**, *8*, 6131–6140. [CrossRef] [PubMed]
79. Macchieraldo, R.; Esser, L.; Elfgen, R.; Voepel, P.; Zahn, S.; Smarsly, B.M.; Kirchner, B. Hydrophilic Ionic Liquid Mixtures of Weakly and Strongly Coordinating Anions with and without Water. *ACS Omega* **2018**, *3*, 8567–8582. [CrossRef] [PubMed]
80. Ohto, T.; Dodia, M.; Imoto, S.; Nagata, Y. Structure and Dynamics of Water at the Water–Air Interface Using First-Principles Molecular Dynamics Simulations within Generalized Gradient Approximation. *J. Chem. Theory Comput.* **2019**, *15*, 595–602. [CrossRef] [PubMed]
81. Ohto, T.; Dodia, M.; Xu, J.; Imoto, S.; Tang, F.; Zysk, F.; Kühne, T.D.; Shigeta, Y.; Bonn, M.; Wu, X.; et al. Accessing the Accuracy of Density Functional Theory through Structure and Dynamics of the Water–Air Interface. *J. Chem. Theory Comput.* **2019**, *10*, 4914–4919. [CrossRef]
82. Lippert, G.; Hutter, J.; Parrinello, M. A hybrid Gaussian and plane wave density functional scheme. *Mol. Phys.* **1997**, *92*, 477–487. [CrossRef]

83. VandeVondele, J.; Hutter, J. Gaussian basis sets for accurate calculations on molecular systems in gas and condensed phases. *J. Chem. Phys.* **2007**, *127*, 114105. [CrossRef] [PubMed]
84. Galib, M.; Duignan, T.T.; Misteli, Y.; Baer, M.D.; Schenter, G.K.; Hutter, J.; Mundy, C.J. Mass density fluctuations in quantum and classical descriptions of liquid water. *J. Chem. Phys.* **2017**, *146*, 244501. [CrossRef] [PubMed]
85. Goedecker, S.; Teter, M.; Hutter, J. Separable dual-space Gaussian pseudopotentials. *Phys. Rev. B* **1996**, *54*, 1703–1710. [CrossRef] [PubMed]
86. Jonchiere, R.; Seitsonen, A.P.; Ferlat, G.; Saitta, A.M.; Vuilleumier, R. Van der Waals effects in *ab initio* water at ambient and supercritical conditions. *J. Chem. Phys.* **2011**, *135*, 154503. [CrossRef]
87. Grimme, S.; Antony, J.; Ehrlich, S.; Krieg, H. A consistent and accurate *ab initio* parametrization of density functional dispersion correction (DFT-D) for the 94 elements H-Pu. *J. Chem. Phys.* **2010**, *132*, 154104. [CrossRef] [PubMed]
88. Durrant, T.R.; Murphy, S.T.; Watkins, M.B.; Shluger, A.L. Relation between image charge and potential alignment corrections for charged defects in periodic boundary conditions. *J. Chem. Phys.* **2018**, *149*, 024103. [CrossRef]
89. Bussi, G.; Donadio, D.; Parrinello, M. Canonical sampling through velocity rescaling. *J. Chem. Phys.* **2007**, *126*, 014101. [CrossRef]
90. Marzari, N.; Vanderbilt, D. Maximally localized generalized Wannier functions for composite energy bands. *Phys. Rev. B* **1997**, *56*, 12847–12865. [CrossRef]
91. Humphrey, W.; Dalke, A.; Schulten, K. VMD: Visual Molecular Dynamics. *J. Mol. Graph.* **1996**, *14*, 33–38. [CrossRef]
92. Racine, J. gnuplot 4.0: A portable interactive plotting utility. *J. Appl. Econom.* **2006**, *21*, 133–141. [CrossRef]
93. Brancato, G.; Barone, V. Free Energy Landscapes of Ion Coordination in Aqueous Solution. *J. Phys. Chem. B* **2011**, *115*, 12875–12878. [CrossRef] [PubMed]
94. Cooper, T.E.; Carl, D.R.; Armentrout, P.B. Hydration Energies of Zinc(II): Threshold Collision-Induced Dissociation Experiments and Theoretical Studies. *J. Phys. Chem. A* **2009**, *113*, 13727–13741. [CrossRef] [PubMed]
95. Marcus, Y. Ionic Radii in Aqueous Solutions. *Chem. Rev.* **1988**, *88*, 1475–1498. [CrossRef]
96. Pestana, L.R.; Mardirossian, N.; Head-Gordon, M.; Head-Gordon, T. Ab initio molecular dynamics simulations of liquid water using high quality meta-GGA functionals. *Chem. Sci.* **2017**, *8*, 3554–3565. [CrossRef]
97. Pestana, L.R.; Marsalek, O.; Markland, T.E.; Head-Gordon, T. The Quest for Accurate Liquid Water Properties from First Principles. *J. Phys. Chem. Lett.* **2018**, *9*, 5009–5016. [CrossRef]
98. Atta-Fynn, R.; Bylaska, E.J.; de Jong, W.A. Free energies and mechanisms of water exchange around Uranyl from first principles molecular dynamics. *Mater. Res. Soc. Symp. Proc.* **2012**, *1383*, 113–118. [CrossRef]

liquids

Article

Solvation Structure and Ion–Solvent Hydrogen Bonding of Hydrated Fluoride, Chloride and Bromide—A Comparative QM/MM MD Simulation Study

Thomas S. Hofer

Theoretical Chemistry Division, Institute of General, Inorganic and Theoretical Chemistry, Center for Chemistry and Biomedicine, University of Innsbruck, Innrain 80-82, A-6020 Innsbruck, Austria; t.hofer@uibk.ac.at; Tel.: +43-512-507-57111; Fax: +43-512-507-57199

Abstract: In this study, the correlated resolution-of-identity Møller–Plesset perturbation theory of second order (RIMP2) ab initio level of theory has been combined with the newly parameterised, flexible SPC-mTR2 water model to formulate an advanced QM/MM MD simulation protocol to study the solvation properties of the solutes F^-, Cl^- and Br^- in aqueous solution. After the identification of suitable ion–water Lennard–Jones parameters for the QM/MM coupling, a total simulation period of 10 ps (equilibration) plus 25 ps (sampling) could be achieved for each target system at QM/MM conditions. The resulting simulation data enable an in-depth analysis of the respective hydration structure, the first shell ligand exchange characteristics and the impact of solute–solvent hydrogen bonding on the structural properties of first shell water molecules. While a rather unexpected tailing of the first shell ion–oxygen peak renders the identification of a suitable QM boundary region challenging, the presented simulation results provide a valuable primer for more advanced simulation approaches focused on the determination of single-ion thermodynamical properties.

Keywords: anionic hydration; hydrogen bonding; aqueous solution; ab initio; molecular dynamics; QM/MM

1. Introduction

Liquids and solutions are omnipresent states of matter in nature [1,2] and are oftentimes regarded as one of the most versatile tools available in chemistry [3–5]. Processes occurring in the liquid state can be steered in numerous ways by adjusting parameters of the reaction environment. The latter can be achieved for instance by comparably simple measures such as variations in temperature and pressure, or by more complex strategies such as the addition of cosolvents [6,7]. In particular, the addition of salt to a polar solvent may have a strong impact on the overall properties of the liquid [8–10], mainly due to the long-ranged nature of the comparatively strong Coulomb interactions [11], which is linked to a number of prevalent problems in physical and chemical sciences [12,13].

In particular, the requirement of electroneutrality makes it impossible to independently investigate the preferential solvation of cat- or anionic species via experimental means [13,14]. As a consequence, the structural and dynamical properties of any ionic species cannot be investigated without the impact of their respective counter ions. This limitation is extended to the field of ion thermodynamics [13], i.e., it is not possible to separate measured thermodynamic properties such as the solvation free energy and the associated enthalpic and entropic contributions of a dissolved salt into the respective cationic and anionic portion. While this is a known limitation often associated to calorimetric measurements, it can be shown that this shortcoming is directly linked to the potential of an electrochemical cell [15]. Also in this case, the electroneutrality constraint prevents an experimental determination of absolute single-electrode potentials.

An increasingly successful alternative to experimental approaches are theoretical calculations [16], foremost chemical simulation methods based on the Monte-Carlo (MC) and molecular dynamics (MD) frameworks [17–19]. These approaches aim at the description of a target system by collecting a large number of individual configurations associated to a thermodynamic ensemble [20,21]. In contrast to experimental measurements, the investigation of a single ionic species in the absence of counter ions is perfectly feasible using these simulation approaches [22,23]. In particular, simulation protocols based on a quantum chemical [24–27] description of the ion–solvent interactions have proven as a versatile approach to study the manifold properties of solvated ionic systems [28–31].

In addition to the determination of the respective structure and dynamics of the solvation complex, suitable protocols to also investigate thermodynamic quantities have been formulated [23,32–39]. These methods have been applied with remarkable success to characterise the hydration properties of monovalent cations including a consistent evaluation of the associated single-ion solvation free energy. The latter can then be directly related to the absolute single-electrode potential of the associated half-cell reactions [13]. Based on these results, it possible to anchor the entire series of electrochemical potentials without the need of a predefined reference such as the standard hydrogen electrode.

Despite the consistent results achieved in previous studies focused on cationic hydration [38,40], further validation of these results is required. Specifically, the determination of solvation free energies for anionic species appears as a crucial requirement to supplement the data obtained in the cationic case. However, the interaction between anions and the solvent water via hydrogen bonding to the anionic solute is much more intricate in the theoretical description for a number of reasons. First, hydrogen bonded interactions as observed in ion–solvent binding are more challenging in their description compared to the conceptually simpler charge–dipole interaction prevalent in the cationic case. Moreover, an anionic solute introduces excess electron density into the system which typically is more sensitive to shortcomings in the theoretical description than cations with a similar (but opposite) net charge. While the majority of theoretical calculations rely on methods based on density functional theory (DFT) [26,27], certain shortcomings such as the calibration to empirical data have been shown to lead to inaccurate results even in case of simpler cationic hydrates [41–43] and even pure water [44,45]. While the application of empirical dispersion corrections [46–49] may compensate some of these inaccuracies [50], the combination of a quantum mechanical (QM) calculation method with forcefield-like correction terms has also been criticised as combing the worst of both worlds, i.e., the demanding computational effort of QM approaches with the dependence on empirical parameters inherent to a forcefield (FF) description.

While being computationally much more expensive, methods based on perturbative approaches such as the Møller–Plesset perturbation theory of second order (MP2) [51] proved as an accurate and versatile alternative approach to describe the molecular interactions in QM-based studies. Especially in conjunction with the resolution-of-identity (RI) framework [52–54] capable of greatly accelerating the calculations with only a very minor reduction in accuracy, RIMP2-based hybrid quantum mechanical/molecular mechanical (QM/MM) simulations became feasible on modern computational infrastructure. Pioneered by the Nobel laureates Martin Karplus, Michael Levitt and Arieh Warshel [55] in the regime of biomolecular simulations [56–60], hybrid QM/MM MD simulations [56,58,61–63] have been applied with large success in the description of solvated ionic systems [28,29,64,65]. In this approach, the chemically most relevant part, comprised of the solute and solvent molecules in its immediate surroundings, is treated via a suitable quantum mechanical (QM) approach. In contrast, the interactions in the remainder of the simulation system are treated based on efficient molecular mechanical (MM) interaction models. That way, QM/MM approaches exploit the accuracy of a QM description inherently accounting for many-body contributions such as polarisation and charge-transfer in conjunction with the high efficiency of MM interaction potentials.

A large variety of MM potential models have been developed in the past, which is particularly true in case of the solvent water [66–68]. Based on arguments provided by statistical thermodynamics and quantum mechanics, the majority of these solvent models employ a rigid description of water, i.e., the individual water molecules can only perform translational and rotational motion while the respective intramolecular degrees fo freedom are subject to holonomic constraints [69]. While a number of highly successful water models have been formulated based on this strategy, a rigid-body description of the solvent does not provide an ideal basis to represent the complex hydrogen bonding pattern associated to anionic hydration. This is due to the fact that the latter induces a symmetry break in the water molecules, leading to variations in the O–H bond lengths and oftentimes to an adjustment of the intramolecular H–O–H angle. The latter can only be described if a fully flexible description accounting for explicit hydrogen motion is employed.

Although a number of water models accounting for molecular flexibility are available in the literature, the majority of them are not suitable for application in a QM/MM MD simulation study of an anionic solute. Recently, an improved flexible water potential providing a reliable representation of the intramolecular water geometry along with improved vibrational and dielectric properties has been presented [70]. In the present study, this new solvent model is applied for the first time in a RIMP2-based QM/MM MD simulation to investigate the solvation properties of fluoride, chloride and bromide in aqueous solution.

2. Methodology

In this section, the simulation strategy applied in the study of aqueous F^-, Cl^- and Br^- is outlined. The setup and simulation protocols are largely inspired by the previous investigations of Li^+, Na^+ and K^+ in aqueous solution [38,40]. In a QM/MM simulation setting, the chemically most relevant region of the system, in this case the ionic solute and the coordinating solvent molecules of the first hydration layer, is treated using the chosen QM method, while classical MM interaction potentials are considered sufficiently accurate to model the remainder of the simulation system. Prior to the execution of the QM/MM MD simulation, suitable interaction potentials for the solute–solute and solute–solvent interaction beyond the QM region have to be defined. While a large number of adequate solvent–solvent potentials are available in the literature, suitable ion–water potentials describing the interaction between the QM solute and water molecules compatible with the QM treatment have to be identified. In this work, new interaction parameters for the ion–water interaction have been derived, which is discussed in the following subsection. Next, the simulation protocol is outlined followed by a description of the applied analysis strategy.

2.1. Ion–Water Interaction Potentials

Although in a QM/MM study, the immediate interaction between the ionic solute and solvent molecules included in the QM region is described based on the selected quantum chemical calculation method, it is still required to provide an ion–solvent potential to account for interactions between the ionic solute located in the center of the QM region and the classically treated water molecules assigned to the MM zone. While it is possible to avoid the non-coulombic part of this ion–solvent potential if an enlarged QM region encompassing at least two full layers of hydration is employed [71,72], such a simulation setup is prohibitively expensive when aiming at the application of a correlated ab initio method as done in this work.

A key requirement in the selection of the ion–water potential model is a qualitatively correct description of the hydration complex compatible with that of the applied QM method. This is, to some extent, a conceptual dilemma since the structural description resulting from the QM/MM description is only known after the respective simulation is completed, while a suitable ion–solvent interaction potential has to be provided for the execution of said QM/MM simulation. To avoid this situation, suitable potential data are obtained from a series of classical simulations with varying potential settings, that are compared to data of previous QM/MM [73,74] and Car-Parrinello molecular dynamics

(CPMD) simulations [75–79]. In addition, results obtained from experimental investigations such as X-ray absorption fine structures (XAFS) and neutron diffraction with hydrogen isotope substitution (NDIS) also serve as a suitable reference [76,80–82]. While a number of experimental studies are reported in the literature, the solvation properties of halogen ions seem to be quite sensitive to concentration effects (see for instance Wallen et al. [83]). In addition, many experimental studies of salt solutions employing X-ray diffraction report data measured at comparably large concentrations [84–89], highlighting even the formation of solvent-separated ion pairs [90]. However, in the context of this study, treating a single solute ion solvated by 2000 water molecules, reference data for dilute solutions are preferred.

Since the Cl^- ion is an important solute in the context of biomolecular simulations, a large number of potentials are reported in the literature. In contrast, F^- and Br^- are less relevant in this context and, hence, only a smaller number of potential parameters are available. To treat all considered solutes on the same basis, new interaction parameters have been generated for the three ions by performing a series of purely classical MD simulations of a negatively charged ion ($q = -1.0e$) in aqueous solution. A total of 42 individual MD runs have been executed, each with a different setting of the ion–O Lennard–Jones (LJ) parameters [91,92] as depicted in Figure 1a.

The simulation systems in this preliminary potential evaluation were composed of the solute plus 1000 SPC/E water molecules [93] contained in a cubic, periodic simulation cell with a side length of approx. 31.1 Å corresponding to a target density of 0.997 kg dm^{-3}. Long-range coulombic contributions have been accounted for via the reaction-field approach [94] employing a relative static permittivity of 78, with the general cutoff distance set to 12.5 Å. The equations of motion have been integrated via the velocity-Verlet algorithm [95]. To achieve a time step of 1.0 fs, the M-SHAKE/M-RATTLE [69,96] algorithms have been applied. Thermal control was realised via the Berendsen algorithm [97] with the target temperature and relaxation time set to 298.15 K and 0.1 ps. For each tested LJ parameter pair, an equilibration for 30,000 MD steps (30 ps) has been carried out, followed by 50,000 steps of sampling (50 ps) collecting data every 25th MD step.

By comparing the location of the resulting first shell peak maximum in the ion-O radial distribution function and the average coordination number with reference data from the literature [73–83], the best potential parameters for the F^-, Cl^- and Br^- could be identified (see Figure 1b and Table 1).

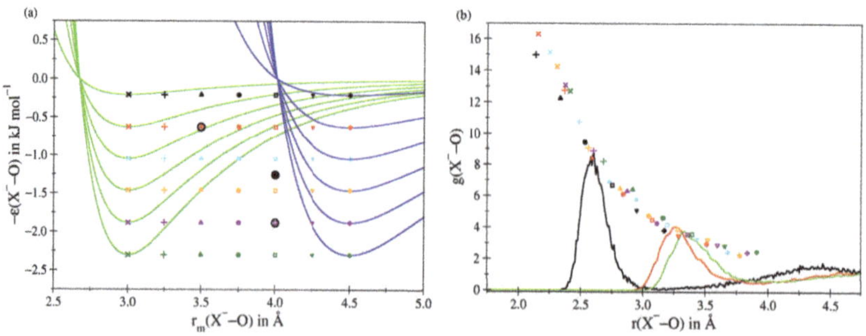

Figure 1. (**a**) Overview of the 42 ion–oxygen Lennard–Jones parameter combinations employed to identify suitable interaction potentials for the description of aqueous F^-, Cl^- and Br^-. (**b**) Maxima of the first shell peak in the ion–oxygen RDF determined for a hydrated ion with charge $-1.0e$ in combination with the 42 Lennard–Jones parameters. The RDFs obtained for F^- (black), Cl^- (red) and Br^- (green) have been selected by comparing the resulting ion–oxygen RDFs and the the coordination number in the first hydration shell to data provided in the literature. The respective parameter sets are highlighted via the black circles and are listed in Table 1. In the case of Cl^- an additional simulation employing an ε-value of 1.2552 kJ mol^{-1} proved necessary.

Table 1. Lennard–Jones potential parameters $\varepsilon(X^-–O)$ in kJ mol^{-1} and $\sigma(X^-–O)$ in Å of the ion–oxygen potential describing the interaction between the QM solute and water molecules in the MM zone. The respective minimum distance $r_m(X^-–O)$ in Å is listed as well (see also Figure 1).

	$\varepsilon(X^-–O)$	$\sigma(X^-–O)$	$r_m(X^-–O)$
F^-	0.6276	3.1181	3.5
Cl^-	1.2552	3.5636	4.0
Br^-	1.8828	3.5636	4.0

2.2. *QM/MM MD Simulation Protocol*

The QM/MM MD simulations carried out in this work are following the general setup of previous successful simulation studies of solvated Li^+, Na^+ and K^+ [38,40]. The QM/MM treatment is based on the highly successful quantum mechanical charge field molecular dynamics (QMCF MD) framework [71,72]. In this approach, the partial charges of the MM atoms are included in the Hamiltonian of the QM calculation in form as an external potential [62,98,99]. It was shown that this framework referred to as electrostatic embedding greatly improves the treatment of the QM zone over calculations in an artificial in vacuo environment, thereby enhancing for instance the description of hydrogen bonds passing through the QM/MM interface [100,101]. In addition, polarization and charge transfer effects occurring in the QM treatment are incorporated into the QM/MM coulombic coupling by employing QM-derived partial charges that are updated in every simulation step [71,72]. In order to ensure compatibility with the partial charges of the employed MM model [70], Mulliken populations [102,103] proved as an adequate choice [100,101] despite their general criticism of showing a pronounced basis set dependence. To avoid the latter, it is crucial to employ balanced basis sets, i.e., assigning diffuse and polarization functions in an equal matter to all atoms in the systems including the oftentimes neglected hydrogen atoms.

In order to account for the distortion of the molecular geometries of water molecules in the vicinity of the solute, the flexible SPC-mTR2 water model [70] has been employed. The latter is a re-parametrization of the SPC-mTR (simple point charge modified Toukan–Rahman) water potential [104], in which the charges, the equilibrium bond length and angle, and further parameters have been adjusted to provide a more reliable representation of the intramolecular water geometry while at the same time the vibrational and dielectric properties were improved.

An enlarged simulation cell containing the ion and 2000 water molecules in the isobaric-isothermal ensemble (NPT) under periodic boundary conditions (side length approx. 39.4 Å) was employed in each case. Also in case of the QM/MM MD simulations, the long-range electrostatic contributions were treated via the reaction-field approach [94] utilizing an extended cutoff radius of 18.0 Å. The MD time step in the velocity-Verlet integration [95] was set to 0.5 fs to describe explicit hydrogen motion. The Berendsen thermostat and manostat algorithms [97] have been employed to maintain the target conditions of 298.15 K and 1.013 bar, the associated relaxation times were set to 0.1 and 0.5 ps, respectively.

As in the previous studies of Li^+, Na^+ and K^+, resolution-of-identity Møller–Plesset perturbation theory of second order (RIMP2) [51–54] has been applied to describe the interactions inside the QM region as implemented in the TURBOMOLE package [105]. Since, in contrast to cationic solutes, anions display an excess of electrons, the use of larger basis sets is recommended. In this study, 6-31++G(d,p) bases [106–109] were assigned to the atoms of the water molecules, while the larger TZVP basis set [110] was employed for the description of the anions. In addition, suitable auxiliary basis sets [53] for the resolution-of-identity (RI) treatment were applied.

The size of the QM region is set to a radius to include the solute plus its entire first solvation layer based on the center of mass of the individual water molecules. Since the ideal setting is unknown prior to starting the simulation study, the radius is typically selected based on the minimum distance separating the first and second solvation shell in the ion–O RDF obtained from purely classical MD simulation. This strategy proved adequate in a number of similar QM/MM studies. The radius determining the size of the QM region was set to 3.2 Å in the case of F^-, while for both Cl^- and Br^- a value of 4.1 Å was determined from the classical simulation. The reason for the similar QM radii in case of the latter two systems can be explained by the fact that the respective σ-values in the ion–O Lennard–Jones potential are similar, while the ε-parameters representing the depth of potential are different. This not only influences the attractive part of the potential, but also increases the steepness of the repulsive branch, thereby leading to a larger ion–O distance in the Br^- case. In order to enable a smooth transition of molecules between the QM and MM zone, a small buffer region of 0.2 Å thickness is considered in which the forces are continuously changed from the QM to the MM contribution and vice versa.

The simulation systems for each solute were equilibrated in the NPT ensemble via classical MD for a total of 100,000 MD steps (50 ps) at target conditions. Considering the much larger size of the MM zone compared to the QM region, all molecules associated to the MM treatment are well-equilibrated after this procedure. Next, a re-equilibration for a minimum of 20,000 steps (10 ps) was carried out using the QM/MM setup, followed by 50,000 MD steps of sampling, resulting in a total simulation time of 25 ps. Since, in the construction of the ion–oxygen Lennard–Jones potentials special care was given to replicate the ideal ion–water distances and the associated coordination numbers, the only required equilibration is associated to the intramolecular degrees of freedom. The impact of polarization and many-body effects then has an influence on the first shell ligand exchange rate. Since monovalent ions are known to quickly exchange the first shell ligands within the picosecond timescale [111], the re-equilibration period of 10 ps can be considered as sufficient.

2.3. Analysis

The solute–solvent interactions were analysed in terms of ion-water radial distribution functions (RDFs), thereby also comparing the differences observed between the classical and QM/MM-based simulations. In order to provide further information about the composition of the first solvation layer, the respective coordination number distributions (CNDs), local density corrected three-body correlations g^3 [112] as well as the oxygen–ion–oxygen angular distributions functions (ADFs) have been analysed.

Typically, the coordination number distribution and the associated average coordination number (CN) of a given solvation shell are determined based on a radial cutoff criterion coinciding with minima determined from the ion–oxygen radial distribution function $g(r)$. However, such a simple $g(r)$-based cutoff (GC) criterion is not always an adequate choice. First, the determination of a shell boundary can be to some extent ambiguous, since it may be influenced by noise and depends inter alia on the type of normalisation applied in the RDF analysis [113]. Moreover, a strictly radial criterion may exclude ligands, that should be considered as part of the first coordination sphere, while on the other hand molecules inside the cutoff should not be included as they have to be considered "blocked" by other, closer ligands.

The relative angular distance (RAD) algorithm [113,114] provides a simple and effective framework to classify ligands in the first solvation shell that is not dependent on any predefined cutoff radius. Instead, the list of potential first shell ligands is evaluated based on their increasing distance from the solute i. A molecule j is considered as part of the first solvation layer if it remains unblocked by all previously identified closer ligands k according to the criterion

$$\frac{1}{r_{ij}^2} > \frac{1}{r_{ik}^2} \cos\left(\theta_{jik}\right) \tag{1}$$

with r_{ij} and r_{ik} being the respective ion–ligand pair distances and θ_{jik} is the associated solvent–solute–solvent angle. As soon as the first ligand j' violates the RAD criterion, it is considered as blocked and the search is ended. Thus, all unblocked ligands closer than molecule j' are considered as part of the first solvation shell in this respective configuration. Averaging over all configurations of the simulation trajectory then yields CN_{RAD}. In the original work, the molecules' centres of mass have been considered [113]. However, in this article the analysis was based on the oxygen atoms in order to make the results for CN_{RAD} directly comparable with those of the radial cutoff CN_{GC}. Since the oxygen atom of a water molecule is always in very close vicinity to the respective center of mass, this approximation is not expected to have any profound impact on the determination of the coordination numbers and their distributions.

The structural properties of the first shell water molecules are of particular interest as well. To analyse the deviation from the ideal geometry, the two hydrogen atoms of a water molecule have been classified based on their orientation in the solvation complex. The H-atom directly coordinated to the solute via hydrogen bonding is considered as proximal (H_1) while the one located at larger distance is referred to as distal (H_2). Based on this classification, the associated O–H_1 and O–H_2 radial distributions as well as the corresponding ion$\cdots H_1$–O and ion$\cdots H_2$–O angular distributions have been analysed.

The mean ligand residence time τ (MRT) within the first shell has been characterised based on a direct measure of the exchange events [111]. In this approach, the MRT-value of a specific solvation shell is determined as

$$\tau_{0.5} = \frac{t_{\text{sim}} \cdot \text{CN}_{\text{av}}}{N_{0.5}} \tag{2}$$

where t_{sim} is the total simulation time, CN_{av} is the average coordination number of the respective solvation layer and $N_{0.5}$ corresponds to the total number of border crossing event to/from this solvation shell lasting for a minimum excursion time $t^* \geq 0.5$ ps [111]. The associated rate coefficient R_{ex} is given as the ratio between the number of all border crossing attempts $N_{0.0}$ for $t^* = 0$ and $N_{0.5}$.

3. Results

In the following, the results obtained for the target systems via classical and QM/MM MD simulations are discussed. The key quantities are listed in Table 2, including a comparison to other theoretical and experimental investigations. Overall, a broad variation in the determined quantities can be identified, clearly highlighting the challenging nature associated to the characterisation of anionic solvation. The experimental studies by Wallen et al. [83] and Soper et al. [81] also point towards a quite sensitive concentration dependence, in particular with respect to the observed coordination number. This, on the other hand, might prove to be a limiting factor when comparing to theoretical results based on the Car-Parrinello MD framework, which especially in the past were typically limited to small system sizes, often encompassing in addition to the solute only about 30 to 128 water molecules. This and the oftentimes limited accuracy of DFT functionals at the GGA (generalised gradient approximation) level make it difficult to judge the quality of the results obtained via CPMD simulations. The latter is particularly true whenever hydrogen bonded systems are to be investigated. Simulation data on pure water have shown that the results can be notably improved if a corrective treatment of dispersive contributions is considered [44,45,50]. The latter is confirmed in a study by Bankura and coworkers comparing CPMD simulation data [78] with and without the application of the Tkatchenko–Scheffler van-der-Waals (TS-vdW) correction [115]. The same study also highlighted that the simulation results obtained using the Perdew–Burke–Enzerhofer (PBE) GGA-type functional can be notably improved if the more accurate yet more demanding hybrid variant PBE0 is applied.

Table 2. Maximum, avarage and minimum distances r^1_M, $\langle r^1_M \rangle$ and r^1_m of the first solvation shell in the ion–O RDF in Å, average first shell coordination number determined via a g(r)-based cutoff CN^1_{GC} as well as the relative angular distance approach CN^1_{RAD}, first shell mean ligand residence time τ_1 in ps, number of registered ligand exchange events $N^1_{0.5}$ ($t^* \geq 0.5$ ps) and associated rate coefficient R^1_{ex} obtained for aqueous F^-, Cl^- and Br^- via classical (MM) and RIMP2-based QM/MM MD simulations in comparison to data reported in the literature.

		r^1_M	$\langle r^1_M \rangle$	r^1_m	CN^1_{GC}	CN^1_{RAD}	τ_1	$N^1_{0.5}$	R^1_{ex}	
F^-	MM MD	2.59	2.63	3.25	6.2	6.3	15.4	10	5.4	this work
	RIMP2/MM MD	2.46	2.68	3.36	4.9	5.2	1.1	115	5.2	this work
	MM MD	2.53			5.8 ± 0.1					Ref. [73]
	HF/MM MD	2.68			4.6 ± 0.2					Ref. [73]
	BLYP CPMD	2.66			5.1					Ref. [75]
	BLYP CPMD		2.7	3.4						Ref. [77]
	NDIS KF/D_2O 1.2:100	2.54		3.27	6.9					Ref. [81]
Cl^-	MM MD	3.26	3.35	3.93	7.6	7.8	4.2	46	6.7	this work
	RIMP2/MM MD	3.23	3.48	4.16	8.1	7.5	1.6	178	4.5	this work
	MM MD	3.15			5.9 ± 0.1					Ref. [73]
	HF/MM MD	3.24			5.9 ± 0.1					Ref. [73]
	HF/MM MD	3.25		3.9	6.8		2.0		4.6	Ref. [74]
	PBE-D3 CPMD	3.14		3.78	6.0					Ref. [79]
	PBE0-D3 CPMD	3.17		3.85	6.1					Ref. [79]
	SCAN CPMD	3.17		3.85	6.7					Ref. [79]
	PBE CPMD	3.11		3.64	5.5 ± 0.2					Ref. [78]
	PBE+TS-vdW CPMD	3.14		3.78	6.3 ± 0.9					Ref. [78]
	PBE0 CPMD	3.14		3.72	5.8 ± 0.7					Ref. [78]
	PBE0+TS-vdW CPMD	3.16		3.73	6.3 ± 0.8					Ref. [78]
	EXAFS NaCl 40 mM	2.91/3.11			4+3					Ref. [82]
	NDIS KCl/D_2O 1.2:100	3.14		3.78	7.1					Ref. [81]
Br^-	MM MD	3.33	3.45	4.05	8.1	8.1	3.0	313	4.7	this work
	RIMP2/MM MD	3.31	3.68	4.30	9.1	7.4	0.9	390	2.9	this work
	MM MD	3.27		3.9	7.6 ± 0.5		2.6			Ref. [76]
	BLYP CPMD	3.33		3.9	6.5 ± 0.3		5.7			Ref. [76]
	XAFS/MM MC YBr_3 0.3M		3.44 ± 0.07		6 ± 0.5					Ref. [80]
	XAFS RbBr 0.2M	3.35			7.1 ± 1.5					Ref. [83]
	XAFS RbBr 1.5M	3.36			7.2 ± 0.4					Ref. [83]
	XAFS RbCl 0.5 mM	3.26			10					Ref. [82]
	NDIS KBr/D_2O 1.2/100	3.32		3.90	6.7					Ref. [81]

The above-mentioned shortcomings (i.e., the small number of solvent molecules and a quantum chemical treatment typically limited to GGA-DFT level) can be overcome using QM/MM MD type protocols. However, in the existing literature, only simulation studies employing the Hartree–Fock (HF) level of theory have been presented for aqueous F^- and Cl^- [73]. While being in a strict sense an ab initio method, HF theory suffers from the inability to take electron correlation into account, which especially in the case of hydrogen bonded systems might prove as a limiting factor. The QM/MM MD simulation protocol applied in this study aims at overcoming the above-mentioned limitations by treating the target systems at a correlated ab initio level while at the same time providing a sufficient number of solvent molecules represented via the newly parameterised SPC-mTR2 water model [70].

A number of key features of the hydration complexes can be directly extracted from the respective ion–solvent radial distributions functions depicted in Figure 2. In addition to the quantification of the average coordination number and ion–ligand distances, the differences between the purely classical and QM/MM-based simulations proved to be of particular interest. Overall, the structural properties of the solvation complexes are as expected with the main peak observed in the ion–O interaction being located in between two separate peaks in the corresponding ion–H RDF. Based on the separation of the first peaks, the hydrogen bonded nature of the solute–solvent interaction can be directly identified. While, in principle, patterns with similar ion–water distances are observed when comparing the MM and QM/MM simulation results, a number of notable differences can be identified, foremost being a dramatic decrease in peak intensity when changing from the MM to the QM/MM description by a factor of approx. 3 for F^- to about 2 in case of Br^-. This implies that the MM description is notably overstructured despite showing similar ion–H and ion–O distances. As outlined above, great attention has been given to identify suitable potential parameters resulting in an adequate description of the system. When comparing the peak maxima observed in the ion–oxygen RDFs determined via classical MD simulations employing the different Lennard–Jones potentials (see Figure 1b), it can be seen that it is not possible to identify a parameter set that results in an adequate ion–O distance while at the same time resulting in a reduced peak intensity. This demonstrates that the dramatic overstructuring is an inherent shortcoming of the simplified, pairwise-additive description. This trend has already been observed in previous simulation studies of the monovalent cations Li^+, Na^+ and K^+, albeit to a much lesser extent. Not surprisingly, subtle effects associated to the (re)distribution of the electron density such as polarisation, charge-transfer and many-body contributions are much more dominant in the anionic case.

The latter not only leads to the observed reduction in the RDF peak intensities, but also to a very pronounced tailing of the first shell ion-O peaks towards larger distances. This again demonstrates the difficulty in determining a suitable boundary for the QM treatment, which commonly is derived based on preceding classical simulations. Also in this case, simulations of cationic solutes show a much better agreement between the MM and QM/MM case.

Simulation data depicted in the form of RDF plots provide direct access to the most populated distances coinciding with the maxima of the respective RDF peaks. In contrast, experimentally determined ion–oxygen distances can be assumed to measure the associated average over the entire solvation shells of a large number of complexes present in solution as for instance explicitly highlighted by Merkling and coworkers [80]. While the difference between the latter is oftentimes negligible when characterising cationic hydrates, the strong tailing of the first shell peak may result in a notable increase of the average distance. In order to characterise the latter, the respective maximum distances r^1_{M} are compared to their respective average $\langle r^1 \rangle$ determined as weighted average up to the border of the first solvation shell given as r^1_{m}:

$$\langle r^1 \rangle = \frac{\sum\limits_{r=0}^{r^1_{\mathrm{m}}} r \cdot g(r)}{\sum\limits_{r=0}^{r^1_{\mathrm{m}}} g(r)} \tag{3}$$

The resulting values obtained for the MM and QM/MM MD simulations are listed in Table 2. While in case of the classical simulations, $\langle r^1 \rangle$ is only marginally increased compared to the respective maximum distance r^1_{M}, a notable shift to larger values can be observed in the QM/MM simulations. In the latter case, the respective differences between the maximum and average first shell distance amount to 0.22, 0.25 and 0.37 Å for F^-, Cl^- and Br^-, respectively, while in case of the classical simulations, much smaller deviations of 0.04, 0.09 and 0.12 Å have been determined.

The different interaction characteristics also have a notable impact on the first shell coordination numbers determined using the first shell boundary r^1_{m} as cutoff criterion yielding $\mathrm{CN_{GC}}$. While in the case of fluoride a decrease in $\mathrm{CN_{GC}}$ from 6.2 to 4.9 has been observed when changing from the MM to the QM/MM description, in both the chloride and bromide system an increase from 7.6 to 8.1 and from 8.1 to 9.1 was observed, respectively. The non-integer coordination numbers discussed above result from the averaging over different microstates observed along the simulations. Comparison of the calculated coordination numbers to other experimental and theoretical data reveals that the CN_{GC} values show notable deviations especially in case of Cl^- and Br^-. While the extraction of partial coordination patterns from measured data may prove as particularly challenging, the deviations in the coordination numbers might as well be the results of an ill-defined assignment of first shell ligands based on a radial cutoff criterion. The latter can be expected to also be negatively influenced by the pronounced tailing observed in the ion–water pair distributions resulting from the QM/MM MD simulations.

The relative angular distance (RAD) procedure by Higham et al. provides an alternative characterisation of first shell ligand distributions, which is solely based on the comparison between individual ion–ligand distance contributions weighted by the cosine of the associated solute–solvent–solute angle. The associated coordination numbers CN_{RAD} determined without any radial cutoff criterion are compared to the corresponding CN_{GC} values in Table 2. Although in the case of the MM simulations the RAD procedure resulted in highly similar average CN values, the results of the QM/MM simulations are greatly improved after application of the RAD criterion. While in case of F^- an increase from 4.9 to 5.2 has been observed, the average coordination number decreased from 8.1 to 7.5 and from 9.1 to 7.4 in case of Cl^- and Br^-, respectively. Thus, in all cases, the CN_{RAD} values are in much better agreement with the literature compared to the CN_{GC} results.

To further analyse the composition of the first solvation shell, the corresponding coordination number distributions (CNDs) as obtained from the GC and RAD analysis have been evaluated (see Figure 2). As in the case of the RDF plots, the CNDs show a notable decrease associated to a broadening of the distribution when changing from the MM to the QM/MM description in all cases. Comparison of the QM/MM-based CND plots obtained using the GC- and RAD-based ligand assignment again reveals a trend to lower coordination numbers in the latter case as already observed for the average coordination numbers. In contrast, highly similar CND plots are obtained in the MM case when comparing the two different analysis protocols.

The in-depth characterisation of the first shell coordination numbers and their respective distribution over the course of the individual simulations provides a number of important conclusions: First, the data highlight that a simple radial criterion is not always an adequate choice to determine the coordination number, and improved results can be obtained by the application of more elaborate frameworks such as the RAD criterion. In addition, it was found that the application of the latter only improves the results of the QM/MM MD simulations, while in case of the purely classical simulations, both the average CNs as well as the corresponding CNDs remained highly consistent. Considering the strongly tailed first shell peak of the radial distribution functions in the QM/MM case, it appears that interactions incorporating many-body contributions are prone to errors in a purely cutoff-based coordination number determination.

Comparing the more reliable RAD-based coordination numbers, it can be seen that the MM simulations yield notably higher average CN values than their QM/MM counterpart by approx. 21, 5 and 9% in case of F^-, Cl^- and Br^-, respectively. Thus, specifically in the case of fluoride, a notable shift to higher coordination numbers is observed in the MM case. On the other hand, the CNDs determined from the MM and QM/MM simulations of Cl^- display the best agreement, albeit also in this case a reduction of the most populated contributions (coordination numbers 7 and 8) is observed. One possible explanation for these findings could be that the applied potential parameters are ideal in the Cl^- case, while for the other two systems a refinement of the interaction potentials might lead to improved results.

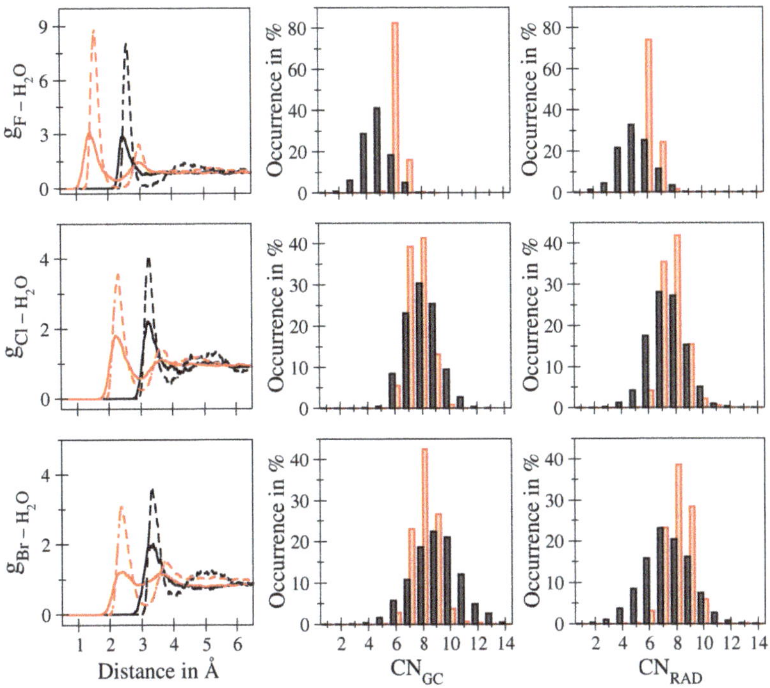

Figure 2. Ion–oxygen (red) and ion–hydrogen (black) radial distribution functions (left column) for aqueous F^- (top row), Cl^- (center row) and Br^- (bottom row) determined from the QM/MM (solid line) and classical (dashed line) MD simulations along with the associated coordination number distributions based on a radial $g(r)$-cutoff criterion (center column) and the RAD analysis (right column) as obtained from the QM/MM (black) and classical (red) simulation trajectory.

Although the latter strategy can be expected to adjust the solvation parameters within a small margin, already great care has been devoted towards the identification of suitable solute–solvent Lennard–Jones potentials as outlined above. However, the pairwise nature of the classical description provides a more reasonable explanation for the observed deviations. Due to the small size of the F^- ion, the hydration complex shows the smallest ion–water distances. This implies that contributions arising from the respective Coulomb interactions are highest in this case. Since the classical description does not account for many-body contributions such as polarisation and charge-transfer, it can be expected that, out of the considered systems, F^- shows the highest susceptibility towards overcoordination. In contrast, the Cl^- and Br^- systems display notably increased first shell ion–water distances, thus reducing the magnitude of the Coulombic contributions and, thereby, the risk for observing much higher coordination numbers. Nevertheless, the high intensities in the coordination number distribution and the ion–water RDFs still points towards an overall too strong solute–solvent interaction that cannot be easily compensated by a variation in the Lennard–Jones parameters (see data in Figure 1b). While similar overcoodination effects were also observed in the case of the previously mentioned cations Li^+, Na^+ and K^+, the increase in the first shell ion–oxygen peak maxima amounts to at most 50%, with an overall very good agreement in the coordination number distributions when comparing the classical and QM/MM MD simulation results. These difficulties clearly demonstrate that simulations of anionic solutes are much more intricate compared to their cationic counterparts. A reduction of the net-charge of the ions in the simulations to non-integer values could offer a potential solution, which has already been discussed by different authors in the past [116–119].

In order to obtain further insight into the composition of the first solvation shell, both local density corrected three-body correlations g^3_{I-O-O} measuring the O–O distance contributions within the first shell as well as oxygen–ion–oxygen angular distribution functions can be employed. While at first sight these analysis schemes appear redundant, the comparison of the different systems provided in Figure 3 demonstrates that the two different methods complement each other. For instance, based on the g^3-plots determined for F^-, a similar O–O distance distribution within the first shell is observed in the MM and QM/MM simulations. However, when comparing the respective angle distributions, it can be seen that, although the O–O distances are nearly identical, the O–F–O angle contributions show remarkable differences, pointing towards a highly increased intra-shell mobility in the QM/MM case. In contrast, similar angular distributions are observed in the case of Cl^-, while notable differences in the three-body correlations are visible. When keeping in mind that Cl^- displayed the best agreement in the coordination number distributions between the MM and QM/MM simulations, the capabilities of the combined g^3 and ADF contributions to provide further details about the solvation structure is clearly demonstrated.

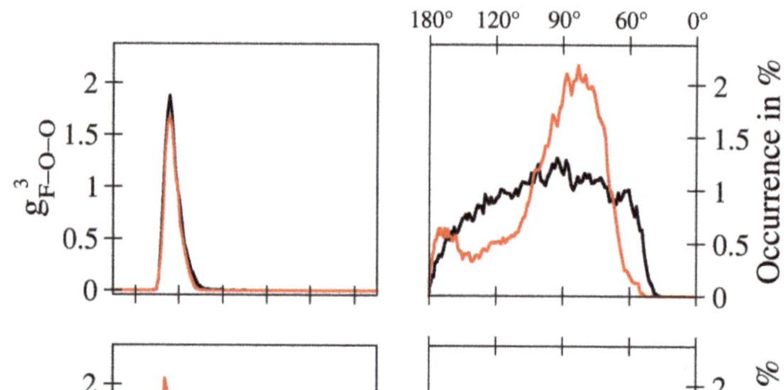

Figure 3. First shell ion–oxygen–oxygen three-body correlation function g^3 (left) and the associated O–ion–O cosine distribution functions (right) obtained for aqueous F^- (top), Cl^- (center) and Br^- (bottom) from the QM/MM (black) and classical (red) MD simulations. The absolute values for the O–ion–O angles follow a non-linear trend and are shown on the secondary *x*-axis.

In addition to highlighting the shortcomings of the classical description, this combined analysis provides manifold information about the differences between the target systems. For instance, in the case of F^- no second neighbour distances are registered in the three-body distributions, which can be attributed to the overall low coordination number permitting only one type of neighbouring molecules. In contrast, a larger number of solvent molecules is present in the first shell of Cl^- and Br^-, giving rise to a notable second neighbour peak located at approx. 4.2 Å. While in the case of the latter two ions, the features of the g^3-distributions appear similar, the respective oxygen–ion–oxygen distributions show a notable broadening especially in the Br^- case, implying that in this system the highest intra-shell mobility of ligands is observed as expected.

In order to further study the ion–ligand binding properties within the first solvation shell, the associated mean ligand residence times τ (MRT) have been determined using the direct method. In this approach, all crossing events of the first shell border r_m^1 lasting for a minimum time of t^* are counted, and the respective MRT value is then determined based on Equation (1). In addition, the respective rate coefficient R_{ex} corresponding to the average number of border crossing events required to achieve one lasting ligand exchange were determined. The respective values obtained from the MM and QM/MM MD simulation trajectories of the target systems are listed in Table 2.

When considering the MRT values calculated from the classical simulations being 15.4, 4.2 and 3.0 ps in the case of F^-, Cl^- and Br^-, respectively, the expected decrease in the solute–solvent interaction is well reflected. This decrease in the mean residence times is well in agreement with the associated ion–O RDF data, specifically the intensity of the minimum separating the first and second solvation shells which gradually increases from the lightest to the heaviest solute. In the case of the QM/MM MD simulations, however, a different picture emerges. First, it can be seen that the MRT values are consistently lower compared to the MM-based estimation. This is not uncommon considering the overstructured hydration discussed above, which is the result of the unpolarised, pairwise nature of the interaction in the classical model. In addition, the MRT values determined in case of F^- are now slightly below that obtained for the Cl^- system. Although at first sight this finding implies that the latter now displays weaker interactions with the solvent, this counterintuitive result is due to the occurrence of the average coordination number in the determination of the MRT value (see also Equation (1)). In contrast, when comparing the actual number of registered ligand exchange events lasting for a minimum excursion time of $t^* \geq 0.5$ ps, only 115 exchanges were registered in the case of F^- over the sampling time of 25 ps, while 178 and 390 exchanges were registered for Cl^- and Br^- over the same time period, respectively. This clearly displays that the solute–solvent interaction follows the expected trend with F^- forming the most stable hydrate, while the Br^-–H_2O interaction is weakest. It should also be noted that MRT values below 1.0 ps have been associated to "structure-breaking" properties in the past [42,120]. The latter is underlined by the associated rate coefficients R_{ex} measuring the average number of border crossing events to achieve one sustainable exchange event lasting for longer than t^*. While in case of F^- on average 5.2 exchange attempts are required, notably smaller R_{ex}-values of 4.5 and 2.9 have been determined for the weaker hydrates of Cl^- and Br^-, respectively. Again, small values in R_{ex} have been associated to "structure breaking" solute–solvent interactions [42,120].

Although it is commonly observed that QM/MM MD simulations display an accelerated first shell ligand exchange compared to MM-based simulations, the difficulties in identifying suitable QM radii discussed above might lead to an additional acceleration of the observed exchange rates. In order to minimise QM/MM transition artifacts, it is a typical strategy to position the QM/MM boundary in a region of low density corresponding to minima in the respective ion–O distributions. As discussed above, the unexpected strong tailing observed in the ion–O RDFs of the QM/MM simulations implies that the MM-derived settings for the QM radii might not be ideal.

While in case of Cl^- and Br^- the ion–O RDFs still show well-defined minima in the QM/MM case, the respective first shell peak in the F^- case displays only a very shallow minimum that directly connects to the bulk without any discernable contribution indicating the presence of a second solvation shell. On the other hand, the good agreement of the first shell distance and coordination number CN_{RAD} with data in the literature implies that the first shell properties are adequately represented. In order to verify that the selected size of the QM region is indeed adequate, it might prove advantageous to execute further QM/MM MD simulations with an increased size of the QM zone to monitor changes in the hydration shells and the associated mean ligand residence times. However, when considering the rather costly execution times of several months of a 10 + 25 ps QM/MM MD simulation at RIMP2 level, such a systematic adjustment of the simulation settings is quite demanding. The costly execution time of a correlated ab initio description in conjunction with the requirement to execute the MD simulations with full consideration of molecular flexibility is a further indication why QM/MM MD simulations of ionic solutes tend to be much more intricate compared to simple (i.e., monoatomic and monovalent) cations in aqueous solution.

In order to underline the requirement of taking molecular flexibility into account, the O–H bond distances observed within the first hydration shell have been analysed, thereby separating the contributions from the proximal H_1 atoms directly interacting with the solute from those of the distal H_2 atoms pointing away from the ion (see sketch in Figure 4). The comparison of the associated RDFs depicted in Figure 4 highlights the different characteristics of the O–H_1 and O–H_2 bonds. In addition to a clear separation of the associated equilibrium distances, the pronounced tailing of the O–H_1 bond to larger distances is visible. This is a direct consequence of the hydrogen bonded nature of the solute–solvent interaction, resulting in elongated bond distances due to the coordination to the negatively charged solute. Simulation protocols employing a rigid-body description of the solvent cannot account for these critical changes in the equilibrium geometry of the first shell ligands. On the contrary, for practical reasons, the same structural constraints have to be applied to molecules in the first solvation shell, whenever such a water model is employed to describe the solvent in the simulation system. While such a strategy might still be sufficient to derive structural and dynamical properties, simulations aiming at the determination of thermodynamic properties will be negatively influenced by the application of artificial constraints employing a reference geometry that is in a strict sense only valid for molecules in the bulk.

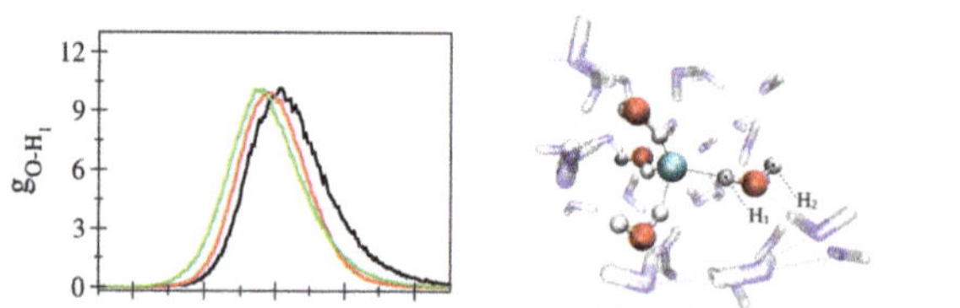

Figure 4. Oxygen-hydrogen radial distribution function of water molecules in the first solvation shell of aqueous F^- (black), Cl^- (red) and Br^- (green) separated into contributions of the H-atoms in proximal (H_1, top left) and distal (H_2, bottom left) position with respect to the solute along with the associated ion$\cdots H_1$–O cosine distribution function (bottom right). The screenshot of aqueous F^- displays the assignment of H_1 and H_2 for an exemplary first shell water molecule, that is re-evaluated in every simulation step.

A further interesting detail linked to the separate consideration of first shell hydrogen atoms is the distribution of the ion$\cdots H_1$–O angle, which is also depicted in Figure 4. While in all cases the highest contributions are observed for angles $\geq 170°$, a pronounced tailing towards lower angles is observed upon increasing size of the solute again pointing towards an increase in intra-shell mobility. In addition, a plateau near 120° is visible for all three solutes that is highest in the Br^- case. This finding could either be (i) a direct consequence of intra-shell hydrogen bonding potentially promoted by the increasing ionic radii and the associated higher coordination numbers, (ii) due to contributions of ligands leaving the second solvation shell or (iii) the result of intra-shell mobility in which the ligands display a strictly dipolar, bifurcated coordination. In order to investigate the origin of this contribution in detail, a two-dimensional histogram representation correlating the ion$\cdots H_1$–O and ion$\cdots H_2$–O angles registered for each first shell molecule in the QM/MM MD simulation of aqueous Br^- has been prepared (see Figure 5). It can be seen that the main contribution is again observed in the range close to 170–180°, which represents the main hydrogen bond angle already visible in the one-dimensional $Br^-\cdots H_1$–O angle distribution (see sketch in Figure 5, top right corner). For lower $Br^-\cdots H_1$–O angles a sideward V-shape is visible, with the main contributions indeed resulting from a bifurcated coordination with the $Br^-\cdots H_2$–O angle averaging at approx. 100° (sketch in Figure 5, top left corner). In addition, a secondary distribution is visible in which the $Br^-\cdots H_1$–O hydrogen bond appears to be broken while the O–H_2 bond is pointing away from the ion. The latter explains the very low angle of approx. 30° (sketch in Figure 5, bottom left). Thus, an overlap of two different configurations contribute to the plateau at 120° visible in the one-dimensional $Br^-\cdots H_1$–O distribution (Figure 4), namely, a strictly polar, bifurcated ion–water coordination plus a less populated contribution arising from configurations with broken ion–solvent hydrogen bonding most likely associated to ligand exchange. Similar distributions can be observed in the case of F^- and Cl^-, although with significantly lower magnitude (data not shown).

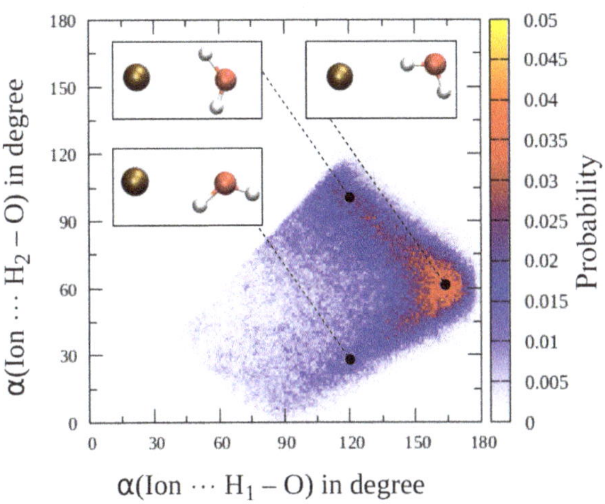

Figure 5. Two-dimensional histogram correlating the ion$\cdots H_1$–O and ion$\cdots H_2$–O angles registered for each first shell ligand of Br^- obtained from the QM/MM MD simulation.

4. Conclusions

The RIMP2-based QM/MM MD simulations presented in this study provide manifold insight into the structural and dynamical properties of the target systems and provide a proof of concept for similar studies combining a correlated ab initio treatment in conjunction with full flexibility in the description of solute–solvent hydrogen bonding. The comparison to a number of previous experimental and theoretical investigations clearly highlights the challenging nature when studying anionic solvation, being dependent on the actual concentration of studied solutions while at the same time showing a notable sensitivity on the applied level of theory. Out of all the presented calculation methods, resolution-of-identity Møller–Plesset perturbation theory of second order proved to be one of the most demanding yet most accurate calculation methods applied to this research question.

However, even in this case a number of potential shortcomings and pitfalls could be identified. In contrast to the majority of investigations available in the literature the improved description of the solute enabled by the RIMP2 framework applied to achieve comparably long sampling times of 25 ps per solute resulted in a rather unexpected tailing of the first shell peak in the ion–oxygen radial distribution functions. Typically, the boundary of the QM-treatment is selected based on similar, purely classical simulations employing suitably adjusted parameters in the applied interaction potentials. While great care was given to providing newly derived, accurate interaction parameters for the solute–solvent QM/MM coupling, the tendency towards elongated ion–solvent distances in conjunction with a notable reduction of the first shell peak intensities makes the choice of suitable QM radii to some extent ambiguous. Based on the presented simulation results, an increase in the QM/MM boundary distance determined via classical simulations appears necessary in future studies.

In addition, the strong tailing observed in the first shell ligand distributions proved problematic in the evaluation of the associated coordination numbers based on a strictly radial cutoff criterion. Application of the relative angular distance (RAD) algorithm enabling the assignment of coordinating ligands without any predefined radial cutoff distance greatly improved the results obtained for the coordination numbers. The latter only applies in case of the QM/MM MD simulations, however, while in case of the pairwise additive MM description, both the cutoff- and RAD-based ligand assignments yielded highly similar results. It can be concluded that the many-body description inherent to the quantum chemical treatment gives rise to the observed first shell tailing which in turn makes an unambiguous assignment of coordination numbers more difficult.

Despite these challenges, a consistent description of the three target systems F^-, Cl^- and Br^- could be achieved, enabling the characterisation of more complex structural properties such as three-body correlations and the distortion of first shell solvent molecules from their ideal, symmetric equilibrium geometry. In addition, the ligand exchange dynamics within the first shell could be characterised based on the simulation data, decreasing with increasing size of the solute as expected.

The presented simulation data serve as a valuable primer for advanced simulations studies, foremost in the determination of single-ion solvation free energies for which a variety of successful simulation protocols have already been implemented in case of alkaline ions. With the availability of both cationic and anionic solvation data, an inherent instrument to compare the simulation data to experimental references values determined for salt solutions becomes available, which in turn provides access to further investigations focused on absolute single-electrode potentials. Both the determination of single ion solvation properties as well as of single-electrode potentials still represents one of the few remaining questions in present day thermodynamics and the advanced QM/MM MD simulation protocol outlined in this study provides an important step towards a generalised workflow to address these prevalent research questions with the help of modern computational techniques.

Funding: This research received no external funding.

Data Availability Statement: The data that support the findings of this study are available from the corresponding author upon reasonable request.

Acknowledgments: The computational results presented have been achieved (in part) using the HPC infrastructure of the University of Innsbruck. Dedicated to the memory of Bernd M. Rode.

Conflicts of Interest: The author declares no conflict of interest.

References

1. Hertz, H.G. *Water: A Comprehensive Treatise*; Plenum Press: New York, NY, USA, 1973.
2. Stumm, W.; Morgan, J.J. *Aqueous Chemistry*, 3rd ed.; John Wiley & Sons: New York, NY, USA, 1996.
3. Reichardt, C. *Solvents and Solvent Effects in Organic Chemistry*, 3rd ed.; Wiley-VCH: Weinheim, Germany, 2003.
4. Buncel, E.; Stairs, R.A.; Wilson, H. *The Role of the Solvent in Chemical Reactions*; Oxford University Press: Cambridge, UK, 2003.
5. Buncel, E.; Stairs, R.A. *Solvent Effects in Chemistry*, 2nd ed.; John Wiley & Sons: New York, NY, USA, 2016.
6. Shimizu, S.; Boon, C.L. The Kirkwood-Buff theory and the effect of cosolvents on biochemical reactions. *J. Chem. Phys.* **2004**, *121*, 9147–9155. [CrossRef] [PubMed]
7. Wypych, G. (Ed.) *Handbook of Solvents*, 2nd ed.; ChemTec Publishing: Oxford, UK, 2014.
8. Barthel, J.M.G.; Krienke, H.; Kunz, W. (Eds.) *Physical Chemistry of Electrolyte Solutions*; Steinkopff: Darmstadt, Germany, 1998.
9. Wright, M.R. (Ed.) *An Introduction to Aqueous Electrolyte Solutions*; John Wiley & Sons: New York, NY, USA, 1998.
10. Marcus, Y. *Ions in Solution and Their Solvation*; John Wiley & Sons: Hoboken, NJ, USA, 2015.
11. Kastenholz, M.A.; Hünenberger, P.H. Computation of methodology-independent ionic solvation free energies from molecular simulations. I. The electrostatic potential in molecular liquids. *J. Chem. Phys.* **2006**, *124*, 124106. [CrossRef] [PubMed]
12. Lee, L.L. (Ed.) *Molecular Thermodynamics of Electrolyte Solutions*; World Scientific: Singapore, 2008.
13. Hünenberger, P.; Reif, M. *Single Ion Solvation*; RSC Publishing: Cambridge, UK, 2011.
14. Feldberg, S.W. On the dilemma of the use of the electroneutrality constraint in electrochemical calculations. *Electrochem. Commun.* **2000**, *2*, 453–456. [CrossRef]
15. Jackowska, K.; Krysiński, P. *Applied Electrochemistry*; De Gruyter: Berlin, Germany, 2020.
16. Tapia, O.; Bertrán, J. (Eds.) *Solvent Effects and Chemical Reactivity*; Springer: Darmstadt, The Netherlands, 2003.
17. Allen, M.P.; Tildesley, D.J. *Computer Simulation of Liquids*; Oxford Science Publications: Oxford, UK, 1990.
18. Frenkel, D.; Smit, B. *Understanding Molecular Simulation*; Academic Press: San Diego, CA, USA; London, UK, 2002.
19. Tuckerman, M.E. *Statistical Mechanics: Theory and Molecular Simulation*; Oxford University Press: New York, NY, USA, 2009.
20. Laurendeau, N.M. *Statistical Thermodynamics: Fundamentals and Applications*; Cambridge University Press: New York, NY, USA, 2005.
21. Schwabl, F. *Statistical Mechanics, 2nd* ed.; Springer: Berlin, Germnay, 2010.
22. Kastenholz, M.A.; Hünenberger, P.H. Computation of methodology-independent ionic solvation free energies from molecular simulations. II. The hydration free energy of the sodium cation. *J. Chem. Phys.* **2006**, *124*, 224501. [CrossRef]
23. Reif, M.M.; Hünenberger, P.H. Computation of methodology-independent single-ion solvation properties from molecular simulations. III. Correction terms for the solvation free energies, enthalpies, entropies, heat capacities, volumes, compressibilities and expansivities of solvated ions. *J. Chem. Phys.* **2011**, *134*, 144103. [CrossRef]
24. Szabo, A.; Ostlund, N. *Modern Quantum Chemistry: Introduction to Advanced Electronic Structure Theory*; Dover Pubn. Inc.: New York, NY, USA, 1996.
25. Helgaker, T.; Jørgensen, P.; Olsen, J. *Molecular Electronic-Structure Theory*; Wiley: Hoboken, NJ, USA, 2000.
26. Koch, W.; Holthausen, M.C. *A Chemist's Guide to Density Functional Theory*, 2nd ed.; Wiley-VCH: Weinheim, Germany, 2002.
27. Sholl, D.S.; Steckel, J.A. *Density Functional Theory—A Practical Introduction*; Wiley: Hoboken, NJ, USA, 2009.
28. Hofer, T.S.; Rode, B.M.; Pribil, A.B.; Randolf, B.R. Simulations of Liquids and Solutions Based on Quantum Mechanical Forces. *Adv. Inorg. Chem.* **2010**, *62*, 143–175.
29. Weiss, A.K.H.; Hofer, T.S. Exploiting the Capabilities of Quantum Chemical Simulations to Characterise the Hydration of Molecular Compounds. *RSC Adv.* **2014**, *3*, 1606–1635. [CrossRef]
30. Hassanali, A.A.; Cuny, J.; Verdolino, V.; Parrinello, M. Aqueous solutions: State of the art in ab initio molecular dynamics. *Phil. Trans. R. Soc. A* **2014**, *372*, 20120482. [CrossRef]
31. Gregory, K.P.; Elliott, G.R.; Wanless, E.J.; Webber, G.B.; Page, A.J. A quantum chemical molecular dynamics repository of solvated ions. *Sci. Data* **2022**, *9*, 430. [CrossRef]
32. Pratt, L.R.; Rempe, S.B. Quasi-chemical theory and implicit solvent models for simulations. *AIP Conf. Proc.* **1999**, *492*, 172–201.
33. Pliego, J.R.; Riveros, J.M. On the Calculation of the Absolute Solvation Free Energy of Ionic Species: Application of the Extrapolation Method to the Hydroxide Ion in Aqueous Solution. *J. Phys. Chem. B* **2000**, *104*, 5155–5160. [CrossRef]
34. Lamoureux, G.; Roux, B. Absolute hydration free energy scale for alkali and halide ions established from simulations with a polarizable force field. *J. Phys. Chem. B* **2006**, *110*, 3308–3322. [CrossRef]
35. Lev, B.; Roux, B.; Noskov, S.Y. Relative Free Energies for Hydration of Monovalent Ions from QM and QM/MM Simulations. *J. Chem. Theory Comput.* **2013**, *9*, 4165–4175. [CrossRef]
36. Vlcek, L.; Chialvo, A.A. Single-ion hydration thermodynamics from clusters to bulk solutions: Recent insights from molecular modeling. *Fluid Phase Equilib.* **2016**, *407*, 58–75. [CrossRef]
37. Duignan, T.T.; Baer, M.D.; Schenter, G.K.; Mundy, C.J. Real single ion solvation free energies with quantum mechanical simulation. *Chem. Sci.* **2017**, *372*, 6131–6140. [CrossRef] [PubMed]
38. Hofer, T.S.; Hünenberger, P.H. Absolute proton hydration free energy, surface potential of water, and redox potential of the hydrogen electrode from first principles: QM/MM MD free-energy simulations of sodium and potassium hydration. *J. Chem. Phys.* **2018**, *148*, 222814. [CrossRef] [PubMed]

39. Shi, Y.; Beck, T.L. Absolute ion hydration free energy scale and the surface potential of water via quantum simulation. *Proc. Natl. Acad. Sci. USA* **2001**, *117*, 30151–30158. [CrossRef]
40. Prasetyo, N.; Hofer, T.S.; Hünenberger, P.H. Single-Ion Thermodynamics from First Principles: Calculation of the Absolute Hydration Free Energy and Single-Electrode Potential of Aqueous Li^+ Using ab Initio Quantum Mechanical/Molecular Mechanical Molecular Dynamics Simulations. *J. Chem. Theory Comput.* **2018**, *14*, 6443–6459. [CrossRef]
41. Schwenk, C.F.; Rode, B.M. Extended ab initio quantum mechanical/molecular mechanical molecular dynamics simulations of hydrated Cu^{2+}. *J. Chem. Phys.* **2003**, *119*, 9523–9530. [CrossRef]
42. Schwenk, C.F.; Hofer, T.S.; Rode, B.M. The Structure Breaking Effect of hydrated Cs(I). *J. Phys. Chem. A* **2004**, *108*, 1509–1514. [CrossRef]
43. Schwenk, C.F.; Löffler, H.H.; Rode, B.M. Molecular dynamics simulations of Ca^{2+} in water: Comparison of a classical simulation including three-body corrections and Born-Oppenheimer ab initio and density functional theory quantum mechanical/molecular mechanics simulations. *J. Chem. Phys.* **2001**, *115*, 10808–10813. [CrossRef]
44. Yoo, S.; Zeng, X.C.; Xantheas, S. On the phase diagram of water with density functional theory potentials: The melting temperature of ice Ih with the Perdew-Ernzerhof and Becke-Lee-Yang-Parr functionals. *J. Chem. Phys.* **2009**, *130*, 221102. [CrossRef] [PubMed]
45. Schmidt, J.; VandeVondele, J.; Kuo, I.; Sebastiani, D.; Siepmann, J.; Hutter, J.; Mundy, C.J. Isobaric-isothermal molecular dynamics simulations utilising density functional theory: An assessment of the structure and density of water at near-ambient conditions. *J. Phys. Chem. B* **2009**, *113*, 11959–11964. [CrossRef] [PubMed]
46. Grimme, S. Semiempirical GGA-type density functional constructed with a long-range dispersion contribution. *J. Comput. Chem.* **2006**, *27*, 1787–1799. [CrossRef] [PubMed]
47. Grimme, S. Semiempirical hybrid density functional with perturbative second-order correlation. *J. Chem. Phys.* **2006**, *124*, 034108. [CrossRef]
48. Zhang, Y.; Xu, X.; Goddard, W.A., III. Doubly hybrid density functional for accurate descriptions of nonbond interactions, thermochemistry, and thermochemical kinetics. *Proc. Natl. Acad. Sci. USA* **2009**, *106*, 4963–4968. [CrossRef]
49. Barone, V.; Biczyskoa, M.; Pavone, M. The role of dispersion correction to DFT for modelling weakly bound molecular complexes in the ground and excited electronic states. *J. Comput. Chem.* **2008**, *346*, 247–256. [CrossRef]
50. Yoo, S.; Xantheas, S.S. The effect of dispersion corrections on the melting temperature of liquid water. *J. Chem. Phys.* **2011**, *134*, 121105. [CrossRef]
51. Møller, C.; Plesset, M.S. Note on an Approximation Treatment for Many-Electron Systems. *Phys. Rev.* **1934**, *46*, 618–622. [CrossRef]
52. Häser, M.; Weigand, F. RI-MP2: First derivatives and global consistency. *Theor. Chem. Acc.* **1997**, *97*, 331–340.
53. Weigend, F.; Häser, M.; Patzelt, H.; Ahlrichs, R. RI-MP2: Optimized auxiliary basis sets and demonstration of efficiency. *Chem. Phys. Lett.* **1998**, *294*, 143–152. [CrossRef]
54. Hättig, A.H.C.; Köhn, A. Distributed memory parallel implementation of energies and gradients for second-order Møller-Plesset perturbation theory with the resolution-of-the-identity approximation. *Phys. Chem. Chem. Phys.* **2006**, *8*, 1159–1169. [CrossRef] [PubMed]
55. The Nobel Prize in Chemistry 2013, NobelPrize.org. Available online: https://www.nobelprize.org/prizes/chemistry/2013/summary/ (accessed on 11 November 2022).
56. Warshel, A.; Levitt, M. Theoretical Studies of Enzymic Reactions: Dielectric, Electrostatic and Steric Stabilization of the Carbenium Ion in the Reaction of Lysozyme. *J. Mol. Biol.* **1976**, *103*, 227–249. [CrossRef] [PubMed]
57. Åqvist, J.; Warshel, A. Simulation of enzyme reactions using valence bond force fields and other hybrid quantum/classical approaches. *Chem. Rev.* **1993**, *93*, 2523–2544. [CrossRef]
58. Field, M.J.; Bash, P.A.; Karplus, M. A Combined Quantum Mechanical and Molecular Mechanical Potential for Molecular Dynamics Simulations. *J. Comput. Chem.* **1990**, *11*, 700–733. [CrossRef]
59. Lyne, P.D.; Hodoscek, M.; Karplus, M. A Hybrid QM-MM Potential Employing Hartree-Fock or Density Functional Methods in the Quantum Region. *J. Phys. Chem. A* **1999**, *103*, 3462–3471. [CrossRef]
60. Warshel, A. Molecular Dynamics Simulations of Biological Reactions. *Acc. Chem. Res.* **2002**, *35*, 385–395. [CrossRef]
61. Senn, H.M.; Thiel, W. QM/MM studies of enzymes. *Curr. Opin. Chem. Biol.* **2007**, *11*, 182–187. [CrossRef]
62. Lin, H.; Truhlar, D.G. QM/MM: What have we learned, where are we, and where do we go from here? *Theor. Chem. Acc.* **2007**, *117*, 185–199. [CrossRef]
63. Bakowies, D.; Thiel, W. Hybrid Models for Combined Quantum Mechanical and Molecular Mechanical Approaches. *J. Phys. Chem.* **1996**, *100*, 10580–10594. [CrossRef]
64. Rode, B.M.; Schwenk, C.F.; Tongraar, A. Structure and dynamics of hydrated ions—New insights through quantum mechanical simulations. *J. Mol. Liq.* **2003**, *110*, 105–122. [CrossRef]
65. Rode, B.M.; Schwenk, C.F.; Hofer, T.S.; Randolf, B.R. Coordination and ligand exchange dynamics of solvated metal ions. *Coord. Chem. Rev.* **2005**, *249*, 2993–3006. [CrossRef]
66. Guillot, B. A reappraisal of what we have learnt during three decades of computer simulations on water. *J. Mol. Liq.* **2002**, *101*, 219–260. [CrossRef]
67. Cisneros, A.G.; Wikfeldt, K.T.; Ojamäe, L.; Lu, J.; Xu, Y.; Torabifard, H.; Bartók, A.P.; Csányi, G.; Molinero, V.; Paesani, F. Modeling Molecular Interactions in Water: From Pairwise to Many-Body Potential Energy Functions. *Chem. Rev.* **2016**, *116*, 7501–7528. [CrossRef]

68. Demerdash, O.; Wang, L.-P.; Head-Gordon, T. Advanced models for water simulations. *WIREs Comput. Mol. Sci.* **2018**, *8*, e1355. [CrossRef]
69. Kräutler, V.; van Gunsteren, W.; Hünenberger, P.H. A fast SHAKE algorithm to solve distance constraint equations for small molecules in molecular dynamics simulations. *J. Comput. Chem.* **2001**, *22*, 501–508. [CrossRef]
70. Hofer, T.S. Balancing the structural, vibrational and dielectric properties of an advanced flexible water model. *Chem. Phys. Lett.* **2021**, *762*, 138172. [CrossRef]
71. Rode, B.M.; Hofer, T.S.; Randolf, B.R.; Schwenk, C.F.; Xenides, D.; Vchirawongkwin, V. Ab initio Quantum Mechanical Charge Field Molecular Dynamics—A QM/MM MD Procedure for Accurate Simulations of Ions and Complexes. *Theor. Chem. Acc.* **2006**, *115*, 77–85. [CrossRef]
72. Hofer, T.S.; Pribil, A.B.; Randolf, B.R.; Rode, B.M. Ab Initio Quantum Mechanical Charge Field Molecular Dynamics: A Nonparametrized First-Principle Approach to Liquids and Solutions. *Adv. Quant. Chem.* **2010**, *59*, 213–246.
73. Tongraar, A.; Rode, B.M. The hydration structures of F^- and Cl^- investigated by ab initio QM/MM molecular dynamics simulations. *Phys. Chem. Chem. Phys.* **2003**, *5*, 357–362. [CrossRef]
74. Wiedemair, M.J.; Weiss, A.K.H.; Rode, B.M. Ab initio quantum mechanical simulations confirm the formation of all postulated species in ionic dissociation. *Phys. Chem. Chem. Phys.* **2014**, *16*, 7368–7376. [CrossRef] [PubMed]
75. Heuft, J.M.; Meijer, E.J. Density functional theory based molecular-dynamics study of aqueous fluoride solvation. *J. Chem. Phys.* **2005**, *122*, 094501. [CrossRef] [PubMed]
76. D'Angelo, P.; Migliorati, V.; Guidoni, L. Hydration Properties of the Bromide Aqua Ion: The Interplay of First Principle and Classical Molecular Dynamics, and X-ray Absorption Spectroscopy. *Inorg. Chem.* **2010**, *49*, 4224–4231. [CrossRef] [PubMed]
77. Choudhuri, J.R.; Yadav, V.K.; Krmakar, A.; Mallik, B.S.; Chandra, A. A first-principles theoretical study of hydrogen-bond dynamics and vibrational spectral diffusion in aqueous ionic solution: Water in the hydration shell of a fluoride ion. *Pure Appl. Chem.* **2013**, *85*, 27–40. [CrossRef]
78. Bankura, A.; Santra, B.; DiStasio, R.A., Jr.; Swartz, C.W.; Klein, M.L.; Wu, X. A systematic study of chloride ion solvation in water using van der Waals inclusive hybrid density functional theory. *Mol. Phys.* **2015**, *113*, 2842–2854. [CrossRef]
79. DelloStritto, M.; Xu, J.; Wu, X.; Klein, M.L. Aqueous solvation of the chloride ion revisited with density functional theory: Impact of correlation and exchange approximations. *Phys. Chem. Chem. Phys.* **2020**, *22*, 10666–10675. [CrossRef]
80. Merkling, P.J.; Ayala, R.; Martínez, J.M.; Pappalardo, R.; Marcos, E.S. Interplay of computer simulations and X-ray absorption spectra in the study of the bromide hydration structure. *J. Chem. Phys.* **2003**, *119*, 6647–6654. [CrossRef]
81. Soper, A.K.; Weckström, K. Ion solvation and water structure in potassium halide aqueous solutions. *Biophys. Chem.* **2006**, *124*, 180–191. [CrossRef]
82. Antalek, M.; Pace, E.; Hedman, B.; Hodgson, K.; Chillemi, G.; Benfatto, M.; Sarangi, R.; Frank, P. Solvation structure of the halides from X-ray absorption spectroscopy. *J. Chem. Phys.* **2016**, *145*, 044318. [CrossRef]
83. Wallen, S.L.; Palmer, B.J.; Pfund, D.M.; Fulton, J.L.; Newville, M.; Ma, Y.; Stern, E.A. Hydration of Bromide Ion in Supercritical Water: An X-ray Absorption Fine Structure and Molecular Dynamics Study. *J. Phys. Chem. A* **1997**, *101*, 9632–9640. [CrossRef]
84. Ohtaki, H.; Radnai, T. Structure and Dynamics of Hydrated Ions. *Chem. Rev.* **1993**, *93*, 1157–1204. [CrossRef]
85. Ohtaki, H. Ionic Solvation in Aqueous and Nonaqueous Solutions. *Monatsh. Chem.* **2001**, *132*, 1237–1268. [CrossRef]
86. Gaspar, A.M.; Alves Marques, M.; Cabaço, M.I.; de Barros Marques, M.I.; Buslaps, T.; Honkimaki, V. X-ray diffraction investigations of concentrated aqueous solutions of calcium halides. *J. Mol. Liq.* **2004**, *110*, 15–22. [CrossRef]
87. Mile, V.; Pusztai, L.; Dominguez, H.; Pizio, O. Understanding the Structure of Aqueous Cesium Chloride Solutions by Combining Diffraction Experiments, Molecular Dynamics Simulations, and Reverse Monte Carlo Modeling. *J. Phys. Chem. B* **2009**, *113*, 10760–10769. [CrossRef]
88. Mile, V.; Gereben, O.; Kohara, S.; Pusztai, L. On the structure of aqueous cesium bromide solutions: Diffraction experiments, molecular dynamics simulations and Reverse Monte Carlo modeling. *J. Mol. Liq.* **2010**, *157*, 36–42. [CrossRef]
89. Mile, V.; Gereben, O.; Kohara, S.; Pusztai, L. On the Structure of Aqueous Cesium Fluoride and Cesium Iodide Solutions: Diffraction Experiments, Molecular Dynamics Simulations, and Reverse Monte Carlo Modeling. *J. Phys. Chem. B* **2012**, *116*, 9758–9767. [CrossRef]
90. Pethes, I.; Bakó, I.; Pusztai, L. Chloride ions as integral parts of hydrogen bonded networks in aqueous salt solutions: The appearance of solvent separated anion pairs. *Phys. Chem. Chem. Phys.* **2020**, *22*, 11038–11044. [CrossRef]
91. Jensen, F. *Introduction to Computational Chemistry*; Wiley: Hoboken, NJ, USA, 1999.
92. Leach, A.R. *Molecular Modelling: Priciples and Applications*; Prentice Hall: Harlow, UK, 2001.
93. Berendsen, H.J.C.; Grigera, J.R.; Straatsma, T.P. The Missing Term in Effective Pair Potentials. *J. Phys. Chem.* **1987**, *91*, 6269–6271. [CrossRef]
94. Barker, J.; Watts, R. Monte Carlo studies of the dielectric properties of water-like models. *Mol. Phys.* **1973**, *26*, 789–792. [CrossRef]
95. Swope, W.C.; Andersen, H.C.; Berens, P.H.; Wilson, K.R. A computer simulation method for the calculation of equilibrium constants for the formation of physical clusters of molecules: Application to small water clusters. *J. Chem. Phys.* **1982**, *76*, 637–649. [CrossRef]
96. Horn, H.; Swope, W.; Pitera, J.; Madura, J.; Dick, T.; Hura, G.; Head-Gordon, T. Development of an improved four-site water model for biomolecular simulations: TIP4P-EW. *J. Chem. Phys.* **2004**, *120*, 9665–9678. [CrossRef] [PubMed]

97. Berendsen, H.J.C.; Postma, J.P.M.; van Gunsteren, W.F.; DiNola, A.; Haak, J.R. Molecular dynamics with coupling to an external bath. *J. Phys. Chem.* **1984**, *81*, 3684–3690. [CrossRef]
98. Laio, A.; VandeVondele, J.; Rothlisberger, U. A Hamiltonian electrostatic coupling scheme for hybrid Car-Parrinello molecular dynamics simulations. *J. Chem. Phys.* **2002**, *116*, 6941–6947. [CrossRef]
99. Voloshina, E.; Gaston, N.; Paulus, B. Embedding procedure for ab initio correlation calculations in group II metals. *J. Chem. Phys.* **2007**, *126*, 134115. [CrossRef]
100. Hofer, T.S.; Randolf, B.R.; Rode, B.M. Solvation Effects on Molecules and Biomolecules. In *Challenges and Advances in Computational Chemistry and Physics*; Canuto, S., Ed.; Chapter Molecular Dynamics Simulation Methods Including Quantum Effects; Springer: Heidelberg, Germany, 2008; Volume 6, pp. 247–278.
101. Hofer, T.S. Perspectives for hybrid ab initio/molecular mechanical simulations of solutions: From complex chemistry to proton-transfer reactions and interfaces. *Pure Appl. Chem.* **2014**, *86*, 105–117. [CrossRef]
102. Mulliken, R.S. Electronic population analysis on LCAO-MO molecular wave functions. I. *J. Chem. Phys.* **1955**, *23*, 1833–1840. [CrossRef]
103. Mulliken, R.S. Electronic population analysis on LCAO-MO molecular wave functions. II. *J. Chem. Phys.* **1955**, *23*, 1841–1846. [CrossRef]
104. Liew, C.C.; Inomata, H.; Arai, K. Flexible molecular models for molecular dynamics study of near and supercritical water. *Fluid Phase Equilibr.* **1998**, *144*, 287–298. [CrossRef]
105. TURBOMOLE V7.2 2017, a Development of University of Karlsruhe and Forschungszentrum Karlsruhe GmbH, 1989–2007, TURBOMOLE GmbH, Since 2007. Available online: http://www.turbomole.com (accessed on 11 November 2022).
106. Ditchfield, R.; Hehre, W.J.; Pople, J.A. Self-Consistent Molecular-Orbital Methods. IX. An Extended Gaussian-Type Basis for Molecular-Orbital Studies of Organic Molecules. *J. Chem. Phys.* **1971**, *54*, 724–728. [CrossRef]
107. Hehre, W.J.; Ditchfield, R.; Pople, J.A. Self-Consistent Molecular Orbital Methods. XII. Further Extensions of Gaussian-Type Basis Sets for Use in Molecular Orbital Studies of Organic Molecules. *J. Chem. Phys.* **1972**, *56*, 2257–2261. [CrossRef]
108. Hariharan, P.C.; Pople, J.A. The influence of polarization functions on molecular orbital hydrogenation energies. *Theor. Chim. Acta* **1973**, *28*, 213–222. [CrossRef]
109. Clark, T.; Chandrasekhar, J.; Spitznagel, G.W.; Schleyer, P.V.R. Efficient diffuse function-augmented basis sets for anion calculations. III. The 3-21+G basis set for first-row elements, Li-F. *J. Comput. Chem.* **1983**, *4*, 294–301. [CrossRef]
110. Weigend, F.; Ahlrichs, R. Balanced basis sets of split valence, triple zeta valence and quadruple zeta valence quality for H to Rn: Design and assessment of accuracy. *Phys. Chem. Chem. Phys.* **2005**, *7*, 3297–3305. [CrossRef]
111. Hofer, T.S.; Tran, H.T.; Schwenk, C.F.; Rode, B.M. Characterisation of Dynamics and Reactivities of Solvated Ions by ab initio Simulations. *J. Comput. Chem.* **2004**, *25*, 211–214. [CrossRef]
112. Bhattacharjee, A.; Hofer, T.; Rode, B. Local density corrected three-body distribution functions for probing local structure reorganization in liquids. *Phys. Chem. Chem. Phys.* **2008**, *10*, 6653–6657. [CrossRef]
113. Higham, J.; Henchman, R.H. Locally adaptive method to define coordination shell. *J. Chem. Phys.* **2016**, *145*, 084108. [CrossRef]
114. Ali, H.S.; Higham, J.; Henchman, R.H. Entropy of simulated liquids using multiscale cell correlation. *Entropy* **2019**, *21*, 750. [CrossRef] [PubMed]
115. Tkatchenko, A.; Scheffler, M. Accurate Molecular Van Der Waals Interactions from Ground-State Electron Density and Free-Atom Reference Data. *Phys. Rev. Lett.* **2009**, *102*, 073005. [CrossRef]
116. Leontyev, I.V.; Stuchebrukhov, A.A. Electronic continuum model for molecular dynamics simulations. *J. Chem. Phys.* **2009**, *130*, 085102. [CrossRef]
117. Leontyev, I.V.; Stuchebrukhov, A.A. Electronic polarizability and the effective pair potentials of water. *J. Chem. Theory Comput.* **2010**, *6*, 3153–3161. [CrossRef]
118. Leontyev, I.V.; Stuchebrukhov, A.A. Accounting for electronic polarization in non-polarizable force fields. *Phys. Chem. Chem. Phys.* **2011**, *13*, 2613–2626. [CrossRef] [PubMed]
119. Kann, Z.; Skinner, J.L. A scaled-ionic-charge simulation model that reproduces enhanced and suppressed water diffusion in aqueous salt solutions. *J. Chem. Phys.* **2014**, *141*, 104507. [CrossRef] [PubMed]
120. Hofer, T.S.; Randolf, B.R.; Rode, B.M. tructure-Breaking Effects of Solvated Rb(I) in Dilute Aqueous Solution—An Ab Initio QM/MM MD Approach. *J. Comput. Chem.* **2005**, *26*, 949–956. [CrossRef] [PubMed]

liquids

MDPI

Article

An Ab Initio Investigation of the Hydration of Tin(II)

Cory C. Pye * and Champika Mahesh Gunasekara

Department of Chemistry, Saint Mary's University, Halifax, NS B3H 3C3, Canada
* Correspondence: cory.pye@smu.ca

Abstract: The structure of tin(II) is not well known in aqueous solution. The energies, structures, and vibrational frequencies of $[Sn(H_2O)_n]^{2+}$ n = 0–9, 18 have been calculated at the Hartree–Fock and second order Møller–Plesset levels of theory using the CEP, LANL2, and SDD effective core potentials in combination with their associated basis sets, or with the 6-31G* and 6-31+G* basis sets. The tin–oxygen distances and totally symmetric stretching frequency of the aquatin(II) ions were compared with each other, and with solution measurements where available.

Keywords: ab initio; tin(II); hydration; symmetry; vibrational spectrum

1. Introduction

The structure of some metal ions in solution remain elusive [1,2]. Their toxicity to man and the environment is dependent on their oxidation state and speciation, which often depends on pH and the presence of counterions that solubilize the metal by complex formation. Computational chemistry can be useful in supporting and rationalizing proposed speciation models. However, for elements of high atomic number, a drawback is that there are typically few all-electron basis sets that can be used, and relativistic effects can play an important role. One workaround is to use effective core potentials, which replace the explicit description of core electrons by a core potential, which are then paired with basis sets describing the outermost electrons. Previously, some common effective core potentials for the aqua complexes of the heavy metals mercury(II) and thallium(III) (valence electron configuration $5d^{10}$) were benchmarked [3]. This work was extended to lead(II) with a valence electron configuration of $6s^2 5d^{10}$ [4]. We expand our work now to tin(II), which has a valence electron configuration of $5s^2 4d^{10}$. The presence of the ns^2 subshell, as with lead(II), will be shown to have a pronounced effect on the structures compared to those without it. *Hemidirected* structures (which tend to be favored at lower coordination numbers) have ligands that are not symmetrically distributed around the central ion, whereas *holodirected* structures have a symmetrical distribution.

In the gas-phase, it was reported that tin(II), lead(II), and mercury(II) easily underwent a proton transfer reaction and that the only species observed in the mass spectra were the deprotonated $MOH^+(H_2O)_{n-1}$ ions, not the $M(H_2O)_n{}^{2+}$ ions. These ions have anomalously high acidity in the gas phase as well as the solution phase. An ab initio study was carried out to rationalize this behavior, with a focus on the pathways to deprotonation [5]. In solution, it is believed that the relevant species are Sn^{2+} (aq) (pH < 2), $SnOH^+$ (aq), $Sn_2(OH)_2{}^{2+}$ (aq), $Sn_3(OH)_4{}^{2+}$ (aq), $Sn(OH)_2{}^0$ (aq) (pH = 5–8), and $Sn(OH)_3{}^-$ (aq) (pH > 10) [2]. The polynuclear species form at higher concentrations, and the water content cannot be determined through potentiometric means.

The tin(II) ion in aqueous solution has been characterized by an X-ray study of a ~3 mol/L solution of the perchlorate salt [6]. The radial distribution curves showed peaks at 1.4 Å (Cl–O), 2.3 Å (O ... O and Sn–O), 2.8 Å (Sn–O), 3.6 Å (Sn–Sn), and 4.2 Å. (Sn ... O). Some hydrolyzed solutions were also examined, and the largest variation was in the 3.6 Å and 4.2 Å peaks, which suggested a greater degree of clustering as the hydrolysis increased. Essentially the same unhydrolyzed solution was studied by EXAFS [7], in which a Sn–O distance of 2.2–2.3 Å with four water molecules was found. They also reanalyzed the data of [6]. Regarding the hydrolysis products, potentiometric titrations suggested the existence of the species $Sn_3(OH)_4^{2+}$(aq), in addition to $Sn_2(OH)_2^{2+}$(aq) and $SnOH^+$(aq) [8]. The crystal structure of the hydrolysis product $Sn_3O(OH)_2SO_4$, which is potentiometrically equivalent to $Sn_3(OH)_4^{2+}$(aq), has been determined and shown to contain the discrete $[Sn_3O(OH)_2]^{2+}$ ion [9,10].

A QM/MM-MD study has also been carried out on the tin(II) ion in aqueous solution [11]. It was found that the Sn–O distance peaked at 2.5 Å, with a shoulder at 2.65 Å. Gaussian fitting indicated peaks at 2.45 Å and 2.75 Å. A coordination number of eight was found. The power spectrum of the Sn–O stretching suggested peaks at 85 and 208 cm^{-1}.

2. Materials and Methods

Calculations were performed using Gaussian 98 [12]. This program version was the first to allow the analytical frequency calculation of molecules in which core electrons are described by effective core potentials (ECPs), and thus, many variants of these were tried. The MP2 calculations use the frozen core approximation. A stepping-stone approach was used for geometry optimization, in which the geometries at the levels HF/CEP-4G, HF/CEP-31G*, HF/CEP-121G*, HF/LANL2MB, HF/LANL2DZ, and HF/SDD were sequentially optimized. For minimum energy structures, the MP2/CEP-31G* and MP2/CEP-121G* calculations were also performed. Calculations were also carried out using the 6-31G* and 6-31+G* basis sets on the atoms of the water molecules (5d) with an effective core potential and basis set on the metal ion (denoted as ECP+6-31G* or 6-31+G*). For shorthand, we denoted the mixed basis sets as follows: CEP-121G* on Sn and 6-31G* on O, H, as basis set A; LANL2DZ on Sn and 6-31G* on O, H, as basis set B; and SDD on Sn and 6-31G* on O, H, as basis set C. The corresponding basis sets with diffuse functions were indicated by adding a "+" to the basis set name. Default optimization specifications were normally used. After each level, where possible, a frequency calculation was performed at the same level, and the resulting Hessian was used in the following optimization. Z-matrix coordinates constrained to the appropriate symmetry were used as required to speed up the optimizations. Since frequency calculations are done at each level, any problems with the Z-matrix coordinates would manifest themselves by giving imaginary frequencies corresponding to modes orthogonal to the spanned Z-matrix space. The Hessian was evaluated at the first geometry (opt = CalcFC) for the first level in a series in order to aid geometry convergence. We note that, for the heavy elements only, the three different CEP basis sets are equivalent (CEP-121G*) but differ for the oxygen and hydrogen atoms. We also note that the choice of core electrons defining the pseudopotential depends on the specific core potential (CEP and LANL2, $[Kr]4d^{10}$; SDD, $[Ar]3d^{10}$). In some cases, Gaussian 03 [13] and Gaussian 16 [14] were used to correct errors and omissions.

In many cases to follow, the symmetry of the minimum-energy complexes was the same as those previously found for bismuth [15]. To confirm these results, starting with high symmetry structures, systematic desymmetrization along the various irreducible representations was carried out [16,17]. We did not employ an implicit solvation model or additional electron correlation treatments for reasons described previously in [4]. The energies of all structures are found in Table S1.

3. Results

3.1. A Survey of Structures

Tin(II), as a lighter element in the same group as lead(II), might be expected to show similar properties. The point group symmetry for mono- through hexaaquatin(II) was usually found to be C_{2v}, C_2, C_3, C_2, C_s, and C_3. The diaquatin(II) species, like lead, ascends in symmetry to a planar C_{2v} structure at HF/LANL2MB. The tetraaquatin(II) species has C_{2v} symmetry at HF/LANL2DZ and C_s symmetry at HF/LANL2MB. The pentaaqua species has C_2 symmetry at HF/LANL2MB. At all levels for the pentaaqua species, if the pentacoordinate [5+0] species exists, it is competitive in energy with the [4+1], and the most stable form is dependent on the level of theory. For the hexaaquatin(II), the C_3 [3+3] form was always more stable than the C_3 [6+0] form, which occasionally had imaginary frequencies or reverted to the [3+3] form. We did not find stable hepta-, octa-, or enneaaquatin(II) structures. At the HF levels, the [6+12] form always contained at least an imaginary *T* mode.

The results of the systematic desymmetrization procedure [16] for aquatin(II) are as follows (see Figures 1 and S1):

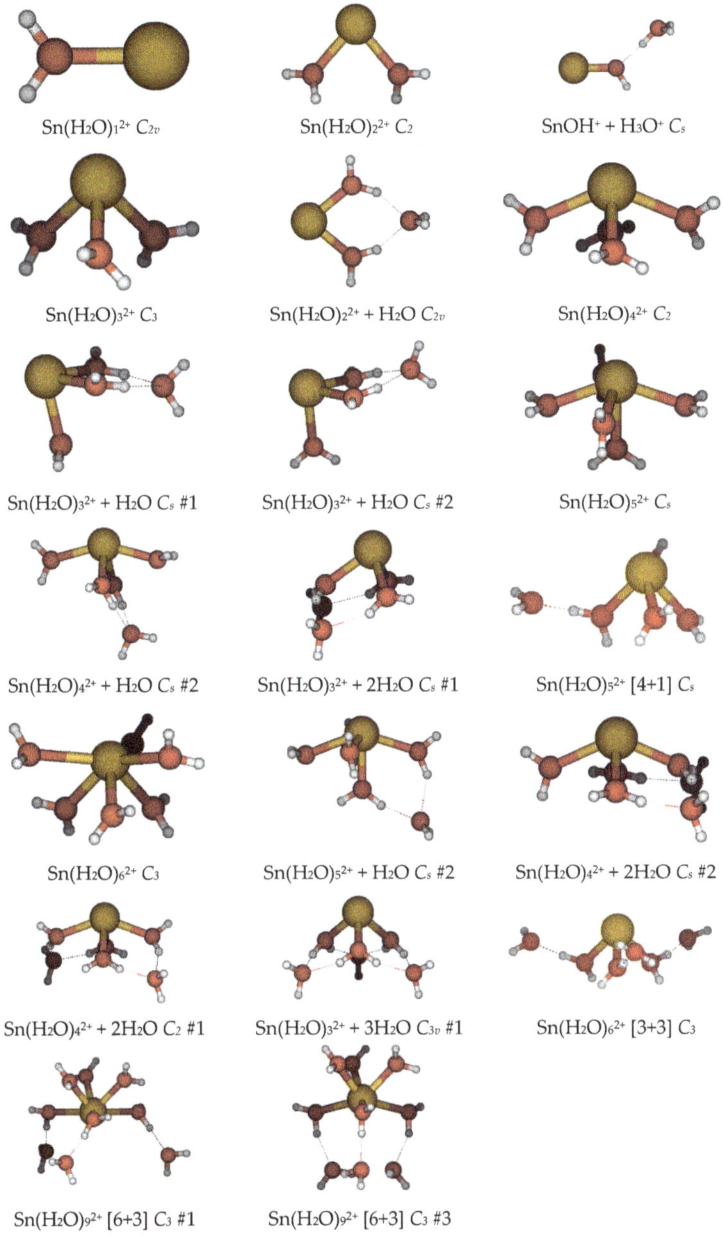

Figure 1. The minimum energy structures of aquatin(II).

- The monoaquatin(II) remained as C_{2v} at all levels;
- The most stable diaquatin(II) remained as the bent C_2 at all levels except HF/LANL2MB (C_{2v} planar). The linear holodirected D_{2d} structure was approximately 50 kJ/mol higher in energy, but the unstable bent C_s structure was only slightly higher in energy (<1 kJ/mol for nonminimal basis sets). All attempts to generate a [1+1] structure instead resulted in proton transfer to give a $SnOH^+$ + H_3O^+ complex, which was 25–40 kJ/mol higher in energy;
- The most stable triaquatin(II) remained as the pyramidal C_3 at all levels. The two pyramidal C_{3v} structures were 12–25 kJ/mol higher in energy, whereas the planar holodirected D_{3h} and D_3 structures were 60–90 kJ/mol higher in energy. The stable [2+1] C_{2v} structure was 25–50 kJ/mol higher in energy;
- The most stable tetraaquatin(II) was usually the see-saw C_2, but it could be the C_{2v} #3 (HF/LANL2DZ, HF/B+) or C_s (HF/LANL2MB). The C_{2v} #3 was slightly higher in energy (<2 kJ/mol), with the other C_{2v} structures being higher (15–25 kJ/mol). The holodirected D_{2d} #1, #2, S_4, and D_2 structures were much higher in energy (50–80 kJ/mol). The C_s #2 [3+1] structure was always competitive in energy, and usually lower, than the tetracoordinate structure;
- The most stable pentaaquatin(II) was the square pyramidal C_s (if it exists), which is closely related to the C_{2v} #1 structure (<3 kJ/mol). The other three C_{2v} structures were ~25 kJ/mol higher in energy. The stable [4+1] and [3+2] structures were competitive in energy, and sometimes lower, depending on level of theory;
- The most stable hexaaquatin(II) was the distorted octahedral C_3 (if it exists). The octahedral T_h structure was ~30 kJ/mol higher in energy;
- Of the 16 different C_{2v} heptaaquatin(II) structures tried, none were stable, and either possessed imaginary modes or dissociated to a [6+1], [5+2], or [4+3] structure. Structures #1–#4, and #11, always dissociated. Structures #5–#8, and #16, usually remained as 7-coordinates. The remaining structures usually dissociated at most levels. Of the 7-coordinate C_{2v} structures, #8 and #16 were the lowest in energy. All of the 7-coordinate structures dissociated at HF/LANL2MB. Upon desymmetrization of the remaining 7-coordinate C_{2v} structures to C_2, nearly all dissociated to [6+1], [5+2], or [4+3] structures. The only exceptions, C_2 #10 and #15 at HF/CEP-31G*, possessed imaginary B modes, of which one corresponds to a water molecule moving to the second hydration shell;
- Of the octaaquatin(II) structures, two D_{4h} and two D_{4d} structures (point group order h = 16) were first examined. Multiple imaginary modes were present;
    - For the D_{4d} #1 and #2 structures, desymmetrization along the A_2 imaginary mode gave the same S_8 structure; along the B_1 imaginary mode, they gave the same D_4 #2 structure; along the B_2 imaginary mode, they gave the C_{4v} #1 and #2 structures; and along the imaginary E_1 mode, they gave the same C_{2v} #1 structure (via C_s);
    - For the D_{4h} #1 and #2 structures, desymmetrization along the A_{1u} imaginary mode gave the D_4 #2 structure found before; along the A_{2g} imaginary mode, they gave the same C_{4h} #1 structure; along the A_{2u} imaginary mode, they gave the C_{4v} #3 [4+4] and #4 structures; along the B_{2g} imaginary mode (D_{4h} #1), the D_{2h} #1 structure ascended in symmetry to D_{4h} #2; along the B_{1g} (D_{4h} #2) imaginary mode, they gave the D_{2h} #2 structure; along the B_{1u} imaginary mode, they gave the same D_{2d} #1 structure; along the B_{2u} mode, they gave the D_{2d} #2 and #4 structures, respectively; and along the E_g and E_u modes, they gave D_{2h} #3 and #4 (via C_{2h} and C_{2v});

  Examination of lower symmetry structures (h = 8) gave the following results:
    - For the S_8 #1 structure, all E_1 and B imaginary modes corresponded to the expulsion of water molecules to the second hydration shell. Desymmetrization along the B mode gave the [4+4] C_4;

- For the D_4 #2 structure, the E mode corresponded to the expulsion of water molecules to the second hydration shell. Desymmetrization along the A_2 mode gave either a [4+4] C_4 structure or a C_4 #2 structure, whereas along the B_1 mode, it gave the D_2 #1 structure;
- For the C_{4h} #1 structure, the imaginary E_u mode corresponded to the expulsion of two water molecules. Desymmetrization along the imaginary A_u mode for the most part gave either a [4+4] C_4 structure or ascended in symmetry to S_8; along the imaginary B_u mode at HF/CEP-4G and HF/LANL2MB, it gave a D_{2d} #5 and S_4 structure, respectively. This D_{2d} #5 structure was then rerun at all levels;
- For the D_{2d} structures, desymmetrization along an A_2 imaginary mode would give an S_4 structure; along a B_1 imaginary mode, they gave a D_2 structure; along a B_2 imaginary mode, they gave a C_{2v} structure; along an E mode, they gave either a C_2 or C_s structure. Along the A_2 mode, an S_4 #2 or #4 structure typically resulted, or ascension in symmetry to the D_{2d} #5; along the B_1 mode, there was usually ascension to D_4 #2; along the B_2 mode, there was dissociation to a [6+2] or [4+4] structure; and along an E mode, dissociation would occur;
- For the D_{2h} structures, desymmetrization along the imaginary A_u mode would give a D_2 structure, and along the imaginary B_{ng} modes, a C_{2h} structure was given. In all cases, these desymmetrized, and most ascended in symmetry to structures already found (D_2 #5, D_4 #2, C_{4h} #1). The B_{nu} modes corresponded to the expulsion of water molecules from the first hydration sphere;

Examination of lower symmetry structures (h = 4) gave the following results:

- For the C_{2v} structures, desymmetrization along the A_2 mode would give a C_2 structure, and along the B_1 or B_2 mode, different C_s structures were given. For the C_{2v} structures, at least one of the imaginary B modes in each structure corresponded to dissociation to a [6+2] structure, whereas desymmetrization along the A_2 mode led to a [4+4] or [4+2+2] structure;
- For the C_4 and S_4 structures, the imaginary E mode corresponded to dissociation to a [6+2] structure, whereas desymmetrization along the B mode to give a C_2 structure resulted in dissociation;
- For the D_2 structures, at least one of the imaginary B_2 or B_3 modes corresponded to dissociation to a [6+2] structure, whereas desymmetrization along the B_1 mode to give a C_2 structure resulted in dissociation to a [6+2] or [4+4];

Based on these results, we must conclude that a stable 8-coordinate octaaquatin(II) ion cannot exist.

- Of the enneaaquatin(II) structures, four D_{3h} structures (point group order h = 12) were first examined. Multiple imaginary modes were present. Desymmetrization along the A_1'' mode would yield D_3 structures; along A_2', C_{3h} structures were given; and along A_2'', C_{3v} structures were given. A common D_3 #1 structure was found for most, and in some cases gave an additional [6+3] structure. Two possible C_{3h} structures were found, and in some cases gave an additional [6+3] structure. Four possible C_{3v} structures were found, and in some cases gave additional [6+3] structures. At least one of the degenerate modes corresponded to the expulsion of water molecule(s) from the inner coordination shell;
  - For the D_3 structure, desymmetrization along the A_2 mode gave a C_3 #1 [6+3] structure. At least one of the E modes corresponded to the loss of water molecules from the first hydration shell.
  - For the C_{3h} structures, desymmetrization along the A'' mode gave either the C_3 #1 [6+3] structure above or a new C_3 #3 [6+3] structure (or occasionally [3+3+3]). At least one of the E modes corresponded to the loss of water molecules from the first hydration shell.
  - For the C_{3v} structures, desymmetrization along the A_2 mode gave one of the C_3 [6+3] structures (or occasionally [3+3+3]) found above. At least one

of the E modes corresponded to the loss of water molecules from the first hydration shell.

Based on these results, we must conclude that a stable 9-coordinate enneaaquatin(II) ion cannot exist.

To summarize these results, by using the systematic desymmetrization procedure, we have found stable structures for the mono- through hexaaquatin(II) complexes, and we have shown that hepta-, octa-, and enneaaqua structures do not exist on the potential energy surface. The hexaaquatin(II) C_3 structure is only stable at HF/CEP-31G*, HF/CEP-121G*, and MP2/CEP-31G*. The pentaaquatin(II) C_s structure is not stable at HF/CEP-4G, HF/LANL2DZ, and HF/SDD. In most cases, for systems with more than three water molecules, the most stable structure on the potential energy surface is tricoordinate, with the remaining water molecules in the second hydration sphere (the main exceptions being HF/CEP-31G* and HF/CEP-121G*). These results suggest that tin(II) would be tricoordinate trigonal pyramidal in an aqueous solution.

3.2. The Sn–O Distance

The average Sn–O distance as a function of coordination number is plotted in Figure 2 for all of the levels studied. The Sn–O distance always lengthened following an increase in the coordination number. We can see some gaps for n = 4–6 at some levels where no local minimum existed. The Sn–O distance using the minimal basis HF/LANL2MB was shorter than the other levels at the same hydration number by 0.1–0.3 Å, which tended to cluster together. The results using CEP-31G* were nearly coincidental with those of CEP-121G*. For all levels, there was a pronounced change in slope at n = 3. Within the cluster noted above, the Sn–O distance using the SDD basis set/pseudopotential on Sn (HF/SDD, HF/C, HF/C+) tended to be the longest (n = 1–5, 2.20–2.50 Å), whereas those using the LANL2DZ basis set/pseudopotential on Sn (HF/LANL2DZ, HF/B, HF/B+) were the shortest. This differed for lead, where the CEP basis set/pseudopotential tended to be the shortest [4]. The effect of the basis set/pseudopotential combination was more important than the presence or absence of correlation (HF vs. MP2). Metal–oxygen bond lengths to those oxygens making a smaller angle to the principal symmetry axis were longer, as was noted previously for aqualead(II) [4].

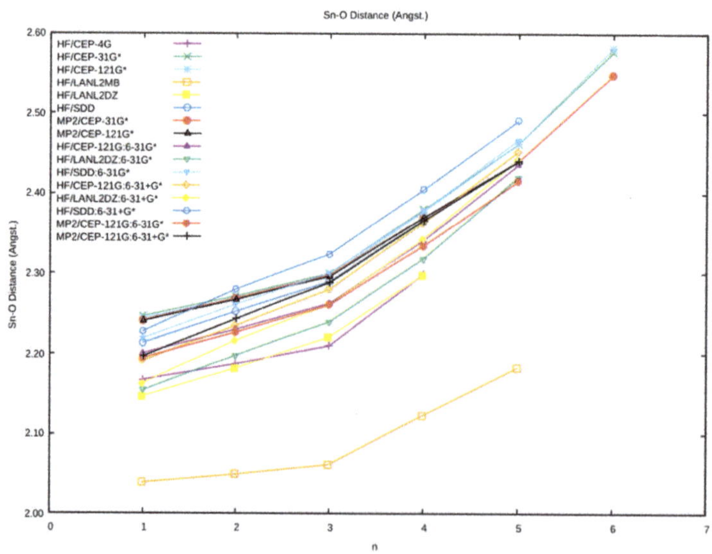

Figure 2. The dependence of the average Sn–O distance on the coordination number n and level of theory.

3.3. The Sn–O Vibrational Frequency

The vibrational frequency of the totally symmetric Sn–O stretch as a function of coordination number is plotted in Figure 3. As expected, the frequencies at the minimal basis HF/LANL2MB were much higher than the other levels which clustered together. For the most part, the vibrational frequency decreased as a function of hydration number. There was a levelling effect upon going from n = 4 to n = 5, because, for the square pyramidal n = 5, the character of the mode changed to be predominantly a Sn–O_{apex} stretch. The results using CEP-31G* were nearly coincidental with those of the CEP-121G*, and the MP2 values were ~15 cm^{-1} higher than the HF values. The addition of diffuse functions (A+ vs. A) lowered the vibrational frequency by ~30 cm^{-1}, and correlation increased it by ~5–15 cm^{-1}. This was also true for the mixed basis set calculations. All other things being equal, the LANL2DZ frequencies were the highest, and the CEP frequencies were the lowest, with the SDD frequencies falling in the middle. The inverse relationship between average bond length and symmetric stretching frequency can be clearly seen.

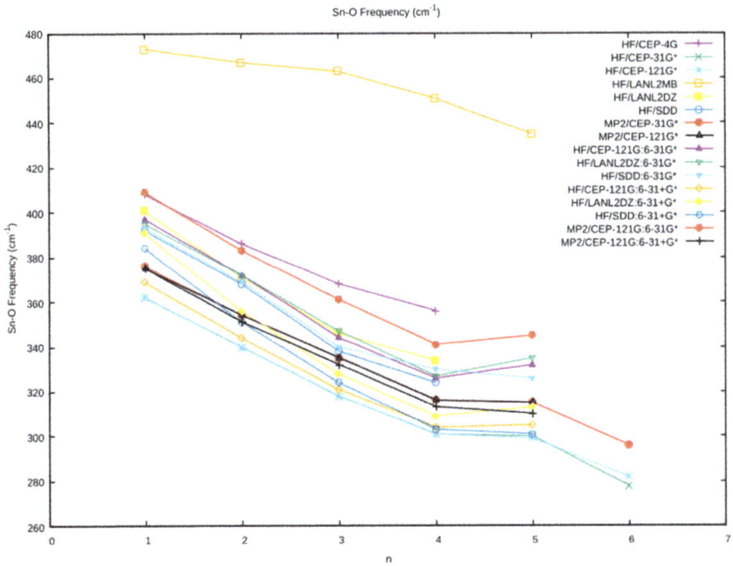

Figure 3. The dependence of the totally symmetric Sn–O frequency on the coordination number n and level of theory.

4. Discussion

Because of the lack of experimental data on aquatin(II) complexes, certainty regarding the structure and vibrational frequencies of the aquatin(II) is lacking. We may compare a series of structures, such as [n+0] and [($n-m$)+m]. For a hydration number of three, the tricoordinate [3+0] is always more stable than the [2+1] structure. For a hydration number of four, the tricoordinate [3+1] is usually more stable than [4+0], except for HF/CEP-31G* and HF/CEP-121G*, but they are competitive in energy. For a hydration number of five, the [4+1] and [3+2] structures are usually more stable than the [5+0], except for HF/CEP-31G* and HF/CEP-121G*, but all are still competitive in energy. For a hydration number of six, the [4+2] structure seems to be the most stable, with the [3+3] structure being competitive in energy. The picture emerging is that of a variable coordination number between three and six, with three (trigonal pyramidal) and four (see-saw) being the most likely.

We recently became aware [18] of a crystal structure determination of an aquatin(II) ion in the compound tin(II) perchlorate trihydrate [19]. The Sn–O distance was reported to be 2.201(7) Å. Examination of Figure 2, at n = 3, and Table 1 suggest that, if crystal packing forces are negligible, then the MP2/A+, HF/LANL2DZ, and HF/B+ levels give excellent agreement with the experiment. This result was confirmed by Persson and co-workers, who obtained 2.208(9) Å and obtained EXAFS results for the crystal and solution Sn–O distance of 2.209(3) and 2.219(3) Å, respectively, which quite importantly shows that the solution structure of tin(II) is the same as the solid. The LAXS measurement gives, at 2.206(2) Å, nearly the same value for the Sn–O distance [20].

Table 1. Bond lengths (Å) of triaquatin(II). The theoretical levels A, B, and C are described in the text. HF = Hartree-Fock, MP2 = second-order Møller-Plesset, Expt. = experiment, XRD = X-ray diffraction, EXAFS = Extended X-ray absorption fine structure, LAXS = Large angle X-ray scattering.

Basis Set/Pseudopotential	HF		MP2
CEP-4G	2.2098		n/c
CEP-31G*	2.3001		2.2978
CEP-121G*	2.2963		2.2964
LANL2MB	2.0616		n/c
LANL2DZ	2.2202		n/c
SDD	2.2898		n/c
A	2.2624		2.2606
B	2.2397		n/c
C	2.3005		n/c
A+	2.2797		2.2887
B+	2.2610		n/c
C+	2.3241		n/c
Expt. XRD [19]		2.201(7)	
Expt. XRD [20]		2.208(9)	
Expt. EXAFS xtal. [20]		2.209(3)	
Expt. EXAFS soln. [20]		2.219(3)	
Expt. LAXS soln. [20]		2.206(2)	

The vibrational frequency for the $n = 3$ structure lay in the range 320–370 cm^{-1}. The effect of a second hydration sphere should be to increase this value somewhat. For most octahedral metal ions that we have previously examined, a rough rule of thumb is that, upon including a second hydration sphere, the vibrational frequency increases by 20q cm^{-1}, where q is the total charge on the octahedral ion. For tin(II), comparison of the vibrational frequencies of $[Sn(H_2O)_3]^{2+}$ C_3, and $[Sn(H_2O)_3]^{2+}{\cdot}(H_2O)_3$ C_{3v} revealed a large ~85 cm^{-1} increase in the vibrational frequency. This suggests that tin(II) should have an observable band in the isotropic Raman spectra corresponding to the totally symmetric stretching motion in the range of 400–450 cm^{-1}.

5. Conclusions

The common CEP, LANL2DZ and SDD pseudopotentials were paired with various basis sets to study the hydrated tin(II) ion. The calculations using minimal basis sets performed poorly. For the most part, the calculated structures were consistent with recent experimental results of a tricoordinate trigonal pyramidal hemidirected aqua complex. The careful use of symmetry can be used to both guide the search for new structures and also to rule out structures.

Supplementary Materials: The following are available online at https://www.mdpi.com/article/10.3390/liquids2040027/s1, Figure S1: Some stationary point (non-minimum) energy structures of aquatin(II); Table S1: Total Energies of Aquatin(II) Species.

Author Contributions: Conceptualization, C.C.P.; methodology, C.C.P.; investigation, C.M.G. and C.C.P.; resources, C.C.P.; data curation, C.C.P.; writing—original draft preparation, C.C.P.; writing—review and editing, C.C.P.; visualization, C.C.P.; supervision, C.C.P.; project administration, C.C.P.; funding acquisition, C.C.P. All authors have read and agreed to the published version of the manuscript.

Funding: This research was funded by the Natural Sciences and Engineering Research Council of Canada.

Data Availability Statement: Data is contained within the article and Supplementary Materials.

Acknowledgments: The authors thank the Department of Astronomy and Physics, at Saint Mary's University (AP-SMU), for providing access to computing facilities, in particular to cygnus, a 10-processor Sun server, which was purchased with assistance from the Canada Foundation for Innovation, Sun Microsystems, the Atlantic Canada Opportunities Agency, and SMU. CCP also thanks ACEnet and Compute Canada for providing access to computers and Gaussian 03/16. CCP acknowledges the former financial support of the NSERC. CMG acknowledges the support of the Department of Economic Development (Government of Nova Scotia), the Office of the Dean of Science, SMU, and the Cooperative Education program (SMU–Work Term 2, Fall 2001 and Work Term 3, Winter 2002).

Conflicts of Interest: The authors declare no conflict of interest. The funders had no role in the design of the study; in the collection, analyses, or interpretation of data; in the writing of the manuscript, or in the decision to publish the results.

References

1. Richens, D.T. *The Chemistry of Aqua Ions*; Wiley: Chichester, UK, 1997.
2. Baes, C.F., Jr.; Mesmer, R.E. *The Hydrolysis of Cations*; Wiley: New York, NY, USA, 1976.
3. Pye, C.C.; Gunasekara, C.M. An Ab Initio Investigation of the Hydration of Thallium(III) and Mercury(II). *J. Solution Chem.* **2020**, *49*, 1419–1429. [CrossRef]
4. Pye, C.C.; Gunasekara, C.M. An Ab Initio Investigation of the Hydration of Lead(II). *Liquids* **2022**, *2*, 39–49. [CrossRef]
5. Cox, H.; Stace, A.J. Molecular View of the Anomalous Acidities of Sn^{2+}, Pb^{2+}, and Hg^{2+}. *J. Am. Chem. Soc.* **2004**, *126*, 3939–3947. [CrossRef]
6. Johansson, G.; Ohtaki, H. An X-Ray Investigation of the Hydrolysis Products of Tin(II) in Solution. *Acta Chem. Scand.* **1973**, *27*, 643–660. [CrossRef]
7. Yamaguchi, T.; Lindqvist, O.; Claeson, T.; Boyce, J.B. EXAFS and X-ray diffraction studies of the hydration structure of stereo-chemically active Sn(II) ions in aqueous solution. *Chem. Phys. Lett.* **1982**, *93*, 528–532. [CrossRef]
8. Tobias, R.S. Studies on the Hydrolysis of Metal Ions. 21. The Hydrolysis of the Tin(II) Ion, Sn^{2+}. *Acta Chem. Scand.* **1958**, *12*, 198–223. [CrossRef]
9. Grimvall, S. On the Crystal Structure of $Sn_3O(OH)_2SO_4$. *Acta Chem. Scand.* **1973**, *27*, 1447. [CrossRef]
10. Davies, C.G.; Donaldson, J.D.; Laughlin, D.R.; Howie, R.A.; Beddoes, R. Crystal structure of tritin(II) dihydroxide oxide sulfate. *J. Chem. Soc. Dalton Trans.* **1975**, 2241–2244. [CrossRef]
11. Hofer, T.S.; Pribil, A.B.; Randolf, B.R.; Rode, B.M. Structure and Dynamics of Solvated Sn(II) in Aqueous Solution: An ab Initio QM/MM MD Approach. *J. Am. Chem. Soc.* **2005**, *127*, 14231–14238. [CrossRef] [PubMed]
12. Frisch, M.J.; Trucks, G.W.; Schlegel, H.B.; Scuseria, G.E.; Robb, M.A.; Cheeseman, J.R.; Zakrzewski, V.G.; Montgomery, J.A., Jr.; Stratmann, R.E.; Burant, J.C.; et al. *Gaussian 98, Revision A.9*; Gaussian, Inc.: Pittsburgh, PA, USA, 1998.
13. Frisch, M.J.; Trucks, G.W.; Schlegel, H.B.; Scuseria, G.E.; Robb, M.A.; Cheeseman, J.R.; Montgomery, J.A., Jr.; Vreven, T.; Kudin, K.N.; Burant, J.C.; et al. *Gaussian 03, Revision D.02*; Gaussian, Inc.: Wallingford, CT, USA, 2004.
14. Frisch, M.J.; Trucks, G.W.; Schlegel, H.B.; Scuseria, G.E.; Robb, M.A.; Cheeseman, J.R.; Scalmani, G.; Barone, V.; Petersson, G.A.; Nakatsuji, H.; et al. *Gaussian 16, Revision A.03*; Gaussian, Inc.: Wallingford, CT, USA, 2016.
15. Pye, C.C.; Gunasekara, C.M.; Rudolph, W.W. An ab initio investigation of bismuth hydration. *Can. J. Chem.* **2007**, *85*, 945–950. [CrossRef]
16. Pye, C.C.; Whynot, D.C.M.; Corbeil, C.R.; Mercer, D.J. Desymmetrization in geometry optimization: Application to an ab initio study of copper(I) hydration. *Pure Appl. Chem.* **2020**, *92*, 1643–1654. [CrossRef]
17. Pye, C.C.; Gilbert, C.R. An ab initio investigation of the second hydration shell of metal cations. *Comput. Appl. Chem.* **2020**, *208*, 395–397. [CrossRef]
18. Persson, I. Structures of Hydrated Metal Ions in Solid State and Aqueous Solution. *Liquids* **2022**, *2*, 210–242. [CrossRef]
19. Hennings, E.; Schmidt, H.; Kohler, M.; Voigt, W. Crystal structure of tin(II) perchlorate trihydrate. *Acta Crystallogr. E* **2014**, *70*, 474–476. [CrossRef] [PubMed]
20. Persson, I.; D'Angelo, P.; Lundberg, D. Hydrated and Solvated Tin(II) Ions in Solution and the Solid State, and a Coordination Chemistry Overview of the $d^{10}s^2$ Metal Ions. *Chem. Eur. J.* **2017**, *22*, 18583–18592. [CrossRef] [PubMed]

liquids

Article

Hydration of Phosphate Ion in Polarizable Water: Effect of Temperature and Concentration

Sandeep Verma [1] and Arup Kumar Pathak [2,3,*]

[1] Radiation and Photochemistry Division, Bhabha Atomic Research Centre, Mumbai 400085, India
[2] Theoretical Chemistry Section, Chemistry Division, Bhabha Atomic Research Centre, Mumbai 400085, India
[3] Homi Bhabha National Institute, Mumbai 400094, India
* Correspondence: akpathak@barc.gov.in

Abstract: The hydration of phosphate ions, an essential component of many biological molecules, is studied using all-atom molecular dynamics (MD) simulation and quantum chemical methods. MD simulations are carried out by employing a mean-field polarizable water model. A good linear correlation between the self-diffusion coefficient and phosphate anion concentration is ascertained from the computed mean-square displacement (MSD) profiles. The HB dynamics of the hydration of the phosphate anion is evaluated from the time-dependent autocorrelation function $C_{HB}(t)$ and is determined to be slightly faster for the phosphate–anion system as compared to that of the water–water system at room temperature. The coordination number (CN) of the phosphate ion is found to be 15.9 at 298 K with 0.05 M phosphate ion concentration. The average CN is also calculated to be 15.6 for the same system by employing non-equilibrium MD simulation, namely, the well-tempered meta-dynamics method. A full geometry optimization of the $PO_4^{3-}\cdot 16H_2O$ cluster is investigated at the ωB97X-D/aug-cc-pVTZ level of theory, and the hydration of the phosphate anion is observed to have both singly and doubly bonded anion–water hydrogen bonds and inter-water hydrogen bonds in a range between 0.169–0.201 nm and 0.192–0.215 nm, respectively. Modified Stokes–Einstein relation is used to calculate the conductivity of the phosphate ion and is found to be in good agreement with the experimentally observed value.

Keywords: hydration; molecular dynamics; electronic structure; mean square displacements; hydrogen bonding; solvent-berg model

1. Introduction

Ion–water interaction is pivotal to many biophysical and biochemical reactivities at the developing stage. Water is unanimously recognized as the universal solvent and solely responsible for diverse chemical and complex biological processes [1,2]. During the process of ion dissolution, the hydrogen-bonded network of pure water is rearranged and attempts to capture the ion in the newly formed hydrogen-bonded network. The solvation of ions in water is dictated by various interactions, e.g., hydrogen bonding, van der Waals interaction, electrostatic interaction, etc. In the hydration process, solvent water forms a network surrounding the ions, commonly termed the hydration layer, sphere of hydration or hydration shell. Thus, the hydration of ions is controlled by the dynamic between the water–water and ion–water interactions. In the literature, the breaking and making of the water network, occurring at a picosecond (ps) time scale, have been attributed to water's physical, chemical and structural properties [3–7]. Numerous experimental techniques, such as X-rays and neutron scattering, in addition to transient spectroscopy detecting in IR to THz frequency region and molecular dynamics (MD) simulations, have been used to depict the static and dynamic structural information of the solvation of ions in water [8–18]. These molecular-level studies have shown how crucial the hydration of ions is in imparting many biological and chemical functions. At the fundamental level, it is known that cations have a simple and rigid hydration structure in comparison to that of anions, as cations bind more strongly with the solvent molecules [1]. On the other hand, a more complex and flexible hydration structure for anions is observed due to the frequent exchange of water molecules among the solvation shells [1,17].

In recent times, the triply negatively charged phosphate anion (PO_4^{3-}) is being considered for detailed investigation, given that it is an indispensable component of many solids, aerosols and electrolyte solutions. Moreover, phosphates are essential components of various biological polymers or molecules, such as DNA, RNA, AMP, ADP, ATP, etc. [18–21]. It is known that the phosphate anion does not exist in isolation, such as in the gas phase, due to Coulomb repulsion causing spontaneous electron loss. In nature, the existence of the PO_4^{3-} anion requires bound water molecules where the ion–water interaction stabilizes against the spontaneous loss of the electron. MD simulations and quantum chemical methods are used to understand the hydration of the PO_4^{3-} anion [22–25]. Existing molecular mechanical models cannot treat electronic polarizability explicitly as two-body additive terms are adopted. Thus, MD simulations on the hydration of PO_4^{3-} are restricted to the non-polarizable water model. The computational cost of MD simulations is increased significantly due to the non-additive force fields in polarizable models. Inconsistency in dealing with intra- and intermolecular polarization is the major drawback of these models [26]. Thus, to accomplish a reasonable level of configurational sampling by including electronic polarizability compared to non-polarizable force fields, the mean-field polarizable (MFP) water model, namely MFP/TIP3P, is developed [27,28]. The MFP/TIP3P model is also effectively used in many biological systems [29–31].

The present study investigates the hydration of PO_4^{3-} by employing classical MD with the MFP/TIP3P water model, and electronic structure calculations. The local solvation environment, microscopic structure, hydration shell, coordination number (CN) and energetics of the hydrated PO_4^{3-} anion are presented in detail. Efforts are also given to building the potential of the mean force (PMF) by considering a suitable collective variable during the hydration process.

2. Computational Methods

Molecular Dynamics Simulation

A 60 Å × 60 Å × 60 Å cubical box is considered to solvate a phosphate anion having a concentration of 0.05 M. Another two cubical boxes of the same size are taken with phosphate anion concentrations of 0.025 M and 0.0125 M, respectively. For electrical charge neutrality, suitable numbers of sodium ions are added to the solvated system in each box. The polarizable water model, MFP/TIP3P, is used to model the solvent water. The local solvation environment in the MFP/TIP3P model is imitated self-consistently and on-the-fly by adjusting the charges on the three atoms of the solvent water molecule. A well-known damping scheme is used to attain the mean-field nature of the local polarization and is parameterized by an average over time. The force field parameters for phosphate are taken from the literature [22]. Presently, two different temperatures, namely, 298 K and 325 K, are considered to study the effect of temperature on the hydration of the phosphate anion. In each simulation box, the temperature is sustained by a Berendsen thermostat with a relaxation time of 2 fs. SHAKE algorithm is used in the present simulation. Initially, the solvent molecules are heated for 500 ps to reach the required temperature, and after that, the whole system is heated for 1 ns to acquire the required temperature. Each system is equilibrated in an NPT ensemble for 5 ns to attain atmospheric pressure. Each system goes through a 10 ns NVT equilibration run before the 100 ns production run in the NVT ensemble using GROMACS [32]. The CN of the phosphate anion is used as a collective variable (CV) to extract the potential of the mean force (PMF) profile at 298 K for an anion concentration of 0.05 M. It is defined as the number of contacts between the PO_4^{3-} ion and water molecules as $CN = \sum_{i,j} CN_{ij}$. A switching function is used to define the individual contribution, CN_{ij} as [33].

$$CN_{ij} = \left\{ \begin{array}{c} 1 \text{ for } r_{ij} \leq 0 \\ \frac{1-\left(\frac{r_{ij}}{r_0}\right)^m}{1-\left(\frac{r_{ij}}{r_0}\right)^n} \text{ for } r_{ij} > 0 \end{array} \right\} \quad (1)$$

Here, $| r_i - r_j | = r_{ij}$ and $| r_i - r_j | - d_0 = r_0$. r_0 and d_0 are the full width at half maxima and the first peak position in the pair distribution function, respectively. The values of m and n are taken as 6 and 12, respectively. The enhanced sampling method, namely, a well-tempered version of meta-dynamics (WT-MtD), is considered to calculate the PMF profiles [34]. The NVT ensemble is used for all the WT-MtD runs, and the simulations are carried out for 50 ns. Gaussian hills height of 0.20 kJ/mol, width = 0.05, a bias factor of 10 and an initial deposition rate of the hills = 5 ps are used in all the WT-MtD simulations. The deposition rate is slightly higher than that used in more complex systems [35,36]. The PMF profiles are generated from the converged WT-MtD run [37,38].

3. Results and Discussion

3.1. Mean Square Displacement

The present study is aimed to investigate the distribution of water molecules around the phosphate anions at various temperatures and phosphate ion concentrations. Atoms in liquids are subject to a deviation from their average coordinates at any particular temperature, i.e., liquid particles move constantly. Although the displacement is very important, the trajectory followed by liquid particles resembles a random walker. Mean square displacement (MSD) is a better quantity than displacement to measure the liquid particles' motion. MSD measures the deviation or "thermal" displacement of the coordinate of a particle for a reference coordinate over time. The MSD of a particle at time t is defined as:

$$MSD = \frac{1}{N}\sum_{i=1}^{N} [r_i(t + \delta t) - r_i(t)]^2, \quad (2)$$

where r_i is the position of a heavy atom in the ith molecule at time t and N is the total number of such atoms. The MSD profiles for water and phosphate ion concentrations of 0.05 M, 0.025 M and 0.0125 M at 298 K and 0.05 M phosphate concentration at 325 K are shown in Figure 1 at 298 K.

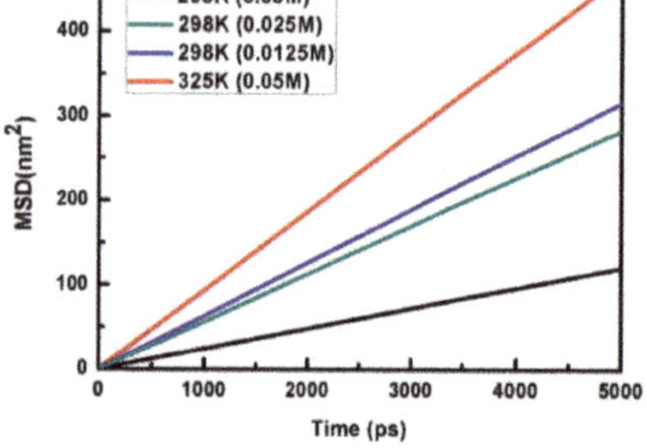

(A)

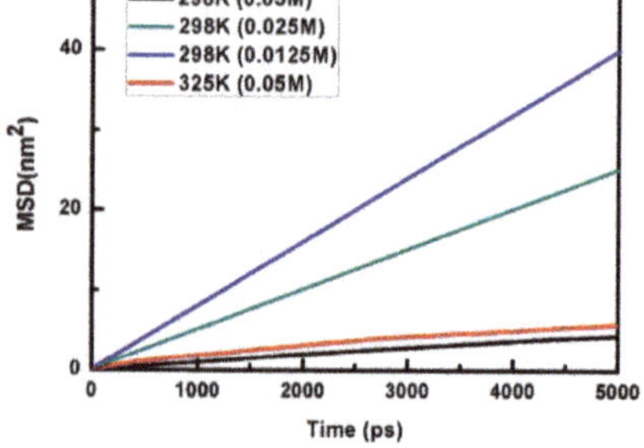

(B)

Figure 1. Plots of MSD (nm^2) vs. time are shown for **(A)** water and **(B)** phosphate anion.

It is known that the information about the diffusion of atoms can also be calculated from the knowledge of MSD. For a solid system, the kinetic energy of the particles needs to be adequate to reach the diffusive behavior and MSD to reach saturation. On the other hand, if the system is liquid, then the MSD of particles will grow linearly with time. For a liquid system, it is possible to explore the behavior of such a system from the knowledge of the slope of the MSD versus the time plot. The self-diffusion coefficient (D) is calculated from the long-time slope of the MSD versus the time plot by applying the Einstein relation [39]:

$$\lim_{t\to\infty}\frac{1}{N}\sum_{i=1}^{N}\left[r_i(t+\delta t)-r_i(t)\right]^2=6Dt \quad (3)$$

The calculated values of the self-diffusion coefficients of the phosphate ion at different concentrations (0.05 M, 0.025 M and 0.0125 M) and temperatures (298 K and 325 K) are shown in Table 1. Interestingly, the self-diffusion coefficient of water is also calculated from a simulation consisting of pure polarizable TIP3P water and is found to be 5.63×10^{-9} $m^2{\cdot}s^{-1}$. It is clear from Table 1 that the self-diffusion coefficient of phosphate is at least one order of magnitude slower than that of water. It is observed that the self-diffusion coefficients of both water and phosphate increase with the rise in temperature. Similarly, self-diffusion coefficient values also increase with a decrease in phosphate ion concentration. The variation of the self-diffusion coefficients of water and phosphate are shown in Figures 2A and 2B, respectively. We observe a good linear correlation between the self-diffusion coefficients and phosphate ion concentration (C) for both water and phosphate. The best-fitted linear expressions are $D(\text{water}) = 5.58 \times 10^{-9} - 31.49C$ and $D(\text{Phosphate}) = 11.18 \times 10^{-10} - 201.6C$. On extrapolating to the infinite dilution ($C \to 0$), the self-diffusion coefficient of water is obtained as 5.58×10^{-9} $m^2{\cdot}s^{-1}$. This value is very close to the calculated self-diffusion coefficient of water in a polarizable TIP3P medium (5.63×10^{-9} $m^2{\cdot}s^{-1}$) in the present study and a non-polarizable TIP3P water medium (5.65×10^{-9} $m^2{\cdot}s^{-1}$) [40]. This justifies the robustness of the linear expressions, and the self-diffusion coefficients can be obtained for both water and phosphate ions at any phosphate ion concentration by using the above linear expressions.

Table 1. The self-diffusion coefficients (D) of phosphate ion and water at different temperatures and phosphate ion concentrations.

Concentration (M)	D of Phosphate ($m^2{\cdot}s^{-1}$)	D of Water ($m^2{\cdot}s^{-1}$)
	298 K	
0.05	9.58×10^{-11}	4.03×10^{-9}
0.025	6.56×10^{-10}	4.70×10^{-9}
0.0125	8.38×10^{-10}	5.24×10^{-9}
	325 K	
0.05	1.92×10^{-10}	7.75×10^{-9}

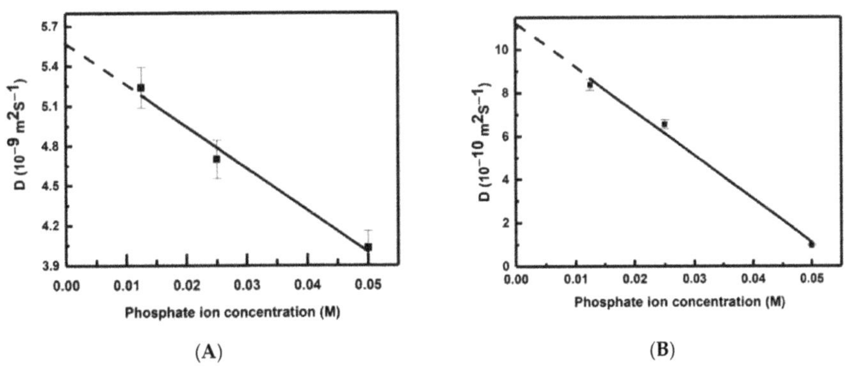

Figure 2. Variation of self-diffusion coefficient (D) of (**A**) water and (**B**) phosphate with concentration at 298 K. Error bars (within ±1.5%) are also shown.

3.2. Hydrogen Bonding

The dynamic balance between water–water and phosphate–water interactions controls the hydration process of the phosphate ions. As the solvent water molecules interact with the phosphate anion by forming hydrogen bonds (SHB and DHB), and the water molecules interact by inter-water hydrogen bonding, the time-dependent behavior of HBs is investigated from MD simulations. The formations of HBs are investigated based on geometric criteria. If the distance between the HB donor (D) and acceptor (A) atom is less than 3.0 Å and the D–H–A angle is lower than 35°, HB is accounted to be formed. The HB dynamics [41,42] are monitored through the HB autocorrelation function, $C_{HB}(t)$, which is defined as:

$$C_{HB}(t) = \sum_{i=1}^{n} \frac{< A(0)A(t) >}{< A >} \quad (4)$$

where n is the all possible hydrogen bond that forms between phosphate and water, $A(0)$ and $A(t)$ are the HB population operators at time, $t = 0$ and $t = \mathrm{t}$, respectively. The value of the *HB* population operators $A(t)$ is equal to 1 when the corresponding *HB* is intact at time t, and it is equal to zero when the associated HB bond breaks down.

The variations of the $C_{HB}(t)$ functions are also shown in Figure 3. The $C_{HB}(t)$ functions are fitted excellently with multi-exponential decay ($\chi^2 \sim 10^{-4}$ and $R > 0.99$), and the fitting parameters, e.g., amplitude, time constants and average lifetime of HB bonds for water–water and phosphate–water are provided in Table 2. All the *HB* correlation functions show tri-exponential decay except for the correlation function at 0.0125 M and 298 K, which shows bi-exponential decay. The information on the fitting parameters of *HB* time correlation functions helps to comprehend the hydration process very effectively in polarizable and non-polarizable water mediums [3,29–31,43,44].

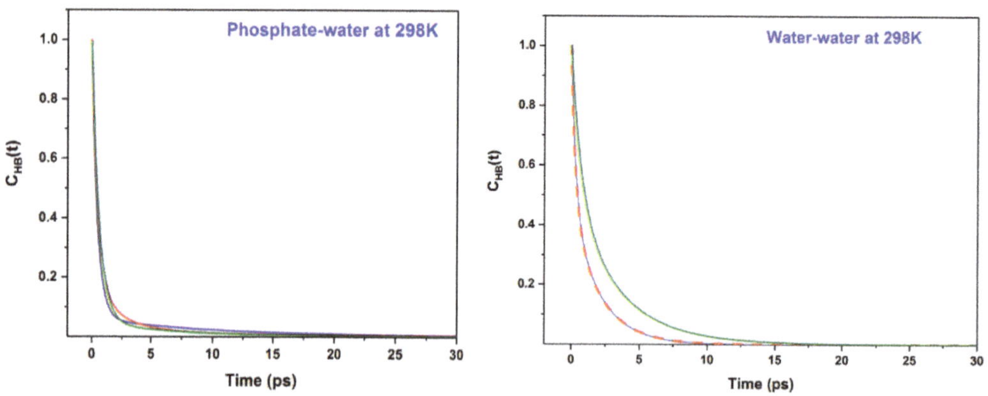

(A) (B)

Figure 3. Time-dependent HB lifetime correlation function, $C_{HB}(t)$ are shown for (**A**) Phosphate–water and (**B**) water–water systems at 298 K. Olive, red and blue color lines, respectively, represent the phosphate ion concentration of 0.05 M, 0.025 M and 0.0125 M.

We do observe a short HB correlation. The lifetime (τ) of each HB bond for water–water and phosphate–water systems is also measured from the knowledge of the *HB* time correlation function, $\tau = \int_0^\infty C_{HB}(t)dt$. The HB lifetimes are also calculated from the $C_{HB}(t)$ versus time plots and are shown in Table 2. The HB lifetime of all phosphate–water systems is observed to be ~1 ps at 298 K. The HB lifetimes of the water–water system are varied from 2 ps to 1.2 ps at 298 K. At 325 K, the observed *HB* lifetimes for phosphate–water and water–water systems are 1.2 ps and 1.1 ps, respectively. It is clear from both Table 2 and Figure 3 that the phosphate–water HB autocorrelation functions decay slightly faster in comparison to that of the water–water HB autocorrelation functions at 298 K.

Table 2. Exponential fitting parameters (time constants and amplitude) of hydrogen bond time correlation functions and average hydrogen bonds lifetime (ps) for various systems. The general expression of the multiexponential fitting is $C_{HB}(t) = \sum_{i=1,2,3,..} A_i[exp(-t/\tau_i)] + C$.

Temperature	Time Constant (ps)	Amplitude (%)	Lifetime (ps)	Time Constant (ps)	Amplitude (%)	Lifetime (ps)
	Phosphate–Water			Water–Water		
			0.05 M			
298 K	0.62 7.2 47.0	95.27% 4.36% 0.36%	1	0.66 3.3 10.0	46.03% 53.56% 0.41 %	2.1
			0.025 M			
	0.5 2.58 22.0	84.12% 14.27% 1.61%	1	0.33 2.02 3.88	52.46% 46.36% 1.18%	1.2
			0.0125 M			
	0.47 10.67	94% 6%	1	0.34 2.04	51.95% 48.05%	1.2
			0.05 M			
325 K	0.6 7.2 40.0	93.47% 6.04% 0.49%	1.2	0.34 1.46 6.0	28.82% 71.04% 0.14%	1.1

3.3. Microscopic Structure

It is known that phosphate ions are rare as an isolated species but are ubiquitous in electrolyte solutions, solids and aerosols. The dynamical balance between phosphate–water and water–water interactions leads to the solvation of the phosphate ion in the presence of water. Thus, it will be interesting to understand the microscopic structure of phosphate in a water medium. At first, the CN of the phosphate ion is calculated from the knowledge of the radial distribution function and is observed as 15.9 at 298 K for an anion concentration of 0.05 M. WT-MtD simulation is also employed to obtain the CN value at 298 K.

It is well understood that the original meta-dynamics (MtD) method is an enhanced sampling method and is very advantageous to construct the PMF profiles along a suitable CV. However, the original version of the MtD simulation method lacks true equilibrium information for a chemical system, and the "slow" build-up limit of Gaussian bias potential is implemented. Thus, due to the inherent non-equilibrium nature and well-known convergence problem, the practicality of the original version of the MtD technique is restricted. The presently implemented method, WT-MtD, has resolved these concerns by rescaling the height of the Gaussian bias potential with the bias grown over time at some fictitious higher temperature. Hence, the PMF profile is generated for the phosphate anion hydration by considering CN as a CV, as defined in Equation (1). The PMF profile is then constructed from the converged WT-MtD simulation at 298 K having 0.05 M phosphate ion concentration, and is shown in Figure 4A. The figure shows that the most probable average CN for a phosphate anion is 15.6. This value is very close to the previously obtained value of ~16 in a non-polarizable water medium [22]. Thus, a triply negatively charged phosphate ion is observed as a strong structure-making ion. In order to find the most preferred candidate of the phosphate ion with 16 water molecules at 298 K, quantum mechanical calculations are used. Full geometry optimization without any symmetry restrictions is carried out for the $PO_4^{3-}\cdot 16H_2O$ cluster at the ωB97X-D/aug-cc-pVTZ level of theory [45]. The most stable structure of the $PO_4^{3-}\cdot 16H_2O$ cluster is shown in Figure 4B. The $PO_4^{3-}\cdot 16H_2O$ cluster consisted of single hydrogen bonding (SHB), having one of the hydrogen atoms of the water molecule bonding with the phosphate moiety, double hydrogen bonding (DHB) with water molecules using both the hydrogen atoms in bonding with the anion, and inter-water hydrogen bonding (WHB). The measured distances of the hydrogen bond (HB) formed between the phosphate anion and water vary from 0.169 to 0.201 nm. The same for the WHB ranges from 0.192 to 0.215 nm. The first peak of g(r) is dominated by ion–water hydrogen bonding (SHB & DHB) and water–water bonding (WHB), and the second and third peaks mainly consist of inter-water hydrogen bonding (see Figure 4B). The average distance of the P–O, P–H and O–O bonds between phosphate and water are observed to be 0.351 nm, 0.290 nm and 0.276 nm, respectively, for the most stable structure. The average POH angle is also observed to be 174°.

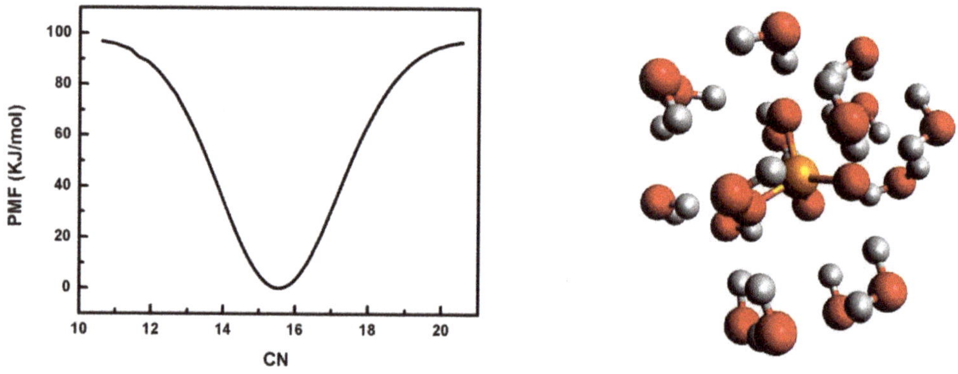

(**A**) (**B**)

Figure 4. (**A**) Plot of potential of mean force, PMF in kJ/mol by employing water CN as a CV in WT-MtD simulation at 298 K. (**B**) Most stable structure of $PO_4^{3-}\cdot 16H_2O$ cluster.

3.4. Solvent-Berg Model

It is observed that the self-diffusion coefficient of phosphate is one order of magnitude slower than that of water, and it increases with decreases in phosphate ion concentration. The solvent-berg model can be invoked to explain this behavior [46]. In this model, the phosphate ion is moving as a solvent-berg, i.e., the ion along with the water molecules of the first solvation shell move together as the diffusing species. It is shown that 16 water molecules can form the first solvation shell of the phosphate ion (see Figure 4B). Thus, the solvent-berg consists of the ion and 16 solvent molecules and is bulkier than the free ion and solvent molecule. This may explain the slow diffusion of phosphate with respect to the water. Similarly, solvent-berg may also have more space to move freely in dilute ion concentration. Thus, the self-diffusion coefficient will increase with decreases in phosphate ion concentration. It will also be interesting to calculate the equivalent conductance of the phosphate ion by employing modified Stokes–Einstein relation. The ratio of the equivalent conductance (Λ_0) of two ions, namely, the phosphate anion and a known reference ion, having the same charge for a fixed solvent and at a given temperature can be expressed as $\frac{(\Lambda_0)_{phosphate}}{(\Lambda_0)_{known}} = \frac{(R_{SB})_{known}}{(R_{SB})_{phosphate}}$. Here, $(\Lambda_0)_{phosphate}$ and $(\Lambda_0)_{known}$ are the equivalent conductance of the phosphate anion and a known reference ion, respectively. Similarly, $(R_{SB})_{phosphate}$ and $(R_{SB})_{known}$ are the solvent-berg radius of the phosphate anion and a known reference ion, respectively. The experimental measured values of the conductivity (70 $S\cdot cm^2\cdot eq^{-1}$) and hydrated radius (4.6 Å) of a ferric ion are used as the known reference system [47,48]. At present, the solvent-berg radius of the phosphate ion is considered as the radius of the phosphate ion with 16 solvent molecules, i.e., for the $PO_4^{3-}\cdot 16H_2O$ cluster. The radius is found to be 4.83 Å. The calculated conductivity of the phosphate ion in water is 67 $S\ cm^2\cdot eq^{-1}$. The present calculated value is very close (within 3%) to that of the experimentally observed value [48].

4. Conclusions

The hydration of the phosphate ion is investigated by employing all-atom molecular dynamics (MD) simulation and quantum mechanical methods. The equilibrium MD simulation is performed using the mean-field polarizable water model. The self-diffusion coefficient of phosphate is observed to be one order of magnitude slower than that of water, and it also increases with decreases in phosphate ion concentration and increases with the solution temperature. Using Einstein's method, a good linear correlation between the self-diffusion coefficient and anion concentration is determined from mean-square displacement (MSD) as a time function. The HB dynamics of the hydration of the phosphate anion are obtained from time-dependent autocorrelation function $C_{HB}(t)$ analysis, and we observed a weak correlation. $C_{HB}(t)$ is calculated to be marginally faster for the phosphate–anion system in comparison to that of the water–water system at 298 K. A maximum computed CN value of 15.9 is observed for 0.05 M phosphate ion concentration at 298 K. We observed a decrease in CN with the rise in the temperature and saturation of CN with the dilution of anion concentration. The value of CN is also calculated to be 15.6 by employing non-equilibrium MD simulation, namely, the well-tempered meta-dynamics method. The structure of the $PO_4^{3-}\cdot 16H_2O$ cluster is optimized with the ωB97X-D/aug-cc-pVTZ approach and is found to consist of both single and double–bonded anion–water HB along with inter-water HB in a range from 0.169 to 0.201 nm and from 0.192 to 0.215 nm, respectively. The conductivity of the phosphate ion is also calculated by employing modified Stokes–Einstein relation, and is found to be in good agreement with the experimentally observed value.

Author Contributions: Conceptualization, A.K.P.; Methodology, A.K.P.; Formal analysis, S.V.; Data curation, S.V.; Writing—original draft, A.K.P. and S.V.; Writing—review & editing, A.K.P. All authors have read and agreed to the published version of the manuscript.

Funding: The work was supported by DAE under project RBA4013.

Data Availability Statement: Data is contained within article.

Acknowledgments: Tyagi, A.K.; Kumar, A.; Bandyopadhyay, T.; Choudhury. N. and Ghosh, H.N. are gratefully acknowledged for their encouragement during the course of the work. BARC computer center is also acknowledged for providing ANUPAM computational facility.

Conflicts of Interest: The authors declare no conflict of interest.

References

1. Ohtaki, H.; Radnai, T. Structure and dynamics of hydrated ions. *Chem. Rev.* **1993**, *93*, 1157–1204. [CrossRef]
2. Bakker, H.J. Structural Dynamics of Aqueous Salt Solutions. *Chem. Rev.* **2008**, *108*, 1456–1473. [CrossRef] [PubMed]
3. Pathak, A.K.; Bandyopadhyay, T. Solvation of arsenate anion: Combined quantum mechanics and molecular dynamics based investigation. *Mol. Phys.* **2016**, *114*, 2029–2036. [CrossRef]
4. Smith, J.D.; Cappa, C.D.; Wilson, K.R.; Messer, B.M.; Cohen, R.C.; Saykally, R.J. Energetics of Hydrogen Bond Network Rearrangements in Liquid Water. *Science* **2004**, *306*, 851–853. [CrossRef] [PubMed]
5. Boero, M.; Terakura, K.; Ikeshoji, T.; Liew, C.C.; Parrinello, M. Hydrogen Bonding and Dipole Moment of Water at Supercritical Conditions: A First-Principles Molecular Dynamics Study. *Phys. Rev. Lett.* **2000**, *85*, 3245–3248. [CrossRef] [PubMed]
6. Eaves, J.D.; Loparo, J.J.; Fecko, C.J.; Roberts, S.T.; Tokmakoff, A.; Geissler, P.L. Hydrogen bonds in liquid water are broken only fleetingly. *Proc. Natl. Acad. Sci. USA* **2005**, *102*, 13019–13022. [CrossRef]
7. Matsumoto, M.; Baba, A.; Ohmine, I. Topological building blocks of hydrogen bond network in water. *J. Chem. Phys.* **2007**, *127*, 134504. [CrossRef] [PubMed]
8. Kim, S.J.; Born, B.; Havenith, M.; Gruebele, M. Real-Time Detection of Protein-Water Dynamics upon Protein Folding by Terahertz Absorption Spectroscopy. *Angew. Chem. Int. Ed.* **2008**, *47*, 6486–6489. [CrossRef]
9. Wernet, P.; Nordlund, D.; Bergmann, U.; Cavalleri, M.; Odelius, M.; Ogasawara, H.; Naslund, L.A.; Hirsch, T.K.; Ojamae, L.; Glatzel, P.; et al. The structure of the first coordination shell in liquid water. *Science* **2004**, *304*, 995–999. [CrossRef]
10. Winter, B.; Aziz, E.F.; Hergenhahn, U.; Faubel, M.; Hertel, I.V. Hydrogen bonds in liquid water studied by photoelectron spectroscopy. *J. Chem. Phys.* **2007**, *126*, 124504. [CrossRef]
11. Omta, A.W.; Kropman, M.F.; Woutersen, S.W.; Bakker, H.J. Faculty Opinions recommendation of Negligible effect of ions on the hydrogen-bond structure in liquid water. *Science* **2003**, *301*, 347–349. [CrossRef] [PubMed]

12. Park, S.; Fayer, M.D. Hydrogen bond dynamics in aqueous NaBr solutions. *Proc. Natl. Acad. Sci. USA* **2007**, *104*, 16731–16738. [CrossRef] [PubMed]
13. Heugen, U.; Schwaab, G.; Bründermann, E.; Heyden, M.; Yu, X.; Leitner, D.M.; Havenith, M. Solute-induced retardation of water dynamics probed directly by terahertz spectroscopy. *Proc. Natl. Acad. Sci. USA* **2006**, *103*, 12301–12306. [CrossRef] [PubMed]
14. Kropman, M.F.; Bakker, H.J. Dynamics of Water Molecules in Aqueous Solvation Shells. *Science* **2001**, *291*, 2118–2120. [CrossRef] [PubMed]
15. Mancinelli, R.; Botti, A.; Bruni, F.; Ricci, M.A.; Soper, A.K. Hydration of Sodium, Potassium, and Chloride Ions in Solution and the Concept of Structure Maker/Breaker. *J. Phys. Chem. B* **2007**, *111*, 13570–13575. [CrossRef] [PubMed]
16. Ebbinghaus, S.; Kim, S.J.; Heyden, M.; Yu, X.; Heugen, U.; Gruebele, M.; Leitner, D.M.; Havenith, M. An extended dynamical hydration shell around proteins. *Proc. Natl. Acad. Sci. USA* **2007**, *104*, 20749–20752. [CrossRef]
17. Schmidt, D.A.; Birer, O.; Funkner, S.; Born, B.P.; Gnanasekaran, R.; Schwaab, G.W.; Leitner, D.M.; Havenith, M. Rattling in the Cage: Ions as Probes of Sub-picosecond Water Network Dynamics. *J. Am. Chem. Soc.* **2009**, *131*, 18512–18517. [CrossRef]
18. Watson, J.D.; Crick, F.H. A Structure for Deoxyribose Nucleic Acid. *Nature* **1953**, *171*, 737–738. [CrossRef]
19. Zoll, J.; Tessari, M.; Van Kuppeveld, F.J.; Melchers, W.J.; Heus, H.A. Breaking pseudo-twofold symmetry in the poliovirus 3′-UTR Y-stem by restoring Watson-Crick base pairs. *RNA* **2007**, *13*, 781–792. [CrossRef]
20. Knowles, J.R. Enzyme-Catalyzed Phosphoryl Transfer Reactions. *Annu. Rev. Biochem.* **1980**, *49*, 877–919. [CrossRef]
21. Törnroth-Horsefield, S.; Neutze, R. Opening and closing the metabolite gate. *Proc. Natl. Acad. Sci. USA* **2008**, *105*, 19565–19566. [CrossRef]
22. Pathal, A.K. Evaluation of the detachment energy of hydrated phosphate anion over a wide range of cluster size and revisiting solvent–berg model: A theoretical study. *Mol. Phys.* **2014**, *112*, 3085–3091. [CrossRef]
23. Pathak, A.K. Stepwise hydration of phosphate anion: A microscopic theory connecting domain of instability and stability. *Int. J. Quantum Chem.* **2015**, *115*, 413–418. [CrossRef]
24. Pye, C.C.; Rudolph, W.W. An ab Initio, Infrared, and Raman Investigation of Phosphate Ion Hydration. *J. Phys. Chem. A* **2003**, *107*, 8746–8755. [CrossRef]
25. VandeVondele, J.; Tröster, P.; Tavan, P.; Mathias, G. Vibrational Spectra of Phosphate Ions in Aqueous Solution Probed by First-Principles Molecular Dynamics. *J. Phys. Chem. A* **2012**, *116*, 2466–2474. [CrossRef]
26. Ponder, P.J.; Case, D.A. Force fields for protein simulations. *Adv. Protein Chem.* **2003**, *66*, 27–85.
27. Leontyev, I.V.; Stuchebrukhov, A.A. Polarizable Mean-Field Model of Water for Biological Simulations with AMBER and CHARMM Force Fields. *J. Chem. Theory Comput.* **2012**, *8*, 3207–3216. [CrossRef] [PubMed]
28. Leontyev, I.V.; Stuchebrukhov, A.A. Polarizable molecular interactions in condensed phase and their equivalent nonpolarizable models. *J. Chem. Phys.* **2014**, *141*, 014103. [CrossRef]
29. Pathak, A.K.; Bandyopadhyay, T. Protein–Drug Interactions with Effective Polarization in Polarizable Water: Oxime Unbinding from AChE Gorge. *J. Phys. Chem. B* **2015**, *119*, 14460–14471. [CrossRef]
30. Pathak, A.K.; Bandyopadhyay, T. Unbinding of fluorinated oxime drug from the AChE gorge in polarizable water: A well-tempered metadynamics study. *Phys. Chem. Chem. Phys.* **2017**, *19*, 5560–5569. [CrossRef] [PubMed]
31. Pathak, A.K.; Bandyopadhyay, T. Dynamic Mechanism of a Fluorinated Oxime Reactivator Unbinding from AChE Gorge in Polarizable Water. *J. Phys. Chem. B* **2018**, *122*, 3876–3888. [CrossRef] [PubMed]
32. Hess, B.; Kutzner, C.; van der Spoel, D.; Lindahl, E. GROMACS 4: Algorithms for Highly Efficient, Load-Balanced, and Scalable Molecular Simulation. *J. Chem. Theory Comput.* **2008**, *4*, 435–447. [CrossRef]
33. Bonomi, M.; Branduardi, D.; Bussi, G.; Camilloni, C.; Provasi, D.; Raiteri, P.; Donadio, D.; Marinelli, F.; Pietrucci, F.; Broglia, R.A.; et al. PLUMED: A portable plugin for free-energy calculations with molecular dynamics. *Comput. Phys. Commun.* **2009**, *180*, 1961–1972. [CrossRef]
34. Barducci, A.; Bussi, G.; Parrinello, M. Well-Tempered Metadynamics: A Smoothly Converging and Tunable Free-Energy Method. *Phys. Rev. Lett.* **2008**, *100*, 020603. [CrossRef] [PubMed]
35. Sun, Z.; Zheng, L.; Wang, K.; Huai, Z.; Liu, Z. Primary vs secondary: Directionalized guest coordination in β-cyclodextrin derivatives. *Carbohydr. Polym.* **2022**, *297*, 120050. [CrossRef]
36. Sun, Z.; Huai, Z.; He, Q.; Liu, Z. A General Picture of Cucurbit[8]uril Host–Guest Binding. *J. Chem. Inf. Model.* **2021**, *61*, 6107–6134. [CrossRef]
37. Tiwary, P.; Parrinello, M. A Time-Independent Free Energy Estimator for Metadynamics. *J. Phys. Chem. B* **2015**, *119*, 736–742. [CrossRef]
38. Bonomi, M.; Barducci, A.; Parrinello, M. Reconstructing the equilibrium Boltzmann distribution from well-tempered metadynamics. *J. Comput. Chem.* **2009**, *30*, 1615–1621. [CrossRef] [PubMed]
39. Allen, M.P.; Tildesley, D.J. *Computer Simulation of Liquids*; Oxford University Press: New York, NY, USA, 1987.
40. Mark, P.; Nilsson, L. Structure and Dynamics of the TIP3P, SPC, and SPC/E Water Models at 298 K. *J. Phys. Chem. A* **2001**, *105*, 9954–9960. [CrossRef]
41. Luzar, A.; Chandler, D. Hydrogen-bond kinetics in liquid water. *Nature* **1996**, *379*, 55–57. [CrossRef]
42. Luzar, A.; Chandler, D. Effect of Environment on Hydrogen Bond Dynamics in Liquid Water. *Phys. Rev. Lett.* **1996**, *76*, 928–931. [CrossRef] [PubMed]

43. Pathak, A.K.; Bandyopadhyay, T. Water isotope effect on the thermostability of a polio viral RNA hairpin: A metadynamics study. *J. Chem. Phys.* **2017**, *146*, 165104. [CrossRef] [PubMed]
44. Mishra, V.; Pathak, A.K.; Bandyopadhyay, T. Binding of human serum albumin with uranyl ion at various pH: An all atom molecular dynamics study. *J. Biomol. Struct. Dyn.* **2022**, 1–11. [CrossRef]
45. Schmidt, M.W.; Baldridge, K.K.; Boatz, J.A.; Elbert, S.T.; Gordon, M.S.; Jensen, J.H.; Koseki, S.; Matsunaga, N.; Nguyen, K.A.; Su, S.; et al. General atomic and molecular electronic structure system. *J. Comput. Chem.* **1993**, *14*, 1347–1363. [CrossRef]
46. Bogatko, S.A. Ion solvation structure and dynamical information via deviations from the solvent-berg diffusion model. *Chem. Phys. Lett.* **2013**, *565*, 148–150. [CrossRef]
47. Kolb, M.; Seubert, M. *Practical Ion Chromatography: An Introduction*; Metrohm Ltd.: Herisau, Switzerland, 2001.
48. Nightingale, E.R. Phenomenological Theory of Ion Solvation. Effective Radii of Hydrated Ions. *J. Chem. Phys.* **1959**, *63*, 1381–1387. [CrossRef]

liquids

MDPI

Article

An X-ray and Neutron Scattering Study of Aqueous $MgCl_2$ Solution in the Gigapascal Pressure Range

Toshio Yamaguchi [1,2,*], Nami Fukuyama [2,†], Koji Yoshida [2], Yoshinori Katayama [3], Shinichi Machida [4] and Takanori Hattori [5]

1 Key Laboratory of Salt Lake Resources Chemistry of Qinghai Province, Qinghai Institute of Salt Lakes, Chinese Academy of Sciences, Xining 810008, China
2 Department of Chemistry, Faculty of Science, Fukuoka University, Jonan 814-0180, Fukuoka, Japan; nami.fukuyama@maedaroad.co.jp (N.F.); kyoshida@fukuoka-u.ac.jp (K.Y.)
3 Synchrotron Radiation Research Center, National Institutes for Quantum and Radiological Science and Technology, Sayo 679-5148, Hyogo, Japan; katayama.yoshinori@qst.go.jp
4 Neutron Science and Technology Center, Comprehensive Research Organization for Science and Society, 162-1, Shirakata 319-1106, Tokai, Japan; s_machida@cross.or.jp
5 J-PARC Center, Japan Atomic Energy Agency, 2–4, Shirakata 319-1195, Tokai, Japan; takanori@post.j-parc.jp
* Correspondence: yamaguch@fukuoka-u.ac.jp
† Present address: Research Institute, Maeda Road, 208 Oohata, Tsuchiura 300-4111, Ibaraki, Japan.

Abstract: The structure of electrolyte solutions under pressure at a molecular level is a crucial issue in the fundamental science of understanding the nature of ion solvation and association and application fields, such as geological processes on the Earth, pressure-induced protein denaturation, and supercritical water technology. We report the structure of an aqueous 2 m (=mol kg^{-1}) $MgCl_2$ solution at pressures from 0.1 MPa to 4 GPa and temperatures from 300 to 500 K revealed by X-ray- and neutron-scattering measurements. The scattering data are analyzed by empirical potential structure refinement (EPSR) modeling to derive the pair distribution functions, coordination number distributions, angle distributions, and spatial density functions (3D structure) as a function of pressure and temperature. Mg^{2+} forms rigid solvation shells extended to the third shell; the first solvation shell of six-fold octahedral coordination with about six water molecules at 0 GPa transforms into about five water molecules and one Cl^- due to the formation of the contact ion pairs in the GPa pressure range. The Cl^- solvation shows a substantial pressure dependence; the coordination number of a water oxygen atom around Cl^- increases from 8 at 0.1 MPa/300 K to 10 at 4 GPa/500 K. The solvent water transforms the tetrahedral network structure at 0.1 MPa/300 K to a densely packed structure in the GPa pressure range; the number of water oxygen atoms around a central water molecule gradually increases from 4.6 at 0.1 MPa/298 K to 8.4 at 4 GPa/500 K.

Keywords: ion solvation; X-ray scattering; neutron scattering

1. Introduction

Understanding the nature of Mg^{2+} solvation is an important issue since Mg^{2+} plays an essential role in various application fields, such as biological systems [1], environmental aerosols [2], and geological processes in the Earth [3]. Mg^{2+} affects many cellular functions, including the transport of K^+ and Ca^{2+}; and modulates signal transduction, energy metabolism, and cell proliferation [1]. Mg^{2+} is the second cation rich in seawater and lake water after Na^+. Thus, Mg^{2+} affects the properties of sea salt aerosols in an atmospheric environment [2]. Most investigations on Mg^{2+} solvation in aqueous solutions have been performed under ambient conditions. When high pressure of gigapascals (GPa) is applied to aqueous solutions, the hydrogen bonding of solvent water is greatly perturbed [4], followed by a possible change in ion solvation. Thus, investigation of Mg^{2+} solvation in the GPa pressure range will provide valuable information for the nature of ion solvation and association.

Furthermore, electrolyte solutions in the GPa pressure range exist in the Earth's upper mantle and play critical roles in geological processes, such as earthquakes, and growing magma, followed by the eruption of a volcano [3]. Due to the development of a high-pressure cell and data acquisition techniques at synchrotron and pulsed neutron sources, high-pressure and high-temperature X-ray- and neutron-scattering measurements in the GPa pressure range became possible. Many studies have been conducted on aqueous solutions of electrolytes, such as NaCl [5–7], RbCl [8], $CaCl_2$ [9,10], and $CeCl_3$ [11].

The Mg^{2+} solvation was investigated at a molecular level under ambient conditions by various techniques, such as X-ray scattering [12,13], neutron scattering [14,15], classical molecular dynamics (MD) simulations [16–21], and ab initio MD simulation [15,22–24]. The results are consistent with the conclusion that Mg^{2+} has a rigid solvation shell of six water molecules arranged in an octahedral geometry. The Mg^{2+}-H_2O distance ranges from 1.92 to 2.12 Å, depending on the methods and pair potentials employed. These features of Mg^{2+} solvation differ from the Ca^{2+} solvation, with seven and eight water molecules in an almost equal population in the first solvation shell at 0.1 MPa [9,10]. When the solution is compressed to 1 GPa/300 K, the distribution of water molecules around Ca^{2+} becomes sharp and has a maximum of eight [10]. The distance of the Ca^{2+}-O_W(I) (2.44 Å) does not change significantly with pressure (O_W(I) is a water oxygen atom in the first solvation shell). However, the Ca^{2+}-H_W(I) distance is shortened from 3.13 Å at 0.1 MPa to 3.10 Å at 4 GPa (H_W(I) is a water hydrogen atom in the first solvation shell). This result suggested that the orientation of the water dipole around Ca^{2+} is affected by pressure. The contact and solvent-shared ion pairs between Ca^{2+} and Cl^- are formed with Ca^{2+}-Cl^- distances of 2.82 and 5.0 Å, respectively [10]. There are no appreciable changes in the coordination number of Ca^{2+}-Cl^- at 0.2–0.3 with pressure. It is interesting to see the behavior of Mg^{2+} solvation and association in the GPa pressure range and compare it with the Ca^{2+} solvation from a biological point of view [1].

In this study, we perform X-ray- and neutron-scattering measurements on a 2 *m* $MgCl_2$ aqueous solution at ambient pressure to 4 GPa and temperatures of 300 and 500 K. An EPSR modeling method is employed to analyze both X-ray- and neutron-scattering data together to extract the pair distribution functions, coordination number distributions, angle distribution (orientational correlation), and spatial density functions (3D structure) in aqueous electrolyte solutions. We show the effects of pressure and temperature on ion solvation and association, and solvent water structure.

2. Materials and Methods

2.1. Sample Solutions

Magnesium chloride ($MgCl_2$, 99.99%, Sigma-Aldrich, St. Louis, MO, USA), commercially available, was used without further purification. $MgCl_2$ powder was dried in a vacuum oven at 393 K for 3 h and then cooled in a desiccator at room temperature. Deuterium oxide (D_2O) (99.8 atom %D) was purchased from Kanto Chemical Ltd., Chuo-ku, Tokyo, Japan. For X-ray-scattering measurements, a required amount of dried $MgCl_2$ powder was weighed and dissolved into degassed H_2O to prepare a 2 *m* solution. For neutron-scattering experiments, a 2 *m* $MgCl_2$ aqueous solution in D_2O was prepared in a similar way to the X-ray sample in a nitrogen-filled glove box to avoid contamination of light water (H_2O) since the incoherent scattering cross-section of H is about 40 times larger than D ($\sigma_i^H = 80.27$ barns, $\sigma_i^D = 2.05$ barns).

2.2. Experimental PT Condition and Density Estimation

Figure 1 shows the phase diagram of water. The red open circles are thermodynamic states for X-ray-scattering measurements, and the solid red arrows show the experimental pressure and temperature path. The black-filled circles show those for neutron scattering experiments, and the black dashed arrows are the experimental *PT* path, which was carefully chosen to avoid the sample blowout by compression at high temperatures. The densities of the sample solutions were measured at 300 K with a vibrational densitometer

(Anton Paar GmbH, Graz, Austria, DMA48). Those at high temperatures and high pressures were estimated from the densities of water in the corresponding thermodynamic states available in the literature [25]. Table 1 summarizes the compositions and densities at the measured pressures and temperatures.

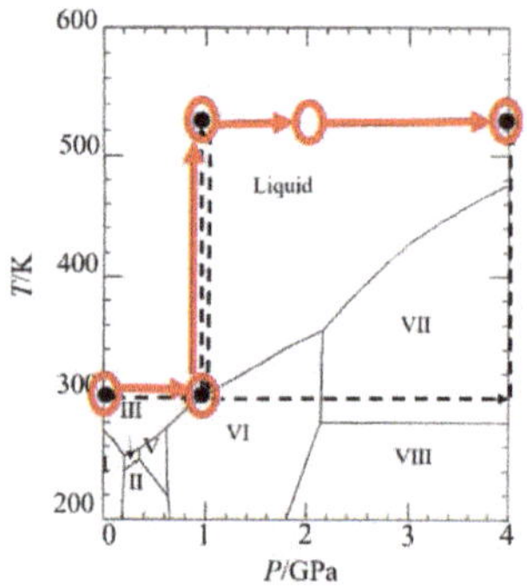

Figure 1. Phase diagram of H_2O [26] and thermodynamic states where the scattering data of a 2 *m* $MgCl_2$ aqueous solution were taken in this study. Red and black circles correspond to *PT* conditions for X-ray- and neutron-scattering measurements. Red solid and black dashed arrows show the experimental pressure and temperature path for the X-ray- and neutron-scattering experiments.

Table 1. Concentration *c*, bulk densities d_X and d_N, the atomic number densities ρ_S of a sample solution, and measured conditions of aqueous $MgCl_2$ solutions. X and N denote X-ray and neutron-scattering, respectively.

Abbreviation	c/mol dm^{-3}	d_X/g cm^{-3}	d_N/g cm^{-3}	ρ_S/atoms Å^{-3}	P_X/GPa	P_N/GPa	T_X/K	T_N/K	Method
0.1 MPa 300 K	2.024	1.113	1.204	0.09577	1.0×10^{-4}	1.0×10^{-4}	300	298	X, N
1 GPa 300 K	2.397	1.377	1.426	0.1134	1.0	0.65	300	298	X, N
1 GPa 500 K	2.125	1.269	1.264	0.1005	1.0	0.63	500	523	X, N
2 GPa 500 K	2.020	1.396	...	0.1386	2.0	...	500	...	X
4 GPa 500 K	2.439	1.459	1.451	0.1388	3.8	4.5	500	523	X, N

2.3. X-ray Scattering Measurements

A sample solution was inserted into a diamond cylindrical cell of 1.5 mm in inner diameter and 1.2 mm in height, which was capped with a thin gold foil and placed in a high-pressure cell assembly, as shown in Figure S1, Supplementary Materials [27]. The incident X-ray beam size was 300 μm (width) and 50 μm (height). The applied pressure was estimated from the equation of state of NaCl of the lattice constant of NaCl powder pellets placed below the diamond cell measured at various thermodynamic states [28]. The sample was heated by applying AC to the graphite heater surrounding the pressure cell. The sample temperature was estimated from the temperature-electric power relation of the power applied to the heater [29]. High-pressure X-ray-scattering experiments were done at 1 GPa/300 and 500 K, 2 GPa/500 K, and 4 GPa/500 K on a high-pressure X-ray diffractometer installed at BL14B1 of SPring-8 [30]. X-ray-scattering data were collected in an energy-dispersive mode with a Ge-SSD over an energy I region of 40~140 keV at the scattering angles (2θ) from 4° to 24° with an increment of 2° to 4°, which corresponded to the magnitude of the scattering vector $Q = (4\pi/hc)E\sin\theta$ (h is the Planck constant, and c is the velocity of light) of 0.5–13.85 Å^{-1}. A set of scattering patterns measured at different scattering angles was empirically merged into the whole Q-range values [27,28]. The experimental scattering data were normalized to the absolute units (electrons) in the usual manner [31] and then corrected for incoherent scattering. The interference functions, $F(Q)$, were obtained by Equation (1).

$$F(Q) = \frac{\left(I^{coh}(Q) - \sum x_i f_i^2(Q)\right)}{\sum x_i f_i^2(Q)} \tag{1}$$

Here, $I^{coh}(Q)$ denotes the coherent scattering intensity per atom, and x_i and f_i are the atomic fraction and atomic-scattering factor of atom i, respectively.

2.4. Neutron Scattering Measurements

A sample solution under ambient conditions was inserted in a cylindrical vanadium cell (2.8 mm in inner diameter, 0.1 mm in thickness, and 30 mm in height) and sealed with an indium wire. Under high-pressure conditions, a sample solution was inserted into a cylindrical Teflon cell (5.5 mm in inner diameter and 6.5 mm in inner height) against corrosion by aqueous salt solutions at high temperatures. The Teflon cell was loaded in a high-pressure cell assembly, as shown in Figure S2. The neutron scattering experiments were carried out on a PLANET high-pressure diffractometer, equipped with the six-axis multi-anvil press ATSUHIME [32], at BL11 of a pulsed neutron facility, J-PARC MLF [33]. The scattered neutrons were collected with 160 ^{3}He position-sensitive detectors placed at the scattering angle (2θ) of 90° for each side with the horizontal and vertical coverage of $90 \pm 11.3°$ and $0 \pm 34.6°$, respectively. The wavelength (λ) range used was 0. 2–12 Å, corresponding to the amplitude of wavevector Q ($=4\pi\sin\theta/\lambda$) of 0.8–40 Å^{-1}. Neutron-scattering measurements were performed at 0.1 MPa/298 K, 0.65 GPa/298 K, 0.63 GPa/523 K, and 4.5 GPa at 523 K according to the pressure and temperature path, as is shown in Figure 1. This path was chosen to avoid the sample blowout by compression at high temperatures. A vanadium rod with the same dimension as the sample was placed in the high-pressure cell and measured at the same thermodynamic states as the sample solution. An empty cell at each pressure point was prepared and measured at ambient conditions to estimate the background. The time for measurements at the ambient condition was 6 h for a sample, 6 h for a vanadium rod, and 3 h for an empty can. For high-pressure, the measurements were 9 h at 1 GPa and 13 h at 4 GPa for the sample, 6 h for the vanadium rod at 1 and 4 GPa, and 6 h for each empty cell.

Small Bragg peaks from a vanadium rod were removed. The corresponding regions were interpolated using data outside of this range. Normalization by proton intensities was applied to the total cross-sections of a sample, a vanadium rod, and an empty cell and binned as an increment of Q = 0.01 Å^{-1}. The scattered data were corrected for absorption by the sample and the cell [34] and multiple scattering [35], and then normalized to the absolute units (barns) by using the data of the vanadium rod. After subtraction of the incoherent scattering, the structure factor $S(Q)$ was calculated by Equation (2). All the data treatments were performed with a program nvaSq [36].

$$S(Q) = \frac{\left(\frac{d\sigma}{d\Omega}\right) - \left\{\left(\sum x_i b_i^2\right) - \left(\sum x_i b_i\right)^2\right\}}{\left(\sum x_i b_i\right)^2} \tag{2}$$

where $\left(\frac{d\sigma}{d\Omega}\right)$, x_i, and b_i are the differential scattering cross-section, atomic fractions, and coherent-scattering length of atom i, respectively. The coherent-scattering length, absorption, and incoherent cross-sections of all atoms except for D were taken from the literature [37]. The absorption and incoherent cross-sections for D were calculated by a least-squares fitting procedure using the experimentally obtained total cross-section of D_2O [38]. The inelastic correction was made by the Kameda method [39] and, subsequently, a third-order polynomial equation. The $S(Q)$ data not available in the experiments below $Q < 0.2$ Å^{-1} were obtained by extrapolation.

The radial distribution function $G(r)$ were obtained by the Fourier transform of $S(Q)$ by Equation (3).

$$G(r) = 1 + \frac{1}{2\pi^2 r \rho_0} \int_{Q_{min}}^{Q_{max}} Q[S(Q) - 1] \sin(Qr) dQ \quad (3)$$

Unphysical ripples below 0.7 Å in the radial distribution functions were corrected in a usual manner [36]. The structure factors $S(Q)$ were transformed to the interference functions $F(Q)$ used in the EPSR calculations by Equation (4).

$$F(Q) = [(\sum x_i b_i)^2 (S(Q) - 1)] \ / \ \sum x_i b_i^2(Q) \quad (4)$$

An overall scaling factor of $F(Q)$ was finally obtained by a least-squares fitting procedure applied to the experimental $F(Q)$ values over 10–30 $Å^{-1}$ compared with the theoretical ones of the intramolecular O–D and D–D interactions in a D_2O molecule. An overall scaling factor, the intramolecular distances, and the root mean square amplitudes were treated as independent variables. The OriginPro software 2022b was used for the calculations [40].

2.5. EPSR Modeling

EPSR calculations were performed in a cubic cell, including 40 Mg^{2+}, 80 Cl^-, and 1000 D_2O, to reproduce the composition of a 2 *m* $MgCl_2$ aqueous solution. Monte Carlo calculations were carried out using the seed potentials to equilibrate the system. The parameter values of the Lennard-Jones potential in this study were taken from the literature for Mg^{2+} [20], Cl^- [41], and water from the SPC/E model [42]. Various parameter values were proposed for Mg^{2+} to reproduce the experimental values, such as the Mg^{2+}-H_2O distance, solvation energy, and ion association [13,16–21]. In this study, we employed those in Ref. [20] to reproduce the solvation energy and contact ion pair formation to meet our aims. The simulated interference function $F^{sim}(Q)$ is defined as Equation (5).

$$F^{sim}(Q) = \sum_i \sum_j (2 - \delta_{ij}) x_i x_j b_i b_j s_{ij}(Q) \quad (5)$$

Here, parameter $\delta_{ij} = 0$ for $i \neq j$ and =1 for $i = j$ and $sij(Q)$ the partial structure factors of atom pair *i* and *j*. According to Equation (5), the contributions of all-atom pairs to X-ray- and neutron-scattering intensities were calculated and shown in Figure S2. The O–O and O–H of solvent water and Cl–O and Mg–O pairs due to ion solvation are distinct in X-ray scattering, whereas the D–D and O–D of solvent water are major contributions, and the water O–O. Moreover, Cl–H pairs of Cl^- solvation are significant in neutron scattering. Thus, the complementary use of X-ray- and neutron-scattering data is an advantage for EPSR calculations. In the present study, X-ray- and neutron-scattering data were employed for EPSR calculations except for 2 GPa/500 K, where only X-ray-scattering data were available. Unfortunately, the pressure and temperature values at which X-ray- and neutron-scattering data were taken were slightly different for 1 and 4 GPa data. In this study, no correction was applied to the $S(Q)$ and $G(r)$ shapes or the number densities to avoid artificial modifications. The number density difference estimated from the pressure dependence of $\rho(d\rho/dP)$ and the pressure difference (ΔP) were 5.4%, 2.0%, and 3.2%, at 1.0 GPa/500 K, 1.0 GPa/500 K, and 4.0 GPa/500 K, respectively, which would affect the coordination number. The pair potentials were empirically optimized from the initial ones by comparing the simulated and experimental data. The above calculations were continued until good agreements were obtained between the simulated and experimental interference functions. Then, we calculate the pair distribution functions (pdfs), coordination number (*CN*) distributions, angle distributions (orientational correlation), and spatial density functions (SDF, 3D structure). The *CN* of atom *j* around atom *i* was calculated from the corresponding pdf $g_{ij}(r)$ by Equation (6),

$$CN_{ij} = 4\pi\rho_j \int_{r_{min}}^{r_{max}} g_{ij}(r) r^2 dr \quad (6)$$

where ρ_j is the number density of atom j, and r_{min} and r_{max} are the lower and upper limits to define the coordination shell. The present study took r_{max} as the first minimum of pdf. The parameters and references of each element for EPSR calculations are given in Table 2. The detailed procedure in EPSR modeling has been described in Refs. [43–45].

Table 2. Potential parameter values for each element.

Atoms	ε/kJ mol^{-1}	σ/Å	Atomic Mass	Coulomb Charge/e	Refs.
Mg^{2+}	0.0040	2.63	24.31	2	[20]
Cl^-	0.4187	4.40	35	−1	[41]
Ow	0.65	3.16	16.00	−0.8476	[42]
Hw	0.00	0.00	2.00	0.4238	[42]

3. Results

3.1. Interference Functions and Radial Distribution Functions

Figure 2 shows the X-ray and neutron experimental interference functions and corresponding radial distribution functions obtained by the experiments (black dots) and EPSR modeling (red lines) of an aqueous 2 *m* $MgCl_2$ solution as a function of pressure and temperature. Symbols X and N denote X-ray and neutron scattering, respectively. In the neutron interference functions, a hump observed at $Q \sim 4$ Å^{-1} at 0.1 MPa/300 K gradually decreases and becomes a plateau at 4 GPa/500 K. On the other hand, the X-ray interference function has first double peaks at 2–3 Å^{-1} at 0.1 MPa/300 K, merging into a single sharp peak at 4 GPa/500 K. These characteristic features suggest the disruption of the tetrahedral hydrogen-bonded network in water with increasing pressure and temperature [4,46]. In the X-ray radial distribution functions, there are peaks at 1.0, 2.0, and 3.0 Å, which are ascribed to the intramolecular O–H bonds within water molecules, Mg^{2+}–O(H_2O) bonds due to Mg^{2+} solvation, and an overlap of the first neighbor O–O bonds of solvent water and Cl—O(H_2O) bonds of Cl^- solvation. In the neutron radial distribution functions, peaks at 0.96 Å and 1.55Å are assigned to the intramolecular O–D and D–D interactions within a D_2O molecule, respectively. The broad peak is seen at about 3.3 Å at 0.1 MPa/300 K and shifts to about 3.0 Å with new broad peaks evolved at 5.9 and 8 Å in the GPa pressure range, indicating the structure change in solvent water because of its dominant contributions to the neutron data as seen in Figure S3.

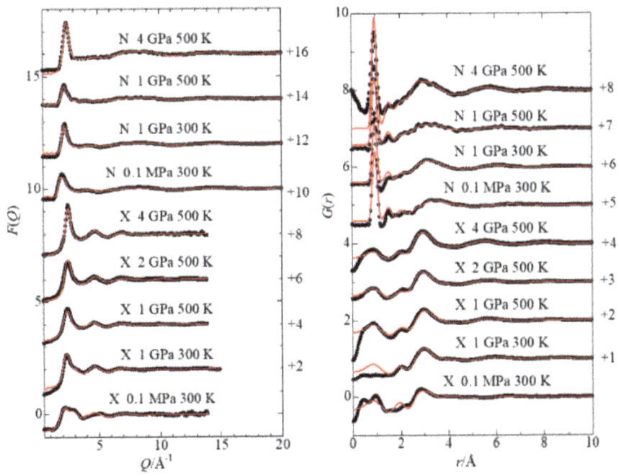

Figure 2. X-ray and neutron interference functions (**left** frame) and the radial distribution functions (**right** frame) obtained by experiments (black dots) and EPSR modeling (red lines) of 2 *m* $MgCl_2$ aqueous solutions as a function of pressure and temperature. X and N denote X-ray and neutron, respectively.

3.2. Mg^{2+} *Solvation*

The structural information of Mg^{2+} solvation is obtained from the pdfs of Mg–Ow and Mg–Hw pairs shown in Figure 3 (left). The numerical values are given in Table S1 of the Supplementary Materials. Table 3 summarizes the peak positions corresponding to the interatomic distances of the first (I) and second (II) coordination shells of Mg^{2+}. The first neighbor Mg–O_W(I) distance is almost constant at 1.92 Å over the pressure range measured, suggesting a rigid solvation shell. The value for the Mg–O_W(I) distance obtained in this study agrees with a lower limit of the Mg–O_W(I) distances from 1.92 to 2.14 Å reported in the literature [12–24]. Despite the constant Mg–O_W(I) distance, the Mg–H_W(I) distance is sensitive to pressure and temperature. At 300 K, the compression to 1 GPa does not significantly affect the Mg–H_W(I) distance at 2.58 Å. However, it becomes longer to 2.61 Å while elevating the temperature to 500 K. The second neighbor Mg–O_W(II) distance shows an appreciable change with pressure and temperature. As the pressure increases from 0.1 MPa to 1 GPa at 300 K, the Mg-O_W(II) distance is lengthened from 3.78 to 4.53 Å but is shortened to 4.14 Å while elevating the temperature to 500 K. At 500 K, the compression from 1 to 4 GPa shortens the Mg-O_W(II) distance from 4.14 to 3.99 Å.

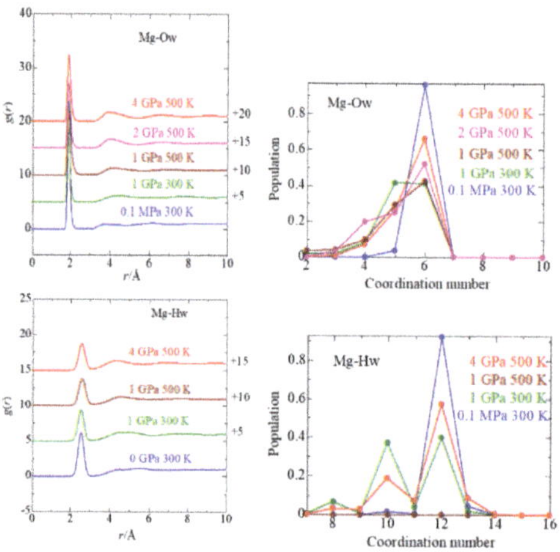

Figure 3. Pdfs (**left** frames) and *CN* distributions (**right** frames) of Mg–Ow and Mg–Hw pairs in 2 *m* $MgCl_2$ aqueous solutions in different thermodynamic states obtained by EPSR modeling.

Figure 3 (right) shows the *CN* distribution of Mg–O_W(I) and Mg–H_W(I) pairs. Table 3 summarizes the mean coordination number calculated by Equation (6). The Mg–O_W(I) pair has a single peak at six at 0.1 MPa/300 K. With increasing pressure and temperature, the population of the six-fold coordination decreases, and five- and four-fold coordination gradually increases. This decrease in *CN* of Mg–O_W(I) is compensated by an increase in *CN* of Mg^{2+}–Cl^- due to the formation of contact ion pairs, as discussed later. The *CN* for Mg–H_W(I) is twice that of Mg–O_W(I). These findings demonstrate a rigid solvation shell of six-fold coordination of Mg^{2+}.

Figure 4 shows the angle distributions of $\angle O_W$–Mg–O_W and $\angle$Mg–O_W–H_W. The $\angle O_W$–Mg–O_W angle has maxima at 90° and 180°, which correspond to *cis*- and *trans*-positions of solvated water molecules of an octahedral geometry. The $\angle$Mg–O_W–H_W angle shows a maximum of 126°, leading to the tilt angle of the water dipole from the Mg–Ow direction of 26°. This tilt angle does not agree with the value (0°) estimated by an MD simulation [16]. Since the $\angle$Mg–O_W–H_W angle is widely distributed around 126°, the tilt angle of the water dipole would be widely distributed around 26°. With increasing pressure and temperature, the distributions at 90° and 180° slightly broaden, much less than those for Cl^- solvation and solvent water shown later. These features again support a rigid octahedral solvation shell of Mg^{2+} over the pressure range measured. The present result contrasts the relatively flexible Ca^{2+} solvation at high pressures in an aqueous $CaCl_2$ solution [9,10]. This rigid solvation shell of Mg^{2+} could be responsible for many cellular functions in biological systems.

Table 3. Interatomic distances (r) and coordination numbers (CN) of the individual atom pairs in a 2 m $MgCl_2$ aqueous solution at various pressures and temperatures. I and II denote the first- and second-coordination shells, respectively. r_{max} is the upper limit of integration in Equation (5) to define the coordination shell. The uncertainties in the interatomic distance are estimated as 0.02 and 0.04 Å for the first and outer shells, respectively.

P, T	Parameter	Mg-Ow(I)	Mg-Hw(I)	Mg-Ow(II)	Cl-Ow(I)	Cl-Hw(I)	Cl-Ow(II)	Mg-Cl(I)	Mg-Cl(II)
0.1 MPa/ 300 K	r/Å	1.92	2.58	3.78	3.18	2.32	5.76	2.37	4.32
	C.N.	6.0 ± 0.2	12.0 ± 0.4		8.2 ± 1.2	6.7 ± 1.1		0.01 ± 0.12	
	r_{max}/Å	2.40	3.15		4.05	3.00		3.21	
1 GPa/ 300 K	r/Å	1.92	2.58	4.53	3.14	2.24	4.41, 5.75	2.33	4.57
	C.N.	5.0 ± 1.3	10.2 ± 2.6		8.2 ± 1.8	6.2 ± 2.0		0.78 ± 0.90	
	r_{max}/Å	2.37	3.09		3.87	2.91		3.15	
1 GPa/ 500 K	r/Å	1.92	2.61	4.14	3.21	2.31	5.86	2.19	4.50
	C.N.	4.2 ± 2.0	8.6 ± 4.0		8.4 ± 2.2	4.0 ± 2.0		1.12 ± 1.27	
	r_{max}/Å	2.58	3.18		4.17	2.88		2.85	
2 GPa/ 500 K	r/Å	1.92	···	4.03	3.12	···	5.70	2.35	4.34
	C.N.	5.3 ± 0.9	···		14.2 ± 1.5	···		0.65 ± 0.75	
	r_{max}/Å	2.32	···		4.19	···		3.00	
4 GPa/500 K	r/Å	1.93	2.61	3.99	3.14	2.18	5.73	2.36	4.34
	C.N.	5.6 ± 0.7	11.4 ± 1.2		14.0 ± 1.4	6.2 ± 1.6		0.33 ± 0.54	
	r_{max}/Å	2.31	3.07		4.15	2.72		3.30	

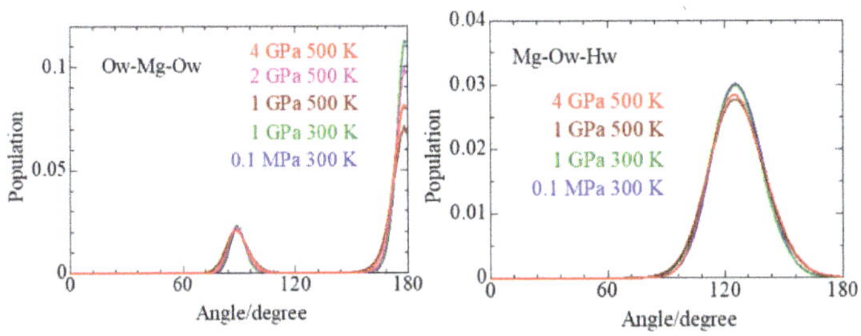

Figure 4. Angle distributions of ∠ Ow–Mg–Ow and ∠ Mg–Ow–Hw for the Mg^{2+} solvation in 2 m $MgCl_2$ aqueous solutions in different thermodynamic states obtained by EPSR modeling.

3.3. Cl^- Solvation

Figure 5 shows the pdfs (left frames) and the corresponding CN distributions (right frames) of Cl–Ow and Cl–Hw pairs, respectively. The numerical values are given in Table S1 of Supplementary Materials. The pdf of Cl–Ow shows the first peak at 3.1–3.2 Å due to Cl^- solvation. The broad peak appears at 5.67 Å, corresponding to the second solvation shell of Cl^-. It should be noted that the third peak evolved at 8 Å at 4 GPa/500 K. With elevating temperature and pressure, the first peak gradually becomes more asymmetrical toward the longer distance, suggesting more water molecules entering into the solvation shell of Cl^-. In the pdfs of Cl–H_W pairs, the first peak is observed at 2.2–2.3 Å, ascribed to the first-neighbor hydrogen-bonded Cl··· H-OH interactions, whereas the second broad peak at 3.6 Å corresponds to the interatomic distance between Cl- and another water hydrogen atom. With increasing temperature and pressure, the height of both peaks gradually decreases, and the peak becomes asymmetric toward the long-distance side, which is consistent with the changes of Cl–O_W pairs with temperature and pressure. The Cl–O_W (I) and Cl–H_W (I) distances follow an expected change with pressure and temperature, i.e., both distances decrease upon compression and increase with elevating temperature.

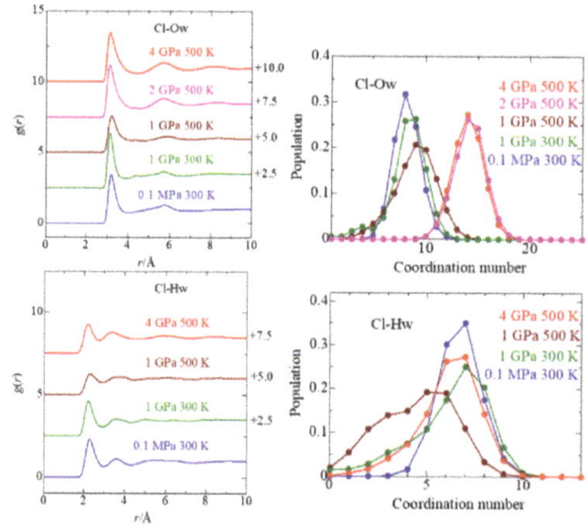

Figure 5. Pdfs (**left** frames) and the *CN* distributions (**right** frames) of Cl–Ow and Cl–Hw pairs in 2 *m* $MgCl_2$ aqueous solutions in different thermodynamic states obtained by EPSR modeling.

Figure 5 (right frames) shows the *CN* distributions of Cl–O_W pairs. The *CN* distributions of Cl–O_W pairs gradually become broader and shift to larger values with increasing pressure and temperature, corresponding to the asymmetry of the first peak in the Cl–Ow pdfs. Table 3 summarizes the mean *CN* of Cl–O_W. The mean *CN* of Cl–O_W changes from 8.2 ± 1.2 at 0.1 MPa/300 K to 14.0 ± 1.4 at 4 GPa/500 K, contrasting sharply with the change in the mean *CN* of Mg–O_W. Here, it is noted that the uncertainties in the coordination number are larger than the estimated errors in the number densities by slightly different thermodynamic states employed in neutron scattering experiments stated in the experimental section. As discussed in the next section, the increase in the *CN* of Cl^- is probably ascribed to an increase in non-hydrogen bonded water molecules at high pressures and temperatures. On the other hand, interestingly, the *CN* of Cl–H_W(I) remains almost unchanged (6.7 ± 1.1 at 0.1 MPa/300 K and 6.2 ± 1.6 at 4 GPa/500 K). This result suggests strong $Cl^- \cdots$ H-OH hydrogen bonds. The discrepancy between the mean *CN*s of Cl–Ow and Cl–Hw in the GPa pressure range indicates that most excess oxygen atoms belong to water molecules not forming hydrogen bonds with Cl^-.

The angle distribution of ∠ Ow–Cl–Ow and ∠ Cl–Hw–Ow are plotted in Figure 6. In the distribution of ∠ Ow–Cl–Ow at 0.1 MPa/300 K, there is a dominant peak at 49° and a subdominant, broad peak at 84°. When the pressure is increased to 1 GPa at 300 K, the two peaks shift to 51° and 77°, respectively. The subdominant peak disappears at 1–4 GPa and 500 K, and a peak is evolved at 94°. The distribution of ∠ Cl–Hw–Ow angle peaks close to 180°, showing the almost linear Cl$\cdots$Hw–Ow hydrogen bonds. With increasing pressure and temperature, the distribution broadens due to the distortion of the hydrogen bonds.

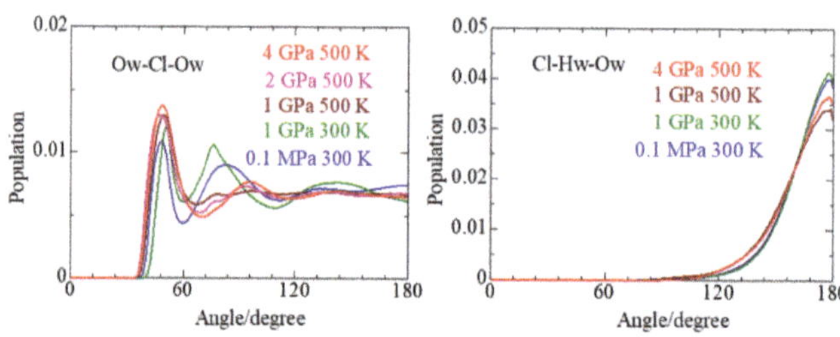

Figure 6. Angle distributions of ∠ Ow–Cl–Ow and ∠ Cl–Hw–Ow for the Cl^-solvation in 2 *m* $MgCl_2$ aqueous solutions in different thermodynamic states.

3.4. Solvent Water

From our previous studies of aqueous 3 *m* NaCl [5,7], 3 *m* RbCl [8], 2 *m* $CaCl_2$ [10], and 1 *m* $CeCl_3$ [11] solutions in the GPa pressure range, we found that the tetrahedral network structure of solvent water at 0.1 MPa/300 K transforms to a random dense-packing structure in the GPa pressure range. As seen in the pdfs of Ow–Ow in Figure 7 (left frame), three peaks are observed at 2.8, 4.1, and 6.8 Å, characteristic for the tetrahedral network structure of water, at 0.1 MPa/300 K. The structure parameters are summarized in Table 4. When the solution is compressed to 1 GPa/300 K, the second and third peaks slightly shift to the shorter distance side. However, the full feature of the O_W–O_W pdf does not change, showing that the tetrahedral water structure remains at 1 GPa/300 K. This finding is in contrast with the case of a 3 *m* NaCl aqueous solution where the tetrahedral structure is completely broken down at 1 GPa/298 K [7]. This structural difference of solvent water is ascribed to the solute concentrations, i.e., more water molecules are present in a 2 *m* $MgCl_2$ solution than in a 3 *m* NaCl solution. A similar change in solvent water's structure has also been observed for a 1 *m* $CeCl_3$ solution [11]. When the solution is heated to 500 K at 1 GPa, the O_W–O_W pdfs show a drastic change in the second and third peaks evolved at 5.6 and 8.2 Å, respectively, showing the transformation to a random dense-packing structure. In addition, the first peak at 2.8 Å becomes broader to the long-distance side, suggesting an increase in non-hydrogen bonded water molecules at the interstitial sites.

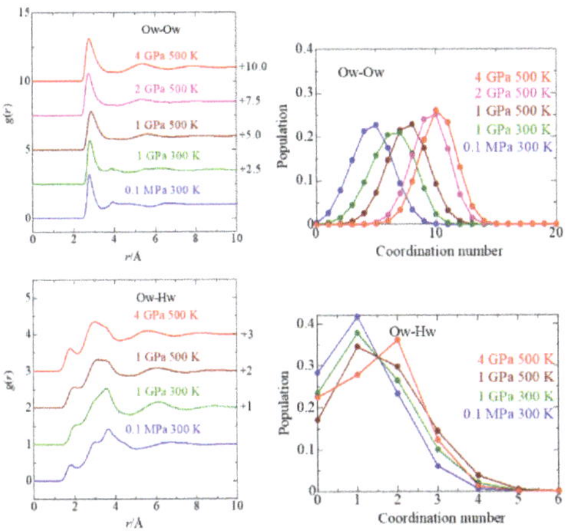

Figure 7. Pair distribution functions (**left** frames) and the coordination number distributions (**right** frames) of Ow–Ow and Ow–Hw pairs for solvent water in 2 *m* $MgCl_2$ aqueous solutions in different thermodynamic states obtained by EPSR modeling.

The detailed picture of hydrogen bonds between solvent water molecules with pressure and temperature is seen in pdfs of O_W–H_W in Figure 7 (left frame). Over the pressure and temperature range measured, the peak at 1.8 Å due to the intermolecular O-H hydron bonds is observed, demonstrating that the hydrogen bonds are preserved even at 4 GPa/500 K. The O_W–H_W distance is lengthened from 1.80 Å at 0.1 MPa/300 K to 2.07 Å at 1 GPa/300 K. Thus, although the tetrahedral network structure retains at 1 GPa/300 K, the hydrogen bonds are largely distorted by compression. When the temperature is elevated to 500 K at 1 GPa, the O_W–H_W distance is shortened; upon compression to 4 GPa, the O_W–H_W distance is further shortened to 1.7 Å. The *CN* distribution of O_W–H_W does not change largely with pressure and temperature, in contrast with the O_W–O_W. This finding suggests that pressure and temperature do not considerably affect the first-neighbor O_W–H_W.

Table 4. Interatomic distances (*r*) and coordination numbers (*CN*) of Ow–Ow and Ow–Dw for solvent water in 2 *m* $MgCl_2$ aqueous solutions under the four thermodynamic conditions. I, II, and III denote the first-, second-, and third-coordination shells, respectively. r_{max} is the upper limit of the integration Equation (5). The uncertainties in the interatomic distance are estimated as 0.01 and 0.03 Å for the first and outer shells, respectively.

P, T	Parameters	Ow–Ow(I)	Ow–Hw(I)	Ow–Ow(II)	Ow–Ow(III)
0.1 MPa/	r/Å	2.76	1.80	3.91–4.80	6.63
300 K	C.N.	4.6 ± 1.7	1.1 ± 0.90		
	r_{max}/Å	3.39	2.04		
1 GPa/	r/Å	2.79	2.07	3.84	6.09
300 K	C.N.	6.3 ± 1.8	1.3 ± 1.0		
	r_{max}/Å	1.8	2.13		
1 GPa/	r/Å	2.85	1.98	5.61	8.22
500 K	C.N.	7.6 ± 1.7	1.5 ± 1.1		
	r_{max}/Å	3.78	2.22		
2 GPa/	r/Å	2.71	......	5.31	7.99
500 K	C.N.	9.5 ± 1.5	......		
	r_{max}/Å	3.63	......		
4 GPa/	r/Å	2.74	1.74	5.38	7.83
500 K	C.N.	8.4 ± 1.6	0.88 ± 0.84		
	r_{max}/Å	3.69	1.98		

The structural change of solvent water with pressure and temperature is again discussed from the distribution of ∠Ow–Ow–Ow and ∠Ow–Hw–Ow angles, as seen in Figure 8. We can see 60° and 90° πεακσ in ∠Ow–Ow–Ow at 0.1 MPa/300 K. The latter angle is close to 109°, showing the presence of the tetrahedral hydrogen-bonded network. The sharp peak at 60° originates from non-hydrogen bonded or interstitial water molecules. Upon compression to 1 GPa at 300 K, both peaks decrease, suggesting the distortion of the network structure. Heating the solution to 500 K at 1 GPa results in the disappearance of the 90° peak and broadening of the 60° peak. Furthermore, very broad peaks evolved at approximately 116 and 180°. The drastic change in ∠Ow–Ow–Ow corresponds to transforming the tetrahedral structure into a dense random-packing arrangement. The results are consistent with 3 *m* and 5 *m* aqueous NaCl solutions [5,6].

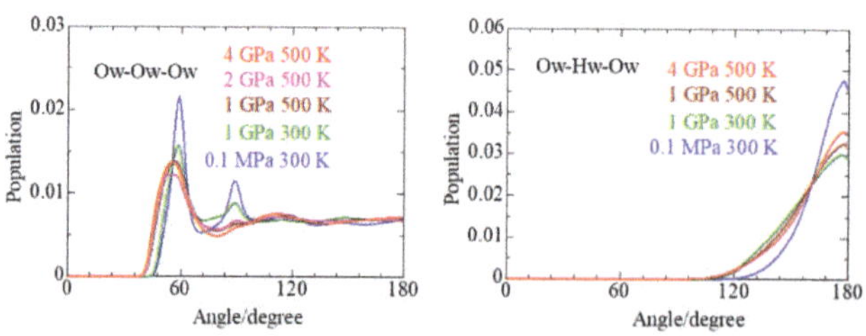

Figure 8. Distributions of ∠ Ow-Ow–Ow and ∠ Ow-Hw–Ow angles of solvent water in 2 *m* $MgCl_2$ aqueous solutions in different thermodynamic states obtained by EPSR modeling.

The geometry of the intermolecular OW–HW hydrogen bonds is seen in the distribution of ∠Ow–Hw–Ow. A unique peak at 180° appears over the pressure and temperature measured. However, in the GPa pressure range, the peak is gradually lowered, and the tail of the distribution is widened to 110°. Thus, the intermolecular O_W–H_W hydrogen bonds are almost linear but distorted in the GPa pressure range.

Figure 9 shows the change in the spatial density functions of the first-, second-, and third-neighbor water oxygen atoms around a central water molecule with pressure and temperature. We can see the tetrahedral hydrogen-bonded network structure at 0.1 MPa/300 K, which remains at 1 GPa/300 K. With increasing pressure and temperature, the water molecules hydrogen bonded to the hydrogen atoms of a central water molecule decrease, showing the distortion of hydrogen bonds. On the other hand, the lobe of water molecules on the lone-pair electron side of a central water molecule expands, and finally, a dense-packing structure of solvent water molecules at 4 GPa/500 K.

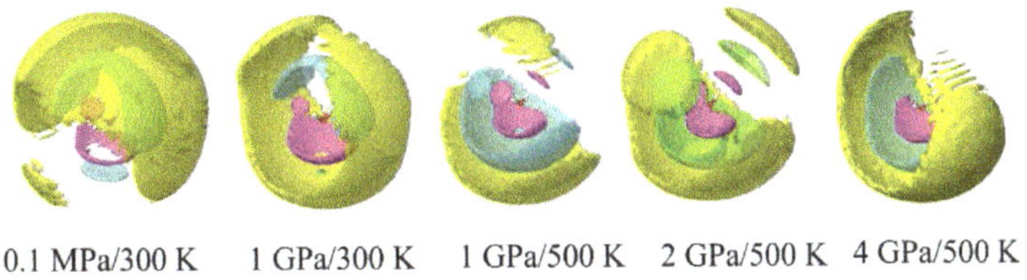

Figure 9. Spatial density functions of the nearest-neighbor water oxygen around a central water molecule for solvent water in 2 *m* $MgCl_2$ aqueous solutions in different thermodynamic states. Red and white balls at the center denote a water molecule's oxygen and hydrogen atoms, respectively. The pink, green, and yellow lobes represent the first, second, and third neighbor water oxygen distributions, respectively.

3.5. Ion Association

Structural information on ion association is obtained from the pdfs of Mg–Cl, as shown in Figure 10. The peak positions are given in Table 3. At 0.1 MPa/300 K, there is a small peak at 2.37 Å and a large, broad peak at 4–5.5 Å. The 2.37 Å peak is ascribed to the contact ion pairs (CIP) Mg^{2+}–Cl^- since the distance is close to the sum of the Shannon effective ionic radii of Mg^{2+} (0.57 Å) and Cl^- (1.81 Å) [47]. The second peak should originate from the solvent-shared ion pairs (SSIP) $Mg^{2+}\cdots OH_2\cdots Cl^-$ estimated from the interatomic distances of Mg–O_W (1.92 Å) and CL–O_W (3.18 Å). Under ambient conditions, SSIP is preferable to CIP since water molecules surround Mg^{2+}. When the pressure is increased to 1 GPa at 300 K, the first peak of CIP is enhanced, and that of SSIP is lowered, showing that CIP is more favorable than SSIP. The formation of CIP is promoted by increasing the temperature to 500 K at 1 GPa. However, with further compression to 2 and 4 GPa at 500 K, CIP decrease. As seen in Table 3, the mean *CN* of Mg–Cl reflects the above change of CIP and SSIP with pressure and temperature. In Figure 10 (right frame), the *CN* distribution of Mg–Cl extends to *CN* = 3, suggesting the formation of higher ion pairs in the GPa pressure range.

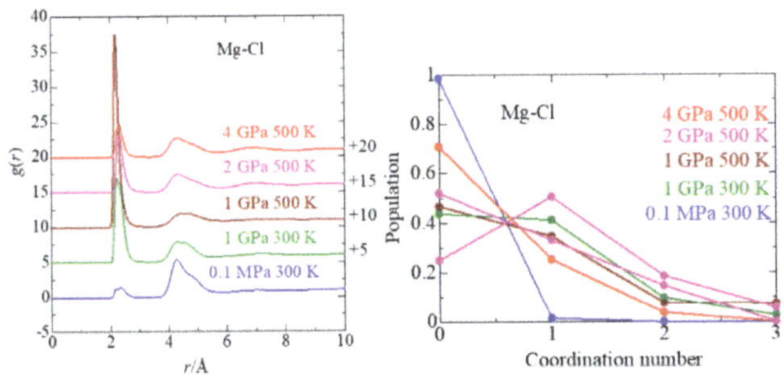

Figure 10. Pair distribution functions (**left** frame) and coordination number distributions (**right** frame) of Mg–Cl pairs for the ion association in different thermodynamic states in 2 *m* $MgCl_2$ aqueous solutions obtained by EPSR modeling.

4. Conclusions

Ion solvation, association, and water structure in a 2 *m* $MgCl_2$ aqueous solution have been revealed in the GPa pressure range by X-ray- and neutron-scattering experiments combined with EPSR modeling. Over pressures from 0.1 MPa to 4 GPa, Mg^{2+} maintain an octahedral solvation shell comprising six water molecules at 0.1 MPa/300 K, one of which is replaced by one Cl^- in the GPa pressure range due to the formation of CIP. The orientational correlation of solvated water dipoles around Mg^{2+} does not change significantly with pressure and temperature, but the distribution of the dipole orientation becomes broadened in the GPa pressure range. On the contrary, the solvation structure of Cl^- is drastically changed upon compression to GPa; the mean *CN* of water oxygen atoms around Cl^- increases from 8.2 at 0.1 MPa/300 K to 10.1 at 4 GPa/500 K. On the other hand, *the CN* of the water hydrogen atom surrounding Cl^- remains at about six regardless of pressure due to linear Cl···H-OH bonds. These results suggest some of the solvated water molecules present slightly distant from Cl^- without forming hydrogen bonds with Cl^-. Compression of solvent water to a GPa pressure range transforms the tetrahedral hydrogen-bonded network structure to a dense random packing arrangement, as evidenced by the mean *CN* change from 4.6 at 0.1 MPa/300 K to 8.4 at 4 GPa/500 K, accompanied by distortion of hydrogen bonds. In a pressure range from 0.1 MPa to 4 GPa, the intermolecular O_W–H_W(I)–O_W hydrogen bonds are linear but distorted in the GPa pressure range. CIP and multiple IPs are promoted with increasing pressure and temperature.

Supplementary Materials: The following supporting information can be downloaded at: https://www.mdpi.com/article/10.3390/liquids3030019/s1. Figure S1: A high-pressure cell assembly for X-ray-scattering experiments; Figure S2: A high-pressure cell assembly for neutron scattering experiments; Figure S3: Comparison of the contributions of atom pairs of a 2 *m* $MgCl_2$ aqueous solution to X-ray- and neutron-scattering intensities. Table S1: Tabulated data of pair distribution functions $g(r)$ of Mg–Ow, Mg–Hw, Cl–Ow, and Cl–Hw in different thermodynamic states. The $g(r)$ values of Mg–Ow and Cl–Ow are given at 2 GPa/500 K for X-ray scattering.

Author Contributions: Conceptualization, writing-original draft preparation-review, and editing: T.Y.; X-ray experiments: T.Y., N.F., K.Y. and Y.K.; X-ray data analysis: T.Y., N.F. and Y.K.; neutron experiments: T.Y., S.M. and T.H.; neutron data analysis: T.Y., writing—review and editing: T.Y., K.Y., Y.K. and T.H. All authors have read and agreed to the published version of the manuscript.

Funding: This work was supported in part by JSPS KAKENHI (No. 19K05551).

Data Availability Statement: The data can be obtained through reasonable demand from the authors.

Acknowledgments: The X-ray-scattering experiments were performed under the Shared Facility Use Program of the Japan Atomic Energy Agency with the approval of the Japan Synchrotron Radiation Research Institute (No. 2012B3623). The neutron scattering experiments at J-PARC's Materials and Life Science Experimental Facility were performed under a user program (Proposal Nos. 2015A0179 and 2017B0179).

Conflicts of Interest: The authors declare no conflict of interest.

References

1. Saris, N.-E.L.; Mervaala, E.; Karppanen, H.; Khawaja, J.A.; Lewenstam, A. Magnesium, An update on physiological, clinical and analytical aspects. *Clin. Chim. Acta* **2000**, *294*, 1–26. [CrossRef] [PubMed]
2. Zhao, L.-J.; Zhang, Y.-H.; Wei, Z.-F.; Cheng, H.; Li, X.-H. Magnesium sulfate aerosols studied by FTIR spectroscopy: Hygroscopic properties, supersaturated structures, and implications for seawater aerosols. *J. Phys. Chem. A* **2006**, *110*, 951–958. [CrossRef] [PubMed]
3. Hirschmann, M.; Kohlstedt, D. Water in Earth's Mantle. *Phys. Today* **2012**, *65*, 40–45. [CrossRef]
4. Yamaguchi, T.; Fujimura, K.; Uchi, K.; Yoshida, K.; Katayama, Y. Structure of water from water to 4 GPa revealed by energy-dispersive X-ray diffraction combined with empirical potential structure refinement modeling. *J. Mol. Liq.* **2012**, *176*, 44–51. [CrossRef]
5. Yamaguchi, T.; Fukuyama, N.; Yoshida, K.; Katayama, Y. Ion solvation and water structure in an aqueous sodium chloride solution in the gigapascal pressure range. *J. Phys. Chem. Lett.* **2021**, *12*, 250–256. [CrossRef]
6. Polidori, A.; Rowlands, R.F.; Zeidler, A.; Salanne, M.; Fischer, H.E.; Annighöfer, B.; Klotz, S.; Salmon, P.S. Structure and dynamics of aqueous NaCl solutions at high temperatures and pressures. *J. Chem. Phys.* **2021**, *155*, 194506. [CrossRef]
7. Yamaguchi, T.; Yoshida, K.; Machida, S.; Hattori, T. Neutron scattering on an aqueous sodium chloride solution in the gigapascal pressure range. *J. Mol. Liq.* **2022**, *365*, 120181. [CrossRef]

8. Zhang, W.-Q.; Yamaguchi, T.; Fang, C.-H.; Yoshida, K.; Zhou, Y.-Q.; Zhu, F.-Y.; Machida, S.; Hattori, T.; Li, W. Structure of an aqueous RbCl solution in the gigapascal pressure range by neutron diffraction combined with empirical potential structure refinement modeling. *J. Mol. Liq.* **2022**, *348*, 118080. [CrossRef]
9. Takamuku, T.; Yagafarov, O.F.; Katayama, Y.; Umeki, T. Structure of aqueous calcium chloride solutions by energy-dispersive X-ray diffraction under high temperatures and high pressures. *Bunseki Kagaku* **2015**, *64*, 203. (In Japanese) [CrossRef]
10. Yamaguchi, T.; Nishino, M.; Yoshida, K.; Takumi, M.; Nagata, K.; Hattori, T. Ion hydration and association in an aqueous calcium chloride solution in the GPa range. *Eur. J. Inorg. Chem.* **2019**, *2019*, 1170–1177. [CrossRef]
11. Yamaguchi, T.; Fukuyama, N.; Yoshida, K.; Katayama, Y. Ion solvation and association and water structure in an aqueous cerium(III) chloride solution in the gigapascal pressure range. *Anal. Sci.* **2022**, *38*, 409–417. [CrossRef] [PubMed]
12. Caminiti, R.; Licheri, G.; Piccaluga, G.; Pinna, G. X-ray diffraction study of $MgCl_2$ aqueous solution. *J. Appl. Crystallogr.* **1979**, *12*, 34–38. [CrossRef]
13. Persson, I. Structures of hydrated metal ions in the solid state and aqueous solution. *Liquids* **2022**, *2*, 210–242. [CrossRef]
14. Bruni, F.; Imberti, S.; Mancinelli, R.; Ricci, M.A. Aqueous solutions of divalent chlorides; ions hydration shell and water structure. *J. Chem. Phys.* **2012**, *136*, 064520. [CrossRef]
15. Duboué-Dijon, E.; Mason, P.E.; Fischer, H.E.; Jungwirth, P. Hydration and ion pairing in aqueous Mg^{2+} and Zn^{2+} solutions; force-field description aided by neutron scattering experiments and ab initio molecular dynamics simulations. *J. Phys. Chem. B* **2018**, *122*, 3296–3306. [CrossRef]
16. Szász, G.I.; Dietz, W.; Heinzinger, K.; Pálinkás, G.; Tadnai, T. On the orientation of water molecules in the hydration shell of the ions in an $MgCl_2$ solution. *Chem. Phys. Lett.* **1982**, *92*, 388–392. [CrossRef]
17. Jiao, D.; King, C.; Grossfield, A.; Darden, T.A.; Ren, P. Simulation of Ca^{2+} and Mg^{2+} solvation using polarizable atomic multipole potential. *J. Phys. Chem. B* **2006**, *110*, 18553–18559. [CrossRef]
18. Callahan, K.M.; Casillas-Ituarte, N.N.; Roeselová, M.; Allen, H.C.; Tobias, D.J. Solvation of magnesium dication; molecular dynamics simulation and vibrational spectroscopic study of magnesium chloride in aqueous solution. *J. Phys. Chem. A* **2010**, *114*, 5141–5148. [CrossRef]
19. Mamatkulov, S.; Fyta, M.; Netz, R.R. Force fields for divalent cations based on single-ion and ion-pair properties. *J. Chem. Phys.* **2013**, *138*, 024505. [CrossRef]
20. Mamatkulov, S.; Scwierz, N. Force fields for monovalent and divalent metal cations in TIP3P water based on thermodynamic and kinetic properties. *J. Chem. Phys.* **2018**, *148*, 074504. [CrossRef]
21. Zeron, I.M.; Abascal, J.L.F.; Vega, C. A force field of Li^+, Na^+, K^+, Mg^{2+}, Ca^{2+}, Cl^-, and SO_4^{2-} in aqueous solution based on the TIP4P/2005 water model and scaled charges for the ions. *J. Chem. Phys.* **2019**, *151*, 134504. [CrossRef]
22. Ikeda, T.; Boero, M.; Terakura, K. Hydration properties of magnesium and calcium ions from constrained first-principles molecular dynamics. *J. Chem. Phys.* **2007**, *127*, 074503. [CrossRef] [PubMed]
23. Tommaso, D.D.; de Leeuw, N.H. Structure and dynamics of the hydrated magnesium ion and of the solvated magnesium carbonates; insights from first principles simulations. *Phys. Chem. Chem. Phys.* **2010**, *12*, 894–901. [CrossRef] [PubMed]
24. Delgado, A.A.A.; Sethio, D.; Kraka, E. Assessing the intrinsic strengths of ion-solvent and solvent0-solvent interactions for hydrated Mg^{2+} clusters. *Inorganics* **2021**, *9*, 31. [CrossRef]
25. Wagner, W.; Pruβ, A. The IAPWS Formulation 1995 for the Thermodynamic Properties of Ordinary Water Substance for General and Scientific Use. *J. Phys. Chem. Ref. Data* **2002**, *31*, 387. [CrossRef]
26. Eisenberg, D.; Kauzmann, W. *The Structure and Properties of Water*; Oxford University Press: Oxford, UK, 2006.
27. Tsuji, K.; Yaoita, K.; Imai, M.; Shimomura, O.; Kikegawa, T. Measurements of X-ray Diffraction for Liquid Metals under High Pressure. *Rev. Sci. Instrum.* **1989**, *60*, 2425. [CrossRef]
28. Decker, D.L. High-Pressure Equation of State for NaCl, KCl, and CsCl. *J. Appl. Phys.* **1971**, *42*, 3239. [CrossRef]
29. Datchi, F.; Loubeyre, P.; LeToullec, R. Extended and Accurate Determination of the Melting Curves of Argon, Helium, Ice (H_2O), and Hydrogen (H_2). *Phys. Rev. B* **2000**, *61*, 6535. [CrossRef]
30. Utsumi, W.; Funakoshi, K.; Katayama, Y.; Yamanaka, M.; Okada, T.; Shimomura, O. High-Pressure Science with a Multi-Anvil Apparatus at SPring-8. *J. Phys. Condens. Matter* **2002**, *14*, 10497–10504. [CrossRef]
31. Kaplow, R.; Strong, S.L.; Averbach, B.L. Radial Density Functions for Liquid Mercury and Lead. *Phys. Rev.* **1965**, *138*, A1336. [CrossRef]
32. Sano-Furukawa, A.; Hattori, T.; Arima, H.; Yamada, A.; Tabata, S.; Kondo, M.; Nakamura, A.; Kagi, H.; Yagi, T. Six-axis multi-anvil press for high-pressure, high-temperature neutron diffraction experiments. *Rev. Sci. Instrum.* **2014**, *85*, 113905. [CrossRef]
33. Hattori, T.; Sano-Furukawa, A.; Arima, H.; Komatsu, K.; Yamada, A.; Inamura, Y.; Nakatani, T.; Seto, Y.; Nagai, T.; Tsumi, W.; et al. Design and performance of high-pressure PLANET beamline at pulsed neutron source at J-PARC. *Nucl. Instrum. Methods Phys. Res.* **2015**, *A780*, 55–67. [CrossRef]
34. Paalman, H.H.; Pings, C.J. Numerical evaluation of X-ray absorption factors for cylindrical samples and annular sample cells. *J. Appl. Phys.* **1962**, *33*, 2635. [CrossRef]
35. Blech, I.A.; Averbach, B.L. Multiple scattering of neutrons in vanadium and copper. *Phys. Rev.* **1965**, *137*, A1113. [CrossRef]
36. NOVA Group, Institute of Materials Structure Science, Tsukuba, Ibaraki 305-080, Japan. Program nvaSq Manual. Available online: https://research.kek.jp/group/hydrogen/reduction/index.html (accessed on 26 June 2023). (In Japanese)
37. Sears, V.F. Neutron scattering lengths and cross sections. *Neutron News* **1992**, *3*, 26. [CrossRef]

38. Granada, J.R.; Gillette, V.H.; Mayer, R.E. Neutron cross sections and thermalization parameters using a synthetic scattering function. II: Applications to H_2O, D_2O, and C_6H_6. *Phys. Rev. A* **1987**, *36*, 5594. [CrossRef] [PubMed]
39. Kameda, Y.; Sasaki, M.; Usuki, T.; Othomo, T.; Itoh, K.; Suzuya, K.; Fukunaga, T. Inelasticity effects on neutron scattering intensities of the null-H_2O. *J. Neutron Res.* **2003**, *11*, 153–163. [CrossRef]
40. OriginPro. *The Ultimate Software for Graphing and Analysis*; OriginLab Corp.: Northampton, MA, USA, 2022.
41. Druchok, M.; Holovko, M. Molecular dynamics study of ion hydration under pressure. *J. Mol. Liq.* **2011**, *159*, 24–30. [CrossRef]
42. Berendsen, H.J.C.; Grigera, J.R.; Straatsma, T.P. The Missing Term in Effective Pair Potentials. *J. Phys. Chem.* **1987**, *91*, 6269. [CrossRef]
43. Soper, A.K. Computer simulation as a tool for the interpretation of total scattering data from glasses and liquids. *Mol. Simul.* **2012**, *38*, 1171–1185. [CrossRef]
44. Soper, A.K. Empirical potential Monte Carlo simulation of fluid structure. *Chem. Phys.* **1996**, *202*, 295–306. [CrossRef]
45. Soper, A.K. Partial structure factors from disordered materials diffraction data: An approach using empirical potential structure refinement. *Phys. Rev. B* **2005**, *72*, 104204. [CrossRef]
46. Amann-Winkel, K.; Bellissent-Funel, M.-C.; Bove, L.E.; Loerting, T.; Nilsson, A.; Paciaroni, A.; Schlesinger, D.; Skinner, L. X-ray and Neutron Scattering of Water. *Chem. Rev.* **2016**, *116*, 7570–7589. [CrossRef] [PubMed]
47. Shannon, R.D. Revised effective ionic radii and systematic studies of interatomic distances in halide and chalcogenides. *Acta Crystallogr.* **1976**, *A32*, 751–767. [CrossRef]

liquids

MDPI

Article

An Ab Initio Investigation of the Hydration of Antimony(III)

Cory C. Pye * and Champika Mahesh Gunasekara

Department of Chemistry, Saint Mary's University, Halifax, NS B3H 3C3, Canada
* Correspondence: cory.pye@smu.ca

Abstract: The energies, structures, and vibrational frequencies of $[Sb(H_2O)_n]^{3+}$, n = 0–9, 18 have been calculated at the Hartree–Fock and second-order Møller–Plesset levels of theory using the CEP, LANL2, and SDD effective core potentials in combination with their associated basis sets, or with the 6-31G* and 6-31+G* basis sets. The metal–oxygen distances and totally symmetric stretching frequency of the aqua ions were compared with each other and with related crystal structure measurements where available.

Keywords: ab initio; antimony(III); hydration; symmetry; vibrational spectrum

1. Introduction

Although the structure of many metal ions in solution is known, some remain elusive [1]. Many are known to be toxic to man, but this is dependent on the oxidation state and speciation, which often depends on pH and the presence of counterions that solubilize the metal by complex formation. While computational chemistry can assist in supporting and rationalizing proposed speciation models, one drawback is that there are typically few all-electron basis sets that can be used. For elements with a high atomic number, relativistic effects can play an important role. Effective core potentials (ECPs) replace the explicit description of core electrons by a potential, and are paired with basis sets describing the outermost electrons. The ECPs represent the scalar relativistic effects only, but the spin–orbital relativistic effects should be small. In a previous work, we benchmarked some common ECPs for the aqua complexes of the heavy metals mercury(II) and thallium(III), both of which have valence electron configuration $5d^{10}$ [2]. It was shown that the ECPs reproduce the known hexacoordination of thallium(III), and supported a hexacoordinate model for mercury(II) over a heptacoordinate model. We also extended this work to lead(II), with valence electron configuration $6s^25d^{10}$ [3]. The presence of an ns^2 subshell can give rise to either *hemidirected* structures (which tend to be favored at lower coordination numbers) with ligands that are not symmetrically distributed around the central ion, or *holodirected* structures with a symmetrical distribution. No consensus exists on the coordination number of lead(II), with predictions ranging from 4 to 9, but our results were most consistent with a hemidirected hexaaqualead(II) species. For the smaller aquatin(II) ion ($5s^24d^{10}$), we found that the preferred hydration mode was a tricoordinate trigonal pyramidal triaquatin(II), agreeing with recent experiments [4]. We extend our work now to antimony(III), which has the same valence electron configuration as tin(II) but a higher charge. The structure of aquaantimony(III) is unknown, so one of our aims is to predict its structure. The presence of the ns^2 subshell, as with lead(II) and tin(II), will be shown to have a pronounced effect on the structures compared to those without it.

The literature on the solution chemistry of antimony(III) is sparse. The solubility of rhombic Sb_2O_3 was examined by Gayer and Garrett over half a century ago, and they proposed that dissolved antimony(III) exists as SbO^+, $Sb(OH)_3$, and SbO_2^- as the pH is increased [5]. This was confirmed spectrophotometrically by Mishra and Gupta, who suggested that the neutral species could be either $Sb(OH)_3$ or $SbO(OH)$ [6]. Potentiometrically, SbO^+ is equivalent to $Sb(OH)_2^+$, SbO_2^- is equivalent to $Sb(OH)_4^-$, and Sb^{3+} is proposed to exist only in strongly acidic solutions [7]. Ahrland and Bovin examined the solution chemistry of antimony(III) oxide in perchloric and nitric acid [8]. At an ionic strength of 5.0 mol/L, maintained by sodium perchlorate or nitrate, the two modifications of $Sb_2O_3(s)$ are only metastable above a perchloric acid concentration of 0.3 mol/L (orthorhombic) or 0.7 mol/L (cubic), and the equilibrium solid phase is $Sb_4O_5(OH)ClO_4{\cdot}1/2H_2O$. Other solid phases exist at higher acid concentrations. In nitric acid at a nitrate concentration of 5.0 mol/L, the equilibrium solid phase is $Sb_4O_4(OH)_2(NO_3)_2$. Analysis suggests that the aqueous speciation at a low pH consists of $Sb(OH)_2{}^+$, and possibly $Sb_2(OH)_2{}^{4+}$. Zakaznova-Herzog and Seward used UV/visible spectroscopy to examine the equilibria of antimonous acid and was able to explain their results using the species H_3SbO_3(aq), $H_4SbO_3{}^+$, and $H_2SbO_3{}^-$(aq) from 25–300 °C, pH = 0.8–12.5 at a total antimony concentration of ~10^{-4} mol/L [9].

We now consider if insight may be gained by examining the solid-state structures of antimony oxides and their hydrates. Antimony(V) oxide, and its mono and trihydrate, were discussed by Natta and Baccaredda [10]. Well-aged (5 years!) monohydrate $Sb_2O_5{\cdot}H_2O$ (also known as antimonic acid, $HSbO_3$) was found to belong to a cubic space group (Schonflies notation, $O_h{}^7$; Hermann–Mauguin notation, $Fd\bar{3}m_2$), with a = 10.23 ± 0.02 Å, Z = 8. The compound Sb_2O_4 of the same space group was formulated as the pyroantimonate $[Sb^{III}{}_2O]^{4+}[Sb^{V}{}_2O_7]^{4-}$. However, Dihlstrom and Westgren suggested that "Sb_2O_4" was actually Sb_3O_6OH, as $Sb^{III}OOH{\cdot}Sb^{V}{}_2O_5$, with a = 10.28 Å [11]. They found the distances Sb^{III}-O, 2.48 Å; Sb^{III}-O(H), 2.23 Å, Sb^{V}-O, 2.02 Å. Another form of antimonic acid, $Sb_2O_5{\cdot}3H_2O$, was found by England et al. to have a pyrochlore ($A_2M_2O_6O'$, $Fd\bar{3}m$) structure via powder diffraction [12]. Other forms, described by Watelet et al. as $[H(H_2O)_n]_{12}Sb_{12}O_{36}$, ($n$ = 0.33, 1), space group $Im3$, a = 9.470(5), and 9.497(3) Å, respectively, have Sb-O distances in the range 1.939–2.006 Å [13]. These solids are proton conductors. Another antimony oxide hydroxide HSb_3O_8 was described by Jager et al. [14] as space group $P2/n$, with cell constants a = 9.456(2), b = 8.691(2), c = 9.909(2) Å, β = 90.25(2)°, Z = 6. The Sb-O distances fall in the range 1.890–2.140 Å. Riviere et al. obtained an $F\underline{d}3\underline{m}$ space group for $H_2Sb_2O_6{\cdot}H_2O$, a = 10.365(2) Å, with Sb-O = 1.971 Å [15]. For antimony(III) oxide, the structure of the cubic α form was refined by Svensson [16], who found a space group $Fd3m$ and a = 11.1519(2) Å. It forms discrete Sb_4O_6 units of symmetry T_d with trigonal pyramidal antimony(III) with an Sb-O distance of 1.977(1) Å and angles of 95.93(8)°. Svensson also examined the orthorhombic β modification of antimony(III) oxide [17], and found a space group $Pccn$, a = 4.911, b = 12.464, c = 5.412 Å. The structure is composed of double infinite chains, with the antimony lone pair perpendicular to the chains. The antimony is trigonal pyramidal, with Sb-O distances in the range 1.977(7)–2.023(4) Å. A novel γ modification was reported by Orosel et al., with space group $P2_12_12_1$, a = 11.6411(1), b = 7.5666(0), c = 7.4771(0) Å, with Sb-O distances in the range 1.87–2.24 Å [18]. In $Sb_4O_4(OH)_2(NO_3)_2$, Bovin found sheets of 3-coordinate trigonal pyramidal and 4-coordinate seesaw antimony(III) separated by nitrate ions [19]. The Sb-O distances in the SbO_3 units are in the range 1.942(7)–2.067(6) Å, and in the SbO_4 units, 2.019(6)–2.265(6) Å. In $Sb_4O_5(OH)ClO_4{\cdot}1/2H_2O$, Bovin found short chains of SbO_n polyhedra forming layers separated by the perchlorate ions and water molecules [20]. The Sb-O distances in the SbO_3 units are in the range 1.969(11)–2.027(8) Å, and in the SbO_4 units, 1.952(9)–2.378(8) Å. In the antimony oxide sulfate $Sb_6O_7(SO_4)_2$, Bovin found that the three unique antimony atoms were trigonally pyramidally coordinated to oxygen atoms, with Sb-O distances between 1.994(17) and 2.207(18) Å, but with a fourth oxygen close 2.327(14)–2.382(18) [21]. Mercier et al. found for the oxonium antimony sulfate, $(H_3O)_2Sb_2(SO_4)_4$, that antimony exists as both a seesaw SbO_4 and square pyramidal SbO_5 polyhedra, with corresponding distances of 2.032(10)–2.359(19) Å and 2.012(15)–2.263(13) Å [22]. The axial distance of SbO_5 is the shortest. To summarize this literature, antimony(V) tends to exist as an octahedrally coordinated moiety in the presence of oxide/hydroxide, whereas antimony(III) can exist as either a trigonal pyramidal SbO_3, seesaw SbO_4, or square pyramidal SbO_5 coordinated moiety in a similar environment.

2. Materials and Methods

Calculations were performed using Gaussian 98 [23]. In this program version, the ability to calculate analytical frequencies of molecules in which core electrons are described by effective core potentials was introduced. Therefore, many variants of these were tried. The MP2 calculations use the frozen core approximation. A stepping-stone approach was used for geometry optimization, in which the geometries at the levels HF/CEP-4G, HF/CEP-31G*, HF/CEP-121G*, HF/LANL2MB, HF/LANL2DZ, and HF/SDD were sequentially optimized. For minimum energy structures, the MP2/CEP-31G* and MP2/CEP-121G* calculations were also performed. Calculations were also carried out using the 6-31G* and 6-31+G* basis sets on the atoms of the water molecules (5*d*) with an effective core potential and basis set on the metal ion (denoted as ECP+6-31G* or 6-31+G*). For shorthand, we denote the mixed basis sets as follows: CEP-121G* on Sb, 6-31G* on O,H, as basis set A; LANL2DZ on Sb, 6-31G* on O, H, as basis set B; SDD on Sb, 6-31G* on O,H, as basis set C; and the corresponding basis sets with diffuse functions are indicated by adding a "+" to the basis set name. Default optimization specifications were used. After each level, where possible, a frequency calculation was performed at the same level and the resulting Hessian was used in the following optimization. Z-matrix coordinates constrained to the appropriate symmetry were used to speed up the optimizations. Because frequency calculations are carried out at each level, any problems with the Z-matrix coordinates would manifest themselves by giving imaginary frequencies corresponding to modes orthogonal to the spanned Z-matrix space. The Hessian was evaluated at the first geometry (opt = CalcFC) for the first level in a series to aid geometry convergence. We note that, for the heavy elements only, the three different CEP basis sets are equivalent (CEP-121G*) but differ for the oxygen and hydrogen atoms. The choice of core electrons defining the pseudopotential depends on the specific core potential (CEP and LANL2, [Kr]$4d^{10}$; SDD, [Ar]$3d^{10}$). Gaussian 03 [24] and Gaussian 16 [25] were used to correct errors and omissions.

In many cases to follow, the symmetry of the minimum energy complexes was the same as those previously found for bismuth [26]. To confirm these results, starting with high symmetry structures, systematic desymmetrization along the various irreducible representations was carried out [27,28]. We did not employ an implicit solvation model for reasons described previously [3].

3. Results

3.1. A Survey of Structures

Antimony(III) might be expected to show similar properties to tin(II), although the higher charge would cause stronger interactions with water molecules. The point group symmetry for mono- through octaaquaantimony(III) was initially usually found to be C_{2v}, C_2, C_3, C_{2v} or C_2, C_{2v}, C_3, C_{2v} [5+2], and S_8, respectively. The diaquaantimony(III) species, like lead and tin, ascended in symmetry to a planar C_{2v} structure at HF/LANL2MB. Initially, all attempts to generate a stable 7-coordinate antimony resulted in two water molecules moving to the second hydration sphere. We initially did not find a stable D_3 enneaaquaantimony(III) structure.

The results of the systematic desymmetrization procedure [27] for aquaantimony(III) are as follows (see Figures 1 and S1):

- The monoaquaantimony(III) remains C_{2v} at all levels.
- The most stable diaquaantimony(III) remains the bent C_2 at all levels except HF/LANL2MB (C_{2v} planar). The linear holodirected D_{2d} structure is approximately 50 kJ/mol higher in energy, but the unstable bent C_s structure is only slightly higher in energy (<1 kJ/mol for nonminimal basis sets). All attempts to generate a [1+1] structure instead result either in proton transfer to give $SbOH^{2+}$ + H_3O^+, which was much lower in energy, or recoordination to give the [2+0] C_s structure.
- The most stable triaquaantimony(III) remains the pyramidal C_3 at all levels. The two pyramidal C_{3v} structures are 10–30 kJ/mol higher in energy, whereas the planar holodirected D_{3h} and D_3 structures are 125–150 kJ/mol higher in energy still. The unstable

[2+1] C_{2v} structure was ~90 kJ/mol higher in energy, and upon desymmetrization, underwent proton transfer to give the much-lower-in-energy $SbOH(H_2O)^{2+}$ + H_3O^+.

- The most stable tetraaquaantimony(III) is usually the seesaw C_{2v} #3 (all mixed basis sets, HF/SDD), but it can be C_2 (all CEP, HF/LANL2DZ) or C_s (HF/LANL2MB). The other C_{2v} structures are higher in energy (10–25 kJ/mol). The holodirected D_{2d} #1, #2, and S_4 structures are much higher in energy (75–125 kJ/mol). The unstable C_s #2 [3+1] structures are up to 50 kJ/mol higher in energy, and upon desymmetrization, undergo a proton transfer to give the much-lower-in-energy $SbOH(H_2O)_2{}^{2+}$ + H_3O^+.
- The most stable pentaaquaantimony(III) is the square pyramidal C_{2v} #1, with the other three C_{2v} structures which are 25–50 kJ/mol higher in energy. The usually unstable [4+1] structures are 15–50 kJ/mol higher in energy. To our surprise, the C_{2v} #1 [4+1] is stable at HF/CEP-121G*, HF/LANL2MB, HF/A, HF/B, HF/C, and HF/A+. At the other levels, there is a B_1 imaginary mode, along which a proton transfer occurs to give $SbOH(H_2O)_3{}^{2+}$ + H_3O^+. At the MP2 levels, desymmetrization along an A_2 mode gives rise to a C_2 structure, which is slightly lower in energy.
- The most stable hexaaquaantimony(III) is the distorted octahedral C_3. The octahedral T_h structure is ~18–50 kJ/mol higher in energy. The two C_s [5+1] structures examined often underwent proton transfer to give $SbOH(H_2O)_4{}^{2+}$ + H_3O^+. However, if the [5+1] structures were stationary points, they were lower in energy than the [6+0] forms by about 25 kJ/mol. In some cases, they were actually minima (HF/CEP-121G*, all HF mixed basis sets). Given how shallow (or in some cases, nonexistent) the barrier to proton transfer is to give the deprotonated forms, it is reasonable to conclude that the aquaantimony(III) species would only exist in an extremely acidic solution. Combined with the results for the pentaaquaantimony(III), this suggests that antimony(III) is actually pentacoordinate square pyramidal.
- Of the 16 different C_{2v} heptaaquaantimony(III) structures tried, none were stable, either possessing imaginary modes or dissociating to a [6+1], [5+2], or [4+3] structure. Structures #1, #2, and #4, nearly always dissociated. The remainder usually remained 7-coordinate, except occasionally dissociating at HF/CEP-4G and HF/LANL2MB. In some cases, the [5+2] structures underwent a double proton transfer and possibly a water elimination to give $Sb(OH)_2(H_2O)_{2/3}{}^+$ + $1/0H_2O$ + $2H_3O^+$. Of the 7-coordinate C_{2v} structures, #16 was the lowest in energy, unless #1 was a stationary point, in which case, it was lower in energy. In all cases, the [5+2] structure was 20–80 kJ/mol lower in energy than the 7-coordinate C_{2v} #16 structure. Upon desymmetrization of the remaining 7-coordinate C_{2v} structures to C_2, many coalesced to C_2 #16, some dissociated to [5+2], but at some levels, C_2 #5, #6, #11, and #13 also exist, with at least one imaginary B mode. When these were desymmetrized to C_1 #1–5, a stable C_1 #3 or #5 was found at some levels, or ligand dissociation to a [6+1] or [5+2] structure took place. Desymmetrization of the C_{2v} structures along the B modes gave one of 27 possible C_s structures. Of these, C_s #1, #5, and either #13 or #20 often coalesce (to C_s #5); C_s #2, #7, #14 and #22 often coalesce (to C_s #7); C_s #12 coalesces to C_s #8; C_s #11, #18, and #26 often coalesce (to C_s #26); C_s #19 and C_s #15 ascend in symmetry to either C_{2v} #10 or #12; C_s #16 and #24 often coalesce to C_s #10 or to each other; C_s #25 coalesces to C_s #21; C_s #27 coalesces to C_s #23; and C_s #17 remains unique. None of these C_s structures are minima, except C_s #13 at HF/A+. The C_s structures desymmetrize to C_1 #6–21. Of these possibilities, for the most part, they coalesced to the previously found C_1 #3 or #5 structures, or underwent ligand dissociation to give [6+1] or [5+2] structures. This exemplifies the power of the systematic desymmetrization procedure in finding minimum energy structures that would otherwise be difficult to locate. In all cases where a seven-coordinate minimum energy structure exists (HF/CEP-31G*, HF/CEP-121G*, MP2/CEP-31G*, MP2/CEP-121G*, HF/C, HF/A+, HF/B+, HF/C+, MP2/A+), it is less stable than a [5+2] or [6+1] structure (or a proton transferred version thereof). The 7-coordinate structures are unlikely to have any significant population in an aqueous solution under ambient conditions.

- For octaaquaantimony(III), two D_{4h} (square prism) and two D_{4d} (square antiprism) structures were first examined. Multiple imaginary modes were present.
- For the D_{4d} #1 and #2 structures, desymmetrization along the A_2 imaginary mode gave the same S_8 #1 structure; along the B_1 imaginary mode, the same D_4 #2 structure; and along the B_2 imaginary mode, the C_{4v} *#1* and #2 structures. For the D_{4h} #1 and #2 structures, desymmetrization along the A_{1u} imaginary mode gave the D_4 #2 structure found before; along the A_{2g} imaginary mode, the same C_{4h} #1 structure; along the A_{2u} imaginary mode, the C_{4v} #3 [4+4] and #4 structures; along the B_{2g} imaginary mode (D_{4h} #1), the D_{2h} #1 structure ascended in symmetry to D_{4h} #2; along the B_{1u} mode, D_{2d} #1 and #3 coalesced; along the B_{2u} mode, D_{2d} #2 and #4, respectively. Another D_{2d} structure (#5) was formed by combining other D_{2d} structures.
- Desymmetrization of the C_{4h} #1 structure along the A_u mode usually gives the S_8 #1 structure (via C_4 #1), and along the B_u mode, D_{2d} #5 (via S_4 #1); the S_8 #1 structure along the B mode at two levels, C_4 #1; the D_4 #1 structure along the A_2 mode, C_4 #1, and along the B_2 mode, D_2 #1; the D_{2d} #1 structure along the A_2 mode, S_4 #2, the B_1 mode, D_2 #1, and the B_2 mode, [4+4] C_{2v} #1; the D_{2d} #2 structure along the A_2 mode, S_4 #2, the B_1 mode, D_2 #1, and the B_2 mode, C_{2v} #2; the D_{2d} #4 structure along the A_2 mode, S_4 #2, the B_1 mode, D_2 #1, and the B_2 mode, C_{2v} #3; the D_{2d} #5 structure along the B_1 mode, D_2 #1, and the B_2 mode, [6+2] C_{2v} #4. All of these structures had at least one imaginary frequency.
- Desymmetrization of the S_4 #2 structure along the B mode, C_2 #1 (which was close in structure to the D_2 #1 structure); the D_2 #1 structure along the B_1 mode gave either the [4+4] or [6+2] C_2 #1, or more usually the stable S_8 #1.
- For enneaaquaantimony(III), four D_{3h} structures were first examined. Desymmetrization along the A_1'' mode led to the common D_3 #1 structure (which was only stable at MP2/CEP-31G*); the A_2' mode led to either the unstable C_{3h} #1 or #2 structures; and the A_2'' mode led to the unstable C_{3v} #1–4 structures. The C_{3v} #2 and 3 structures were usually of [6+3] coordination. When these structures were desymmetrized, they nearly always resulted in expulsion of three water molecules to the second hydration sphere to give the (usually) stable C_3 #1 [6+3]. An unstable [6+3] D_3 structure was also found, which desymmetrized to give the even more stable C_3 #2 [6+3].

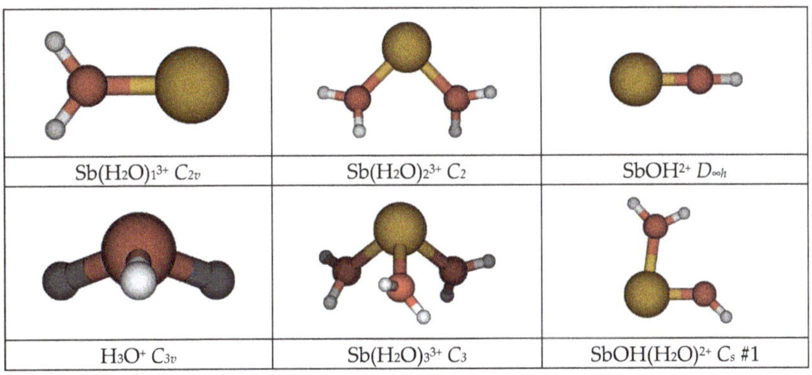

Figure 1. *Cont.*

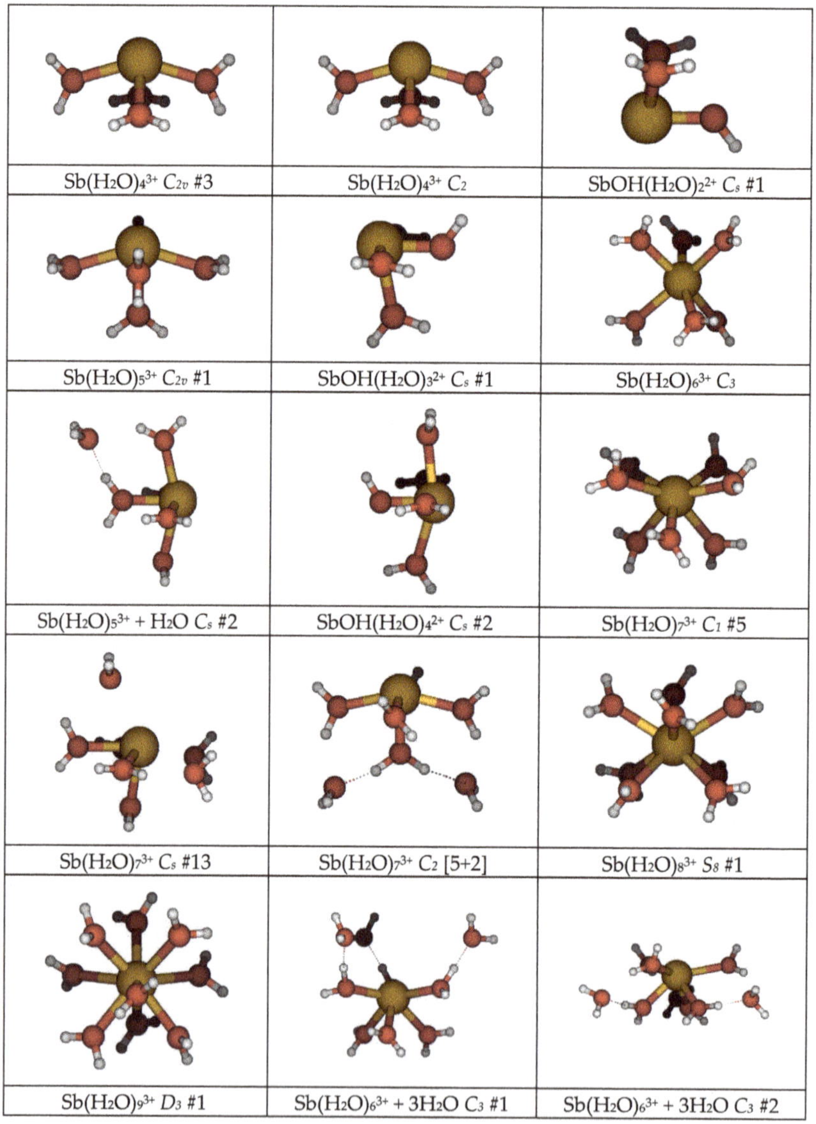

Figure 1. The minimum energy structures of aquaantimony(III) and related species.

To summarize these results, it appears that the square pyramidal pentaaquaantimony(III) ion is the most stable aqua ion. There is a strong propensity to react with second-shell water molecules to form hydroxo complexes. While stable structures with coordination numbers between six and nine do exist at some levels of theory, the corresponding structures in which some waters have moved to the second hydration shell are more stable (in some cases, with proton transfer). These higher-coordination aqua complexes might exist at high pressures. The aqua ion, if it exists, would only exist at a very low pH.

3.2. The Sb-O Distance

In Figure 2, a plot of the dependence of the average Sb-O distance as a function of the coordination number n is given for all of the levels investigated. The Sb-O distance lengthened with the increase in coordination number. With the exception of HF/LANL2MB, the results were fairly uniform (for the most part, within 0.05 Å). The HF/LANL2MB level is the only Hartree–Fock level calculation using a minimal basis set on the valence shell of all atoms, and therefore does not have enough flexibility to fully describe the bonding between antimony(III) and oxygen. The pairs HF/CEP-31G* and HF/CEP-121G*, and MP2/CEP-31G* and MP2/CEP-121G*, were nearly coincident with each other, with the latter pair giving slightly longer bond lengths. The variation in the Sb-O distance with the level of theory was smaller than that of Sn-O [4]. Based on our calculations, if the coordination number is indeed 5, then we would expect an average Sb-O distance of around 2.25 Å.

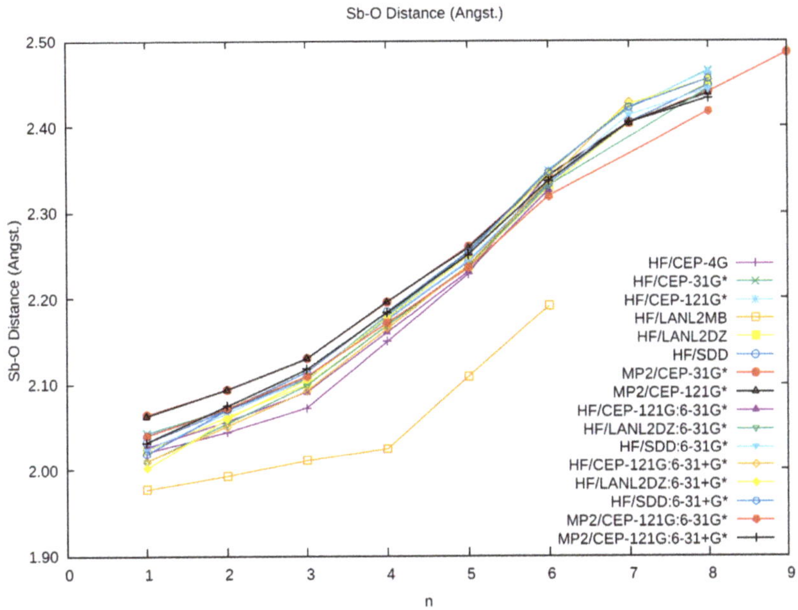

Figure 2. The dependence of the average Sb-O distance (Å) in $Sb(H_2O)_n^{3+}$ on the coordination number n and level of theory.

3.3. The Sb-O Vibrational Frequency

In Figure 3, a plot of the dependence of the frequency of the most intense Raman Sb-O stretching mode as a function of the coordination number n is given for all of the levels investigated. As expected, this frequency drops with hydration number n. The minimal basis set levels HF/LANL2MB, and to some extent, HF/CEP-4G, are different than the others, having higher values. The CEP-4G basis set is a minimal basis set on oxygen and hydrogen, but is actually using the triple-zeta valence basis set (CEP-121G) on the antimony atom. There is a levelling off of the frequency, and then a bigger drop at $n = 6$. For $n = 4$ and 5, the vibrational mode is somewhat more localized, being due predominantly to either the equatorial oxygen motion ($n = 4$) or the apical oxygen motion ($n = 5$). The HF/CEP-31G*, HF/CEP-121G*, MP2/CEP-31G*, MP2/CEP-121G*, and MP2/A+ levels form the group with the lowest vibrational frequencies. The HF calculations using the SDD basis set on Sb form a group with the highest vibrational frequencies. For the levels using split valence basis sets, if the coordination number is indeed 5, then the predictions for the Sb-O symmetric stretching motion center around 450 ± 30 cm^{-1}. In solution, the effect of the second hydration shell would increase this by an estimated 60 cm^{-1}, meaning that our best guess for the Sb-O mode in an aqueous solution would be 510 cm^{-1}.

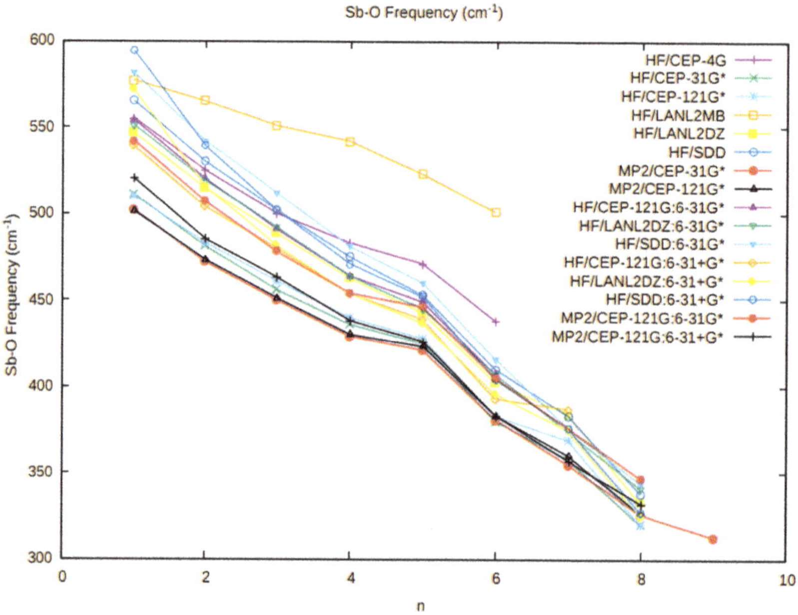

Figure 3. The dependence of the most intense Raman Sb-O frequency (cm^{-1}) in $Sb(H_2O)_n{}^{3+}$ on the coordination number n and level of theory.

4. Discussion

To the best of our knowledge, experimental data on the structure and vibrational spectra of antimony(III) aqua complexes are lacking. Our energetic comparisons suggest that the coordination number should be 5. From this, we have made predictions as to the Sb-O distance (2.25 ± 0.05 Å) and vibrational frequency (510 ± 30 cm^{-1}) that might be observed. We may compare the X-ray crystal structures involving Sb^{3+}-O^{2-} (or OH^-) interactions with the same coordination geometry. Our predictions for trigonal pyramidal (n = 3) lie at around 2.10 Å (cf. 1.977 Å in α-As_4O_6 [16], 1.977–2.023 Å in β-As_4O_6 [16], 1.942–2.067 Å in $Sb_4O_4(OH)_2(NO_3)_2$ [19], 1.969–2.027 Å in $Sb_4O_5(OH)ClO_4 \cdot 1/2H_2O$ [20], 1.994–2.207 Å in $Sb_6O_7(SO_4)_2$ [21]); for seesaw (n = 4), at around 2.17 Å (cf. 2.019–2.265 Å in $Sb_4O_4(OH)_2(NO_3)_2$ [19], 1.952–2.378 Å in $Sb_4O_5(OH)ClO_4 \cdot 1/2H_2O$ [20], 2.032–2.359 Å in $(H_3O)_2Sb_2(SO_4)_4$ [22]); and for square pyramidal (n = 5), at around 2.25 Å (cf. 2.012–2.263 Å in $(H_3O)_2Sb_2(SO_4)_4$ [22]). The separation by distance between the axial and equatorial sites in the seesaw and square pyramid is reproduced. Our predictions for the Sb^{3+}-OH_2 distances are about 0.1 Å longer than the Sb-O distances between the antimony(III) and either the O^{2-} or OH^- moieties in the crystal structures, in which there should be stronger attractive forces.

5. Conclusions

The common CEP, LANL2, and SDD pseudopotentials were paired with various basis sets to study the hydrated antimony(III) ion. Calculations using minimal basis sets (HF/LANL2MB, HF/CEP-4G) performed poorly. The calculated structures of the aqua complexes compared favorably with the crystal structures of the oxides/hydroxides. Symmetry can be used to guide the search for new structures and to rule out structures. For the smaller coordination numbers, the effect of the ns^2 subshell is clearly to form a hemidirected structure in which the electrons act similarly as a ligand or lone pair would.

The factors giving rise to either holodirected or hemidirected structures were thoroughly discussed by Shimony-Livny et al. for Pb(II) and do not need to be repeated here [29]. There is a propensity for the aqua complexes of smaller coordination numbers of type [n + 1] to hydrolyze to form the aquahydroxoantimony(III) complex and a hydronium ion, which is evidence for the strong Lewis acidity of antimony(III). We predict that antimony(III) might exist in extremely acidic solutions with a coordination number of 5, an average Sb-O distance of 2.25 Å, and with a Raman-active Sb-O totally symmetric stretching motion occurring at around 510 cm^{-1}.

Supplementary Materials: The following supporting information can be downloaded at: https://www.mdpi.com/article/10.3390/liquids4020016/s1; Figure S1: Some stationary point (non-minimum) energy structures of aquaantimony(III); Table S1: Total Energies of Aquaantimony(III) Species.

Author Contributions: Conceptualization, C.C.P.; methodology, C.C.P.; validation, C.C.P.; formal analysis, C.C.P.; investigation, C.C.P. and C.M.G.; data curation, C.C.P.; writing—original draft preparation, C.C.P.; writing—review and editing, C.C.P.; visualization, C.C.P. and C.M.G.; supervision, C.C.P.; project administration, C.C.P.; funding acquisition, C.C.P. All authors have read and agreed to the published version of the manuscript.

Funding: This research was funded by the Natural Sciences and Engineering Research Council of Canada and the Department of Economic Development, Government of Nova Scotia.

Data Availability Statement: The data presented in this study are available in the Supplementary Materials.

Acknowledgments: The authors thank the Department of Astronomy and Physics, Saint Mary's University (AP-SMU), for providing access to computing facilities, in particular, to cygnus, a 10-processor Sun server, purchased with assistance from the Canada Foundation for Innovation, Sun Microsystems, the Atlantic Canada Opportunities Agency, and SMU. C.C.P. also thanks ACEnet and Compute Canada for providing access to computers and Gaussian 03/16. C.C.P. acknowledges the former financial support of NSERC. C.M.G. acknowledges the support of the Department of Economic Development (Government of Nova Scotia), the Office of the Dean of Science, SMU, and the Cooperative Education program (SMU—Work Term 2, Fall 2001 and Work Term 3, Winter 2002).

Conflicts of Interest: The authors declare no conflicts of interest. The funders had no role in the design of the study; in the collection, analyses, or interpretation of data; in the writing of the manuscript; or in the decision to publish the results.

References

1. Richens, D.T. *The Chemistry of Aqua Ions*; Wiley: Chichester, UK, 1997.
2. Pye, C.C.; Gunasekara, C.M. An Ab Initio Investigation of the Hydration of Thallium(III) and Mercury(II). *J. Solut. Chem.* **2020**, *49*, 1419–1429. [CrossRef]
3. Pye, C.C.; Gunasekara, C.M. An Ab Initio Investigation of the Hydration of Lead(II). *Liquids* **2022**, *2*, 39–49. [CrossRef]
4. Pye, C.C.; Gunasekara, C.M. An Ab Initio Investigation of the Hydration of Tin(II). *Liquids* **2022**, *2*, 465–473. [CrossRef]
5. Gayer, K.H.; Garrett, A.B. The Equilibria of Antimonous Oxide (Rhombic) in Dilute Solutions of Hydrochloric Acid and Sodium Hydroxide at 25°. *J. Am. Chem. Soc.* **1952**, *74*, 2353–2354. [CrossRef]
6. Mishra, S.K.; Gupta, Y.K. Spectrophotometric Study of the Hydrolytic Equilibrium of Sb(III) in Aqueous Perchloric Acid Solution. *Ind. J. Chem.* **1968**, *6*, 757–758.
7. Baes, C.F., Jr.; Mesmer, R.E. *The Hydrolysis of Cations*; Wiley: New York, NY, USA, 1976.
8. Ahrland, S.; Bovin, J.-O. The Complex Formation of Antimony(III) in Perchloric Acid and Nitric Acid Solutions. A Solubility Study. *Acta Chem. Scand. A* **1974**, *28*, 1089–1100. [CrossRef]
9. Zakaznova-Herzog, V.P.; Seward, T.M. Antimonous acid protonation/deprotonation equilibria in hydrothermal solutions to 300 °C. *Geochim. Cosmochim. Acta* **2006**, *70*, 2298–2310. [CrossRef]
10. Natta, G.; Baccaredda, M. Composti chimici interstiziali. Struttura del pentossido di antimonio idrato e di alcuni antimoniati. [Interstitial chemical compounds. Structure of hydrated antimony pentoxide and some antimoniates]. *Gazz. Chim. Ital.* **1936**, *66*, 308–316. (In Italian)
11. Dihlstrom, K.; Westgren, A. Uber den Bau des sogenannten Antimontetroxyds und der damit isomorphen Verbindung $BiTa_2O_6F$. [About the construction of the so-called antimony tetroxide and the isomorphic compound $BiTa_2O_6F$]. *Z. Anorg. Allg. Chem.* **1937**, *235*, 153–160. (In German) [CrossRef]
12. England, W.A.; Cross, M.G.; Hamnett, A.; Wiseman, P.J.; Goodenough, J.B. Fast Proton Conduction in Inorganic Ion-Exchange Compounds. *Solid State Ion.* **1980**, *1*, 231–249. [CrossRef]

13. Watelet, H.; Picard, J.-P.; Baud, G.; Besse, J.-P. Chevalier, R. Un nouveau conducteur protonique, $[H(H_2O)_n]_{12}Sb_{12}O_{36}$ ($n \leq 1$). [A new protonic conductor, $[H(H_2O)_n]_{12}Sb_{12}O_{36}$ ($n \leq 1$).]. *Mater. Res. Bull.* **1981**, *16*, 1131–1137. (In French) [CrossRef]
14. Jager, G.; Jones, P.G.; Sheldrick, G.M.; Schwarzmann, F. Darstellung und Kristalstruktur von Antimonen(V)oxidhydroxid HSb_3O_8. [Representation and Crystal Structure of Antimony(V) Oxide Hydroxide HSb_3O_8.]. *Z. Naturforschung B* **1983**, *38*, 698–701. (In German) [CrossRef]
15. Riviere, M.; Fourquet, J.L.; Grins, J.; Nygren, M. The cubic pyrochlores $H_{2x}Sb_{2x}W_{2\text{-}2x}O_6{\cdot}nH_2O$: Structural, thermal, and electrical properties. *Mater. Res. Bull.* **1988**, *23*, 965–975. [CrossRef]
16. Svensson, C. Refinement of the Crystal Structure of Cubic Antimony Trioxide, Sb_2O_3. *Acta Cryst. B* **1975**, *31*, 2016–2018. [CrossRef]
17. Svensson, C. The Crystal Structure of Orthorhombic Antimony Trioxide, Sb_2O_3. *Acta Cryst. B* **1974**, *30*, 458–461. [CrossRef]
18. Orosel, D.; Dinnebier, R.E.; Blatov, V.A.; Jansen, M. Structure of a new high-pressure-high-temperature modification of antimony(III) oxide, γ-Sb_2O_3, from high-resolution synchrotron powder diffraction data. *Acta Crystallogr. B* **2012**, *68*, 1–7. [CrossRef] [PubMed]
19. Bovin, J.P. The crystal structure of $Sb_4O_4(OH)_2(NO_3)_2$. *Acta Chem. Scand. A* **1974**, *28*, 267–274. [CrossRef]
20. Bovin, J.P. The crystal structure of $Sb_4O_5(OH)ClO_4{\cdot}1/2\ H_2O$. *Acta Chem. Scand. A* **1974**, *28*, 723–730. [CrossRef]
21. Bovin, J.P. The Crystal Structure of the Antimony(III) Oxide Sulphate $Sb_6O_7(SO_4)_2$. *Acta Cryst. B* **1976**, *32*, 1771–1777. [CrossRef]
22. Mercier, R.; Douglade, J.; Jones, P.G.; Sheldrick, G.M. Structure of an Oxonium Antimony(III) Sulphate, $(H_3O)_2Sb_2(SO_4)_4$. *Acta Cryst. C* **1983**, *39*, 145–148. [CrossRef]
23. Frisch, M.J.; Trucks, G.W.; Schlegel, H.B.; Scuseria, G.E.; Robb, M.A.; Cheeseman, J.R.; Zakrzewski, V.G.; Montgomery, J.A., Jr.; Stratmann, R.E.; Burant, J.C.; et al. *Gaussian 98, Revision A.9*; Gaussian, Inc.: Pittsburgh, PA, USA, 1998.
24. Frisch, M.J.; Trucks, G.W.; Schlegel, H.B.; Scuseria, G.E.; Robb, M.A.; Cheeseman, J.R.; Montgomery, J.A., Jr.; Vreven, T.; Kudin, K.N.; Burant, J.C.; et al. *Gaussian 03, Revision D.02*; Gaussian, Inc.: Wallingford, CT, USA, 2004.
25. Frisch, M.J.; Trucks, G.W.; Schlegel, H.B.; Scuseria, G.E.; Robb, M.A.; Cheeseman, J.R.; Scalmani, G.; Barone, V.; Petersson, G.A.; Nakatsuji, H.; et al. *Gaussian 16, Revision A.03*; Gaussian, Inc.: Wallingford, CT, USA, 2016.
26. Pye, C.C.; Gunasekara, C.M.; Rudolph, W.W. An ab initio investigation of bismuth hydration. *Can. J. Chem.* **2007**, *85*, 945–950. [CrossRef]
27. Pye, C.C.; Whynot, D.C.M.; Corbeil, C.R.; Mercer, D.J. Desymmetrization in geometry optimization: Application to an ab initio study of copper(I) hydration. *Pure Appl. Chem.* **2020**, *92*, 1643–1654. [CrossRef]
28. Pye, C.C.; Gilbert, C.R. An ab initio investigation of the second hydration shell of metal cations. *Comput. Appl. Chem.* **2020**, *37*, 2020. [CrossRef]
29. Shimony-Livny, L.; Glusker, J.P.; Bock, C.W. Lone Pair Functionality in Divalent Lead Compounds. *Inorg. Chem.* **1998**, *37*, 1853–1867. [CrossRef]

liquids MDPI

Article

Ab Initio Investigation of the Hydration of the Tetrahedral d^0 Transition Metal Oxoanions NbO_4^{3-}, TaO_4^{3-}, CrO_4^{2-}, MoO_4^{2-}, WO_4^{2-}, MnO_4^{-}, TcO_4^{-}, ReO_4^{-}, and of FeO_4, RuO_4, and OsO_4

Barbara L. Goodall, Jane P. Ferguson and Cory C. Pye *

Department of Chemistry, Saint Mary's University, Halifax, NS B3H 3C3, Canada
* Correspondence: cory.pye@smu.ca

Abstract: The geometries and vibrational frequencies of various configurations of $XO_4^{m-}(H_2O)_n$, X = Fe, Ru, Os, m = 0; X = Mn, Tc, Re, m = 1; X = Cr, Mo, W, m = 2; and X = Nb, Ta, m = 3; n = 0–6 are calculated at various levels up to MP2/6-31+G* and B3LYP/6-31+G*. These properties are studied as a function of increasing cluster size. The experimental and theoretical bond distances and vibrational spectra are compared where available, and predictions are made where they are not.

Keywords: ab initio; density functional; vibrational spectra; tetrahedral oxoanions; hydration

1. Introduction

The hydration of ions is crucial to understanding the properties of electrolyte solutions. Solution diffraction data using either X-rays or neutrons has established that water molecules arrange themselves in hydration shells around ions [1]. For metal cations, the existence and nature of an aqua ion can often be proven by Raman spectroscopy and/or ab initio calculations [2–7]. For oxoanions, the influence of the first hydration shell of water molecules is much smaller. One of the authors has examined the effect of the first hydration sphere on the main-group tetraoxo anions [8], some of their protonated forms [9–14], as well as several borate species [15–18] by ab initio computational methods. Therein, we have demonstrated that ab initio modeling using restricted Hartree-Fock (HF) theory with modest basis sets gives reasonable structural and vibrational properties, even if a full hydration sphere or an implicit solvation model is not employed. In this paper, we present our studies of naked and explicitly hydrated niobate, tantalate, chromate, molybdate, tungstate, permanganate, pertechnetate, perrhenate; and of iron, ruthenium, and osmium tetroxide, including optimization and frequency calculation up to the MP2/6-31+G* level and with up to twelve water molecules [19].

2. Materials and Methods

Calculations were carried out using Gaussian 03 [20], using the standard 6-31G* and 6-31+G* basis sets in conjunction with the standard HF, MP2, and B3LYP levels of theory. For the atoms of the second and third transition metal series, the standard SDD effective core potential and associated basis set was used in conjunction with the 6-31G(d) and 6-31+G(d) basis sets (O,H). The second-order Moller-Plesset (MP2) calculations use the frozen core approximation. A stepping stone approach was used, where the geometries and molecular orbital coefficients at the levels HF/6-31G*, HF/6-31+G*, MP2/6-31G*, MP2/6-31+G*, B3LYP/6-31G*, and B3LYP/6-31+G* were sequentially optimized (geom = allcheck guess = read). Default optimization specifications were normally used. After each level, a frequency calculation was performed at the same level, and the resulting force constants were used in the following optimization. Z-matrix coordinates constrained to the appropriate symmetry were used to speed up the optimizations and simplify the assignment of vibrational modes (FOpt = z-matrix, ReadFC). The force constants were evaluated at the first geometry as well (FOpt = CalcFC). The quadratic convergence method was applied automatically if the SCF failed to converge (SCF = XQC). Additional options were specified individually or in combination, as needed, to converge the geometry and energy (SCF: NoDIIS and/or IntRep and/or CDIIS; FOpt = CalcAll and/or GDIIS). Additional calculations with Gaussian 16 [21] were carried out to explore the BLYP and PBE functionals, often used for ab initio MD simulations and the effect of an implicit solvation model (CPCM).

3. Results

The XO_4^{m-} ion, or molecule, of T_d symmetry has nine modes of internal vibration spanning the vibrational representation $\Gamma_{vib} = A_1 + E + 2T_2$. All modes are Raman active, whereas only the T_2 modes are IR active. The structures are analogous to those reported previously for perchlorate [8]. In the following subsections, we review the geometries and vibrational frequencies of each molecule or ion, followed by our results.

3.1. Permanganate

The crystal structures of permanganate salts have been examined for several decades. For potassium permanganate (*Pnma*), Mooney determined the average Mn-O distance to be 1.59(9) Å [22]. Refinements include those of Ramaseshan et al. (1.55(1) Å) [23], Palenik (1.629(8) Å) [24], Hoppe (1.60(1)) [25], and Marabello et al. (1.615(1) Å) [26]. Other anhydrous alkali permanganate structure determinations include those of Fischer et al. (Li: neutron diffraction, 1.623(17) Å) [27], Bauchert et al. (Na: 1.610(13) Å) [28], and Hoppe et al. (Rb: 1.603(5) Å; Cs: 1.604(3) Å) [25]. Lithium permanganate can also exist as the trihydrate (1.615(8) Å) [29,30]. The more modern determinations squarely place the Mn-O distance in the range 1.60–1.63 Å.

The resonance Raman spectrum of $KMnO_4$(aq) was measured with the 514.5 and 488 nm argon laser lines by Kiefer and Bernstein [31,32]. The Mn-O stretching bands at 837.5 (ν_1-A_1) and 906.5 cm^{-1} (ν_3-T_2), as well as several overtones $n\nu_1$ and combinations $n\nu_1 + \nu_3$, were observed in light and heavy water. Weinstock, Schulze, and Muller observed, using a He-Ne laser at 633 nm, the normal Raman spectrum of $KMnO_4$(aq) [33] and found bands at 839 ± 1 (ν_1-A_1), 914 ± 5 (ν_3-T_2), 360 ± 5 (ν_2-E), and 430 ± 5 (ν_4-T_2) cm^{-1}. In an argon matrix, the ν_3 mode appears at 896.9 cm^{-1} [34].

The calculated bond lengths and vibrational frequencies for MnO_4^- are given in Table 1. The Hartree-Fock distances are too short compared to the experiment, but the MP2 and especially the DFT distances are in much better agreement. The Hartree-Fock frequencies are overestimated, which is a well-known problem with the theory and can be corrected by an empirical scaling factor. The DFT frequencies are reasonably close to the experiment, with the deformation modes being quite close, although the stretching frequencies are overestimated by up to 100 cm^{-1}. The MP2 stretching frequencies, on the other hand, are nearly twice the experimental values, and this must be regarded as an abysmal failure of the MP2 method. The difficulty that permanganate can present to computational chemistry, especially with regard to electronic transitions, is well known [35–47]. The CPCM solvation model gives rise to slightly smaller Mn-O distances and T_2 frequencies.

The effect of water upon the Mn-O distances (MP2/6-31+G(d)) in $MnO_4^- \bullet nH_2O$ (n = 0–6) is similar to that of the Cl-O distances in the analogous perchlorate [8] (Figure 1). However, the net effect is a very slight lengthening of the Mn-O distance by 0.0007 Å (n = 6), compared with the shortening observed in perchlorate. However, the individual Mn-O distances can vary by up to 0.05 Å (in perchlorate, the variation was only 0.02 Å). This might be reflected in larger bandwidths in the vibrational spectra. Although the Mn-O distances are approximately 0.1 Å longer than the corresponding Cl-O distances, the Mn…O distances are actually about 0.1 Å shorter than the corresponding Cl…O distances (Figure 2). The O…O distances are about 0.2 Å shorter (Figure 3) than the corresponding distances in perchlorate, suggesting that the hydrogen bonding is stronger in permanganate than in perchlorate. This is also seen in the O…H distances (Figure 4). Unlike perchlorate, there is only a slight increase of the hydrogen bonding indicators (O…O, O…H) on the number of water molecules, but the ranges are much larger. This also indicates that the permanganate ion has an unusual solvation dependence. In an aqueous solution, the Mn-O distance is 1.630(5) Å by LAXS, and the Mn…O distance is 4.095(8) Å [48]. While the Mn-O distance is well-reproduced by the hexahydrate calculations, the Mn…O distance is too short because the double-donor water molecules are too close to the central metal. This is rectified in the dodecahydrate model, in which the Mn…O distance lies in the range 3.83–4.09 Å, depending on the level of theory.

Table 1. Bond lengths (Å) and vibrational frequencies (unscaled, cm^{-1}) of permanganate, MnO_4^-.

	d(Mn-O)	ν_1 (A_1)	ν_2 (E)	ν_3 (T_2)	ν_4 (T_2)
HF/6-31G(d)	1.5399	1082	438	1039	490
HF/6-31+G(d)	1.5449	1068	426	1011	480
MP2/6-31G(d)	1.5797	1695	438	1761	446
MP2/6-31+G(d)	1.5864	1694	438	1710	447
B3LYP/6-31G(d)	1.5947	957	379	1018	426
B3LYP/6-31+G(d)	1.6018	936	368	986	417
BLYP/6-31G(d)	1.6232	888	354	959	401
BLYP/6-31+G(d)	1.6315	865	344	927	392
PBE/6-31G(d)	1.6116	909	360	982	406
PBE/6-31+G(d)	1.6194	886	350	949	396
CPCM-HF/6-31G(d)	1.5386	1084	437	1008	480
CPCM-HF/6-31+G(d)	1.5438	1070	426	972	469
CPCM-MP2/6-31G(d)	1.5780	1691	439	1644	454
CPCM-MP2/6-31+G(d)	1.5847	1688	438	1562	453
CPCM-B3LYP/6-31G(d)	1.5927	962	377	1006	419
CPCM-B3LYP/6-31+G(d)	1.5998	941	366	967	409
CPCM-BLYP/6-31G(d)	1.6210	890	352	952	392
CPCM-BLYP/6-31+G(d)	1.6294	866	342	914	382
CPCM-PBE/6-31G(d)	1.6094	912	358	975	398
CPCM-PBE/6-31+G(d)	1.6173	889	348	936	389

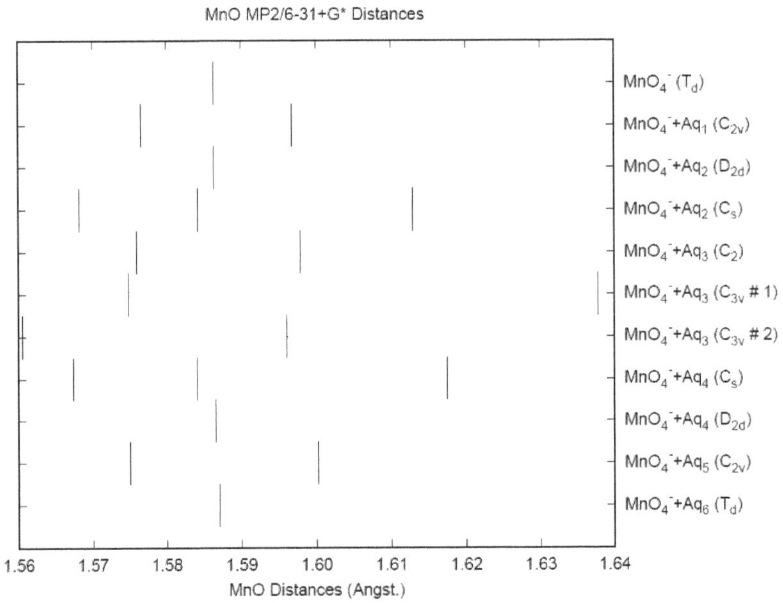

Figure 1. The Mn-O distances (MP2/6-31+G*) in hydrated permanganate.

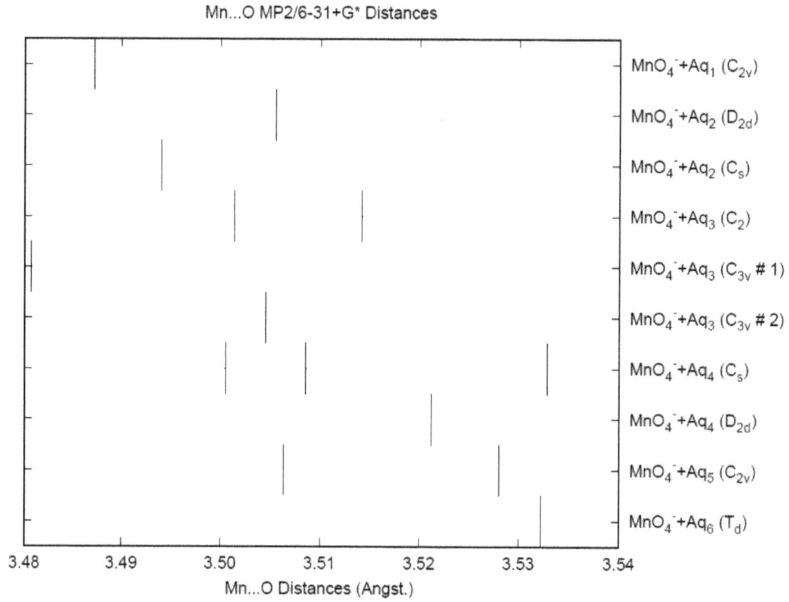

Figure 2. The Mn...O distances (MP2/6-31+G*) in hydrated permanganate.

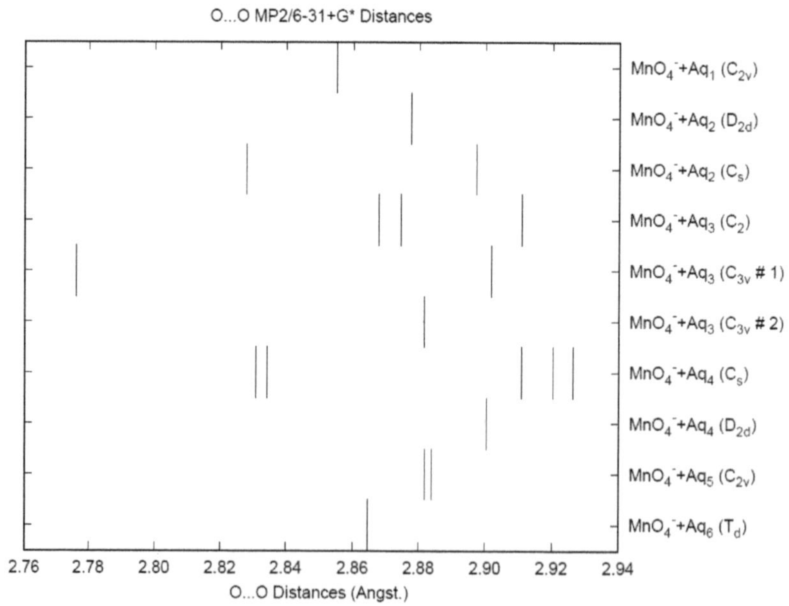

Figure 3. The O...O distances (MP2/6-31+G*) in hydrated permanganate.

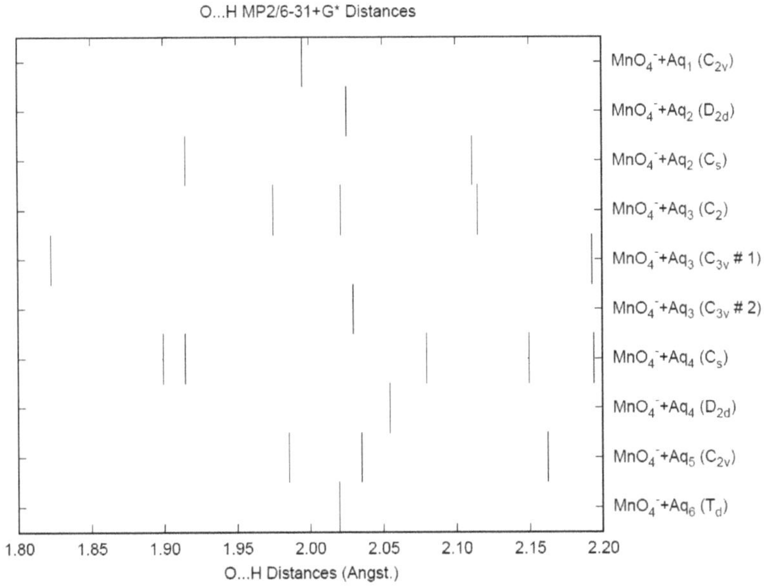

Figure 4. The O...H distance (MP2/6-31+G*) in hydrated permanganate.

The vibrational frequencies (HF/6-31+G*) as a function of hydration number are shown in Figure 5. The deformation frequencies are lower than in perchlorate and are also much closer together. In addition, the HF/6-31+G* level predicts that the asymmetric stretching mode is lower than the symmetric stretching mode for permanganate, whereas the opposite is true in the experiment and also for both the calculated and experimental perchlorate spectrum [8].

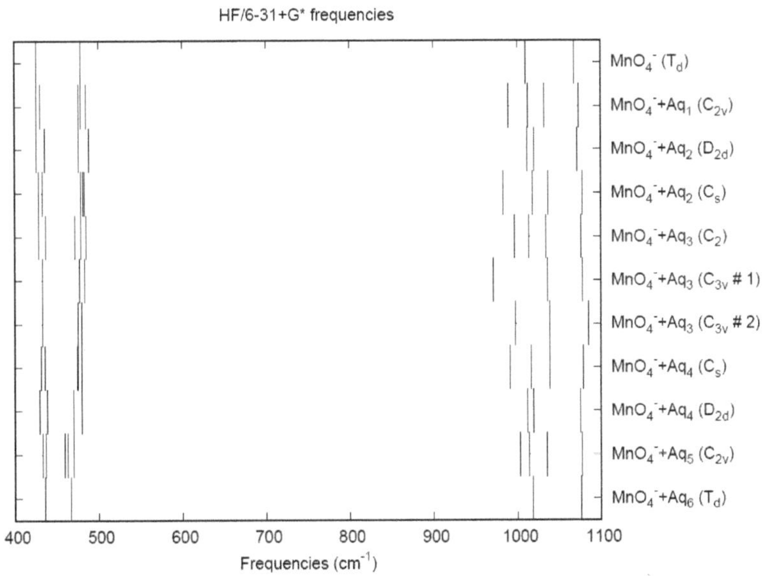

Figure 5. Vibrational frequencies (HF/6-31+G*) of hydrated permanganate.

3.2. Pertechnetate

The determination of the properties of pertechnetate salts requires special care because of the radioactivity of technetium. The crystal structure of potassium pertechnetate gives a Tc-O distance of 1.711(3) Å (1.724 Å corrected) [49]. Other anhydrous pertechnetate crystal structure determinations include those of German et al. (Na: 100 K, 1.7208(3) Å; 296 K, 1.7183(6) Å) [50], Weaver et al. (neutron diffraction: Na, 1.737(4) Å; K, 1.739(14) Å; Rb, 1.723(3) Å; Cs, 1.67(3) Å) [51], and Meyer et al. (Cs: 1.715(50) Å) [52]. The lithium salt crystallizes as the trihydrate (1.717(7) Å) [53]. The average Tc-O distances fall mostly in the high end of the range 1.67–1.74 Å.

Busey and Keller [54] observed Raman bands at 912 and 325 cm^{-1} for 2 mol/L NH_4TcO_4(aq). Weinstock, Schulze, and Muller observed the normal Raman spectrum of $KTcO_4$ (aq) [33] and found bands at 912 ± 1 (ν_1-A_1), 912 ± 4 (ν_3-T_2), 325 ± 2 (ν_2-E), and 336 ± 10 (ν_4-T_2) cm^{-1}. An infrared band was found at 334 ± 10 (ν_4-T_2) cm^{-1}.

The calculated bond lengths and vibrational frequencies for TcO_4^- are given in Table 2. The Hartree-Fock distances are too short compared to the experiment, but the MP2 distances are too long. The DFT distances are in much better agreement. In accordance with the inverse relationship between distance and vibrational frequency, the Hartree-Fock frequencies are overestimated, and the MP2 frequencies are underestimated. In addition, HF places the ν_1-A_1 band above the ν_3-T_2 band by 50–70 cm^{-1}, whereas MP2 reverses this order by 150 cm^{-1}. Only the DFT frequencies are reasonably close to the experiment, predicting the near degeneracy of the two modes. The CPCM solvation model gives rise to slightly larger Tc-O distances and smaller frequencies.

Table 2. Bond lengths (Å) and vibrational frequencies (unscaled, cm^{-1}) of pertechnetate, TcO_4^-.

	d(Tc-O)	ν_1 (A_1)	ν_2 (E)	ν_3 (T_2)	ν_4 (T_2)
HF/6-31G(d)	1.6822	1077	376	1021	400
HF/6-31+G(d)	1.6835	1072	377	1003	396
MP2/6-31G(d)	1.7847	824	291	978	304
MP2/6-31+G(d)	1.7902	813	295	962	304
B3LYP/6-31G(d)	1.7305	935	322	935	342
B3LYP/6-31+G(d)	1.7333	927	325	917	340
BLYP/6-31G(d)	1.7558	875	304	886	325
BLYP/6-31+G(d)	1.7593	865	308	867	324
PBE/6-31G(d)	1.7442	895	310	904	329
PBE/6-31+G(d)	1.7470	886	313	887	328
CPCM-HF/6-31G(d)	1.6831	1068	370	981	345
CPCM-HF/6-31+G(d)	1.6845	1062	371	955	339
CPCM-MP2/6-31G(d)	1.7863	833	291	964	289
CPCM-MP2/6-31+G(d)	1.7929	823	297	941	290
CPCM-B3LYP/6-31G(d)	1.7315	912	284	910	297
CPCM-B3LYP/6-31+G(d)	1.7346	900	280	885	290
CPCM-BLYP/6-31G(d)	1.7575	858	274	869	285
CPCM-BLYP/6-31+G(d)	1.7615	846	276	845	281
CPCM-PBE/6-31G(d)	1.7451	877	277	885	288
CPCM-PBE/6-31+G(d)	1.7484	865	277	863	283

The effect of water upon the Tc-O distances (MP2/6-31+G(d)) in $TcO_4^-\bullet nH_2O$ (n = 0–6) is similar to that of the Cl-O distances in the analogous perchlorate (Figure 6) [8]. The net effect is a slight shortening of the Tc-O distance by 0.005 Å (n = 6), unlike permanganate. The individual Tc-O distances vary by up to 0.01 Å (in perchlorate, the variation was larger at 0.02 Å). The Tc-O distances are approximately 0.3 Å longer than the corresponding Cl-O distances, although the Tc...O distances are about the same as the corresponding Cl...O distances (Figure S1). The O...O (Figure S2) and O...H (Figure S3) distances are about the same as in perchlorate.

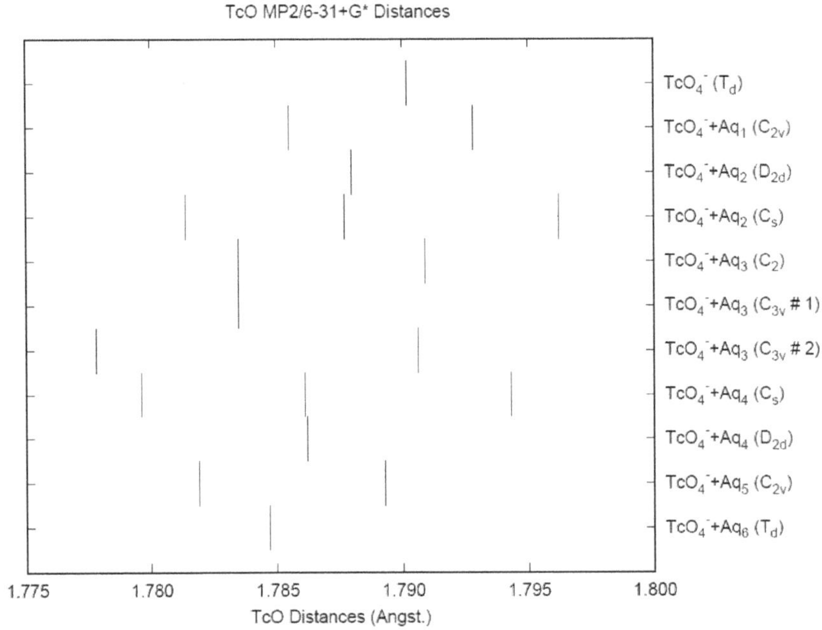

Figure 6. The Tc-O distances (MP2/6-31+G*) in hydrated pertechnetate.

The vibrational frequencies (HF/6-31+G*) as a function of hydration number are shown in Figure 7. The deformation frequencies are lower than in perchlorate and are also much closer together. In addition, the HF/6-31+G* level predicts that the asymmetric stretching mode is lower than the symmetric stretching mode for pertechnetate, whereas experimentally, the modes are degenerate. The effect of hydration is similar to that in perchlorate.

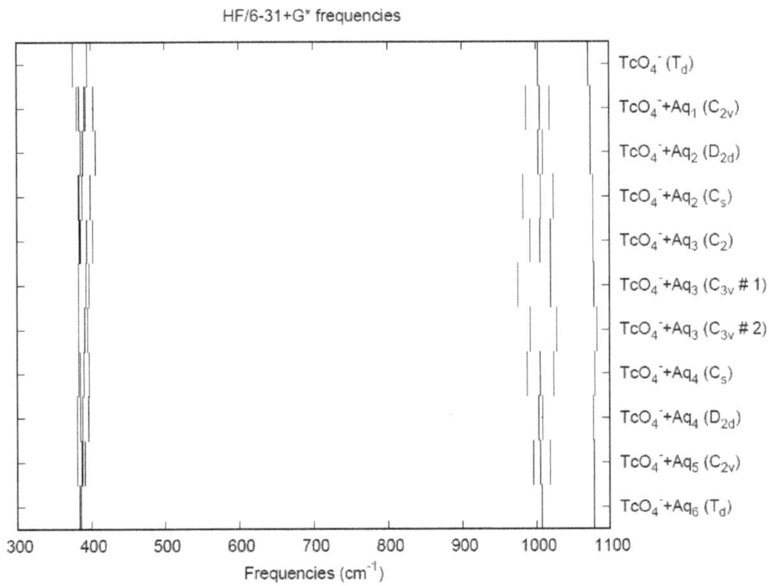

Figure 7. Vibrational frequencies (HF/6-31+G*) of hydrated pertechnetate.

3.3. Perrhenate

The crystal structure of potassium perrhenate was determined by Morrow [55] to give Re-O distances of 1.77(3) Å. The structure was further refined by Lock and Turner [56] (Re-O = 1.723(4) Å), Krebs and Hasse [49] (Re-O = 1.719(5) Å), and Brown, Powell, and Stuart [57] (neutron diffraction, Re-O = 1.739(1) Å, 15 K; 1.736(2), 150 K; 1.731(2) Å, 298 K). Other anhydrous perrhenate determinations include those of sodium perrhenate [58], (Re-O = 1.728(2) Å), rubidium perrhenate [59,60] (Re-O = 1.722(6) Å, 297 K; 1.729(2) Å, 159 K), two forms of cesium perrhenate [61–63] (*Pnma*, Re-O = 1.714(4) Å, 297 K; $I4_1/amd$, 1.683(7) Å, 468 K), and lithium perrhenate [64] (Re-O = 1.715(26) Å). In addition, the crystal structure of some hydrated lithium perrhenate forms are known: $LiReO_4{\bullet}H_2O$ [65] (Re-O = 1.720(12) Å) and $LiReO_4{\bullet}1.5H_2O$ [66] (Re-O = 1.72(1) Å). The more modern determinations squarely place the Re-O distance in the range of 1.683–1.739 Å. A comparison with the nearly identical Tc-O distances is evidence for the well-known lanthanide contraction. Weinstock, Schulze, and Muller observed the normal Raman spectrum of $KReO_4$(aq) [33] and found bands at 971 ± 1 (ν_1-A_1), 920 ± 4 (ν_3-T_2), and 331 ± 2 (ν_2-E) cm^{-1}. An infrared band was found at 322 ± 10 (ν_4-T_2) cm^{-1}.

The calculated bond lengths and vibrational frequencies for ReO_4^- are given in Table 3. The Hartree-Fock distances are slightly too short compared to the experiment, but the MP2 distances are too long. The B3LYP distances are slightly too long. In accordance with the inverse relationship between distance and vibrational frequency, the Hartree-Fock frequencies are overestimated, and the MP2 frequencies are underestimated. In addition, HF places the ν_1-A_1 band above the ν_3-T_2 band by 90–100 cm^{-1}, whereas MP2 reverses this order by 50–60 cm^{-1}. Only the B3LYP frequencies are reasonably close to the experiment. The CPCM solvation model gives rise to slightly larger Re-O distances and usually smaller frequencies.

Table 3. Bond lengths (Å) and vibrational frequencies (unscaled, cm^{-1}) of perrhenate, ReO_4^-.

	d(Re-O)	ν_1 (A_1)	ν_2 (E)	ν_3 (T_2)	ν_4 (T_2)
HF/6-31G(d)	1.7089	1098	375	1009	374
HF/6-31+G(d)	1.7101	1094	275	993	372
MP2/6-31G(d)	1.7885	844	304	906	297
MP2/6-31+G(d)	1.7927	833	308	887	299
B3LYP/6-31G(d)	1.7511	963	325	922	321
B3LYP/6-31+G(d)	1.7535	956	327	905	321
BLYP/6-31G(d)	1.7740	905	308	875	306
BLYP/6-31+G(d)	1.7770	896	312	858	306
PBE/6-31G(d)	1.7635	924	314	892	309
PBE/6-31+G(d)	1.7659	916	317	876	309
CPCM-HF/6-31G(d)	1.7117	1057	322	964	284
CPCM-HF/6-31+G(d)	1.7132	1049	314	940	276
CPCM-MP2/6-31G(d)	1.7890	852	299	907	264
CPCM-MP2/6-31+G(d)	1.7939	841	306	884	268
CPCM-B3LYP/6-31G(d)	1.7548	933	277	899	260
CPCM-B3LYP/6-31+G(d)	1.7578	923	277	876	256
CPCM-BLYP/6-31G(d)	1.7780	889	284	864	257
CPCM-BLYP/6-31+G(d)	1.7817	879	288	843	256
CPCM-PBE/6-31G(d)	1.7667	904	282	878	256
CPCM-PBE/6-31+G(d)	1.7697	894	284	857	254

The effect of water upon the Re-O distances (MP2/6-31+G(d)) in $ReO_4^-{\bullet}nH_2O$ (n = 0–6) is similar to that of the Tc-O distances in the analogous pertechnetate (Figure S4). The net effect is a slight shortening of the Re-O distance by 0.005 Å (n = 6). The individual Re-O distances vary by up to 0.01 Å (in perchlorate, the variation was larger at 0.02 Å). The Re...O, O...O, and O...H distances are about the same as the corresponding distances in the hydrated pertechnetate (Figures S5–S7). The vibrational frequencies (HF/6-31+G*) as a function of hydration number are shown in Figure 8. The deformation frequencies are lower than in perchlorate and are also much closer together. In addition, the HF/6-31+G* level correctly predicts that the asymmetric stretching mode is lower than the symmetric stretching mode for perrhenate. The effect of hydration is similar to that in perchlorate. In an aqueous solution, the Re-O distance is 1.735(2) Å by LAXS, and the Re…O distance is 4.197(7) Å [48]. While the Re-O distance is well-reproduced by the calculations, the Re…O distance is too short in the hexahydrate. This is mostly rectified in the dodecahydrate model, in which the Re…O distance lies in the range 3.87–4.08 Å, depending on the level of theory.

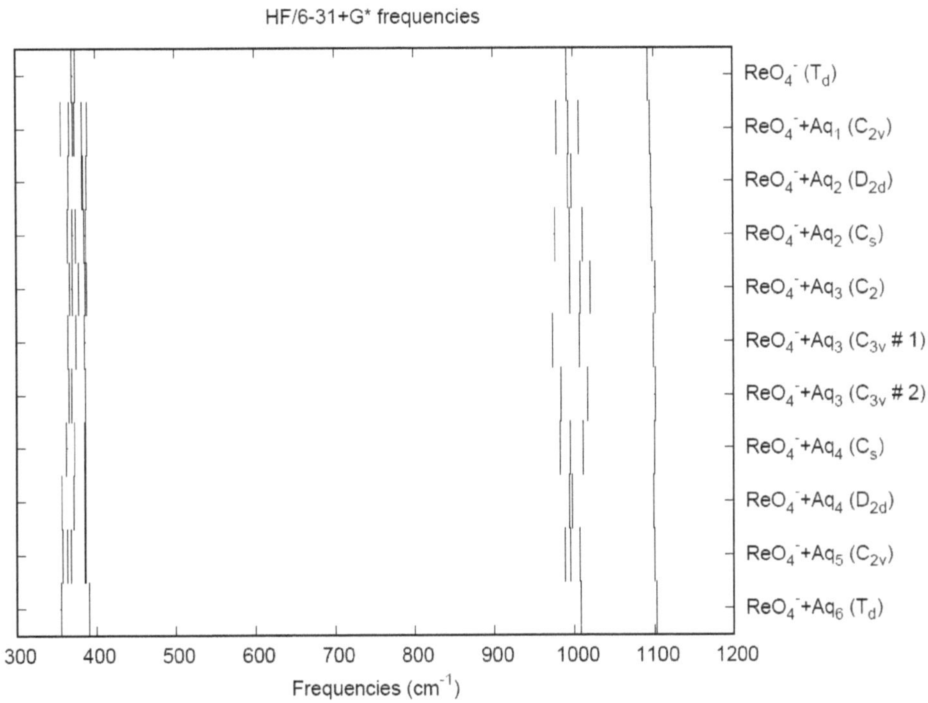

Figure 8. Vibrational frequencies (HF/6-31+G*) of hydrated perrhenate.

3.4. *Iron, Ruthenium, and Osmium Tetroxide*

Iron(VIII) oxide remains unknown. Ruthenium(VIII) oxide occurs in at least two modifications: a cubic form [67] and a monoclinic form [67,68]. The Ru-O distances are 1.696(1) and 1.699(2) Å, respectively. Osmium(VIII) oxide has been studied by both X-ray [69] and (gas phase) electron [70] diffraction. The distances obtained were 1.74(2) and 1.711(3) Å, respectively. The vibrational spectra of these two molecules have been measured by several authors (Table 4). Both molecules contain a tetrahedral MO_4 moiety in the gas and solution phase, consistent with the crystal structure of the solids.

The calculated bond lengths and vibrational frequencies for FeO_4, RuO_4, and OsO_4 are given in Table 5. In all cases, the DFT distances are greater than the HF distances. For RuO_4 and OsO_4, the MP2 distances are longer still, whereas for the hypothetical FeO_4, the MP2 distance is slightly shorter than HF. The DFT levels give the best agreement with the M-O distance and vibrational frequencies for RuO_4 and OsO_4. The MP2 vibrational frequencies are poor in all cases. The CPCM solvation model gives rise to slightly larger M-O distances and usually smaller frequencies.

Table 4. Vibrational Frequencies (cm^{-1}) of RuO_4 and OsO_4.

ν_1 (A_1)	ν_2 (E)	ν_3 (T_2)	ν_4 (T_2)	Type	Ref.
		RuO_4			
		920		IR, gas	[71]
(880) [1]		914		IR, CCl_4 sol.	[71]
(879)		913	330	IR, liquid	[72]
882	323	914	334	R, liquid	[73]
883	318	921	332	R, sat. aq.	[73]
881	obscured	916	obscured	R, CCl_4 sol.	[73]
		OsO_4			
965	335	954	335	R, liquid	[74]
		953	325	IR, solid	[74]
(968) [2]		960		IR, gas	[71]
(967) [3]		954		IR, CCl_4 sol.	[71]
964	338	953	334	R, liquid	[73]
962	333	952	333	R, sat. aq.	[73]
965.5	333.1	960.1	322.7	R, gas	[75]
964.5	335.2	954		R, CCl_4 sol.	[76]
		960.5	329.0	IR, gas	[76]
		955.0	326.0	IR, CCl_4 sol.	[76]

[1] For $RuO_4(CCl_4)$, combination and overtone bands were observed at 1794 cm^{-1} ($\nu_1 + \nu_3$), 1831 cm^{-1} ($2\nu_3$). [2] For OsO_4(g), combination and overtone bands were observed at 1919 cm^{-1} ($\nu_1 + \nu_3$), 1928 cm^{-1} ($2\nu_3$). [3] For $OsO_4(CCl_4)$, combination and overtone bands were observed at 1910 cm^{-1} ($\nu_1 + \nu_3$), 1921 cm^{-1} ($2\nu_3$).

Table 5. Bond lengths (Å) and vibrational frequencies (unscaled, cm^{-1}) of iron tetroxide, FeO_4; ruthenium tetroxide, RuO_4; and osmium tetroxide, OsO_4.

	d(Fe-O)	ν_1 (A_1)	ν_2 (E)	ν_3 (T_2)	ν_4 (T_2)
HF/6-31G(d)	1.5199	927	426	513	446
HF/6-31+G(d)	1.5222	932	419	508	456
MP2/6-31G(d)	1.5023	2493	578	3136	524
MP2/6-31+G(d)	1.5069	2514	575	3143	528
B3LYP/6-31G(d)	1.5711	968	386	1038	437
B3LYP/6-31+G(d)	1.5768	950	377	1015	439
BLYP/6-31G(d)	1.6009	895	361	978	411
BLYP/6-31+G(d)	1.6078	875	352	953	402
PBE/6-31G(d)	1.5891	917	368	1003	418
PBE/6-31+G(d)	1.5955	898	359	978	409
CPCM-HF/6-31G(d)	1.5206	913	413	481	397
CPCM-HF/6-31+G(d)	1.5229	917	408	472	407
CPCM-MP2/6-31G(d)	1.5026	2510	572	3032	517
CPCM-MP2/6-31+G(d)	1.5070	2531	570	3012	522
CPCM-B3LYP/6-31G(d)	1.5722	966	380	1023	422
CPCM-B3LYP/6-31+G(d)	1.5776	951	371	999	414
CPCM-BLYP/6-31G(d)	1.6020	892	347	966	395
CPCM-BLYP/6-31+G(d)	1.6087	869	335	939	385
CPCM-PBE/6-31G(d)	1.5901	918	360	992	404
CPCM-PBE/6-31+G(d)	1.5963	898	349	966	395
	d(Ru-O)	ν_1 (A_1)	ν_2 (E)	ν_3 (T_2)	ν_4 (T_2)
HF/6-31G(d)	1.6394	1116	393	1060	405
HF/6-31+G(d)	1.6396	1115	392	1052	403
MP2/6-31G(d)	1.7631	1080	312	1683	350
MP2/6-31+G(d)	1.7654	1082	315	1698	352
B3LYP/6-31G(d)	1.6977	951	333	974	350
B3LYP/6-31+G(d)	1.6989	947	334	964	348

Table 5. *Cont.*

	d(Ru-O)	ν_1 (A_1)	ν_2 (E)	ν_3 (T_2)	ν_4 (T_2)
BLYP/6-31G(d)	1.7259	883	313	916	332
BLYP/6-31+G(d)	1.7277	877	314	905	331
PBE/6-31G(d)	1.7132	907	320	940	338
PBE/6-31+G(d)	1.7145	903	321	930	337
CPCM-HF/6-31G(d)	1.6422	1089	348	1014	334
CPCM-HF/6-31+G(d)	1.6425	1087	348	1001	331
CPCM-MP2/6-31G(d)	1.7674	1085	286	1660	322
CPCM-MP2/6-31+G(d)	1.7705	1090	291	1675	324
CPCM-B3LYP/6-31G(d)	1.7010	935	328	952	295
CPCM-B3LYP/6-31+G(d)	1.7026	926	319	937	289
CPCM-BLYP/6-31G(d)	1.7308	842	227	891	266
CPCM-BLYP/6-31+G(d)	1.7333	833	224	875	262
CPCM-PBE/6-31G(d)	1.7168	873	257	916	274
CPCM-PBE/6-31+G(d)	1.7186	865	251	902	270
	d(Os-O)	ν_1 (A_1)	ν_2 (E)	ν_3 (T_2)	ν_4 (T_2)
HF/6-31G(d)	1.6672	1152	391	1075	383
HF/6-31+G(d)	1.6668	1154	390	1070	382
MP2/6-31G(d)	1.7863	861	310	1126	311
MP2/6-31+G(d)	1.7867	859	314	1124	314
B3LYP/6-31G(d)	1.7192	991	335	975	330
B3LYP/6-31+G(d)	1.7197	988	335	966	329
BLYP/6-31G(d)	1.7449	925	317	921	313
BLYP/6-31+G(d)	1.7459	920	317	910	313
PBE/6-31G(d)	1.7334	947	324	942	319
PBE/6-31+G(d)	1.7340	944	324	933	318
CPCM-HF/6-31G(d)	1.6715	1131	359	1039	302
CPCM-HF/6-31+G(d)	1.6712	1132	358	1029	299
CPCM-MP2/6-31G(d)	1.7873	870	295	1120	320
CPCM-MP2/6-31+G(d)	1.7885	867	299	1117	324
CPCM-B3LYP/6-31G(d)	1.7228	965	282	957	280
CPCM-B3LYP/6-31+G(d)	1.7238	959	278	943	277
CPCM-BLYP/6-31G(d)	1.7494	896	264	903	292
CPCM-BLYP/6-31+G(d)	1.7511	889	264	888	292
CPCM-PBE/6-31G(d)	1.7370	920	269	925	282
CPCM-PBE/6-31+G(d)	1.7380	914	267	913	281

All attempts to optimize structures of hydrated iron, ruthenium, or osmium(VIII) oxide in which the water was hydrogen-bonded to the oxygen atoms of the metal tetroxide resulted in optimized structures with imaginary modes corresponding to water wagging or partially optimized structures where the water reorients in a wagging motion to approach the metal atom more closely. In the case of iron tetroxide, other structures can form, such as peroxides or ozonides. This is consistent with the supposition that $Fe^{VIII}O_4$ is not the most stable form of iron tetroxide but rather $Fe^{VI}O_2(\mu_2\text{-}O_2)$ [77].

3.5. Chromate

The crystal structures of alkali metal chromate salts are known. Early determinations of the anhydrous salts placed the average Cr-O distance as 1.60 Å (Na [78], K [79], Rb [80], Cs [81]). The redetermination of anhydrous potassium chromate (*Pnam*) [82] gave a Cr-O distance of 1.644 Å (1.670 Å corrected) or 1.646 Å [83]. Other (re)determinations give Li ($R\bar{3}$) [84], 1.653 Å (1.660 Å corr.); Na (*Cmcm*) [85], neutron diffraction 1.645 Å; Rb (*Pnam*) [86], 1.632 Å; and Cs (*Pnma*) [87], neutron diffraction, 1.650(5) Å. Several hydrates of sodium chromate can form, including the tetrahydrate ($P2_1/c$) with a Cr-O distance of 1.64(4) Å [88], decahydrate ($P2_1/c$) and sesquihydrate (*F2dd*) with Cr-O distances of 1.646(7) and 1.652(5) Å, respectively [89]. The crystal structure of the hexahydrate remains unknown, presumably because of its narrow stability range (19.525–25.90 °C) [90,91].

The resonance Raman spectrum of aqueous chromate ion was measured with the 514.5 and 363.8 nm argon laser lines by Kiefer and Bernstein [32]. The Cr-O stretching bands at 848 (ν_1-A_1), 887 (ν_3-T_2), 350 (ν_2-E), and 369 cm^{-1} (ν_4-T_2) were observed with the 514.5 nm line. For the 363.8 nm line, only the stretching modes at 848 (ν_1-A_1) and 890 (ν_3-T_2) were observed, as well as several overtones $n\nu_1$ and combination $\nu_1+\nu_3$, were observed in light and heavy water. Early Raman work by Venkateswaran [92] gave bands at 840–860 (ν_1-A_1), 874–879 (ν_3-T_2), 481–486 (ν_2-E), and 503–513 (ν_4-T_2) cm^{-1}. Stammreich et al. observed aqueous solutions of sodium, potassium, and ammonium chromate using the He 587.56 nm line [93], and found Raman shifts of 847 (ν_1-A_1), 884 (ν_3-T_2), 348 (ν_2-E), and 368 (ν_4-T_2) cm^{-1}. Weinstock et al. observed, using a He-Ne laser at 633 nm, the normal Raman spectrum of K_2CrO_4(aq) [33] and found bands at 846 ± 1 (ν_1-A_1), 890 ± 4 (ν_3-T_2), 349 ± 2 (ν_2-E), and 368 ± 2 (ν_4-T_2) cm^{-1}. In their study of the chromate—dichromate--hydrogen chromate equilibria, the values obtained by Michel and Machiroux [94] were 846, 887, 348, and 371 cm^{-1}. The aqueous IR and EXAFS spectrum were measured by Hoffmann and coworkers [95], who observed the chromate band at 880 ± 1 cm^{-1} at pH 8.5. The EXAFS analyses (pH = 13) gave a Cr-O distance of 1.660(3) and 1.656(3) Å using two models.

The calculated bond lengths and vibrational frequencies for CrO_4^{2-} are given in Table 6. The Hartree-Fock distances are about 0.05 Å too short compared to the experiment, but the MP2 distances are about 0.05 Å too long. The B3LYP distances are very close. The Hartree-Fock frequencies are overestimated. The MP2 stretching frequencies are also overestimated, which is somewhat unusual. In addition, HF places the ν_1-A_1 band above the ν_3-T_2 band by 30–50 cm^{-1}, whereas MP2 reverses this order by 20–70 cm^{-1}. Only the DFT frequencies (and MP2 deformation frequencies) are reasonably close to the experiment. The CPCM solvation model gives rise to smaller Cr-O distances.

Table 6. Bond lengths (Å) and vibrational frequencies (unscaled, cm^{-1}) of chromate, CrO_4^{2-}.

	d(Cr-O)	ν_1 (A_1)	ν_2 (E)	ν_3 (T_2)	ν_4 (T_2)
HF/6-31G(d)	1.5968	1043	410	1010	455
HF/6-31+G(d)	1.6056	1017	395	967	443
MP2/6-31G(d)	1.6868	1029	327	1100	377
MP2/6-31+G(d)	1.6996	998	319	1018	369
B3LYP/6-31G(d)	1.6479	897	348	925	391
B3LYP/6-31+G(d)	1.6572	872	338	887	384
BLYP/6-31G(d)	1.6756	830	323	865	364
BLYP/6-31+G(d)	1.6859	804	316	832	361
PBE/6-31G(d)	1.6642	849	328	885	368
PBE/6-31+G(d)	1.6737	823	321	851	364
CPCM-HF/6-31G(d)	1.5896	1065	410	993	452
CPCM-HF/6-31+G(d)	1.5984	1035	391	934	434
CPCM-MP2/6-31G(d)	1.6802	1023	322	1028	380
CPCM-MP2/6-31+G(d)	1.6920	993	319	916	375
CPCM-B3LYP/6-31G(d)	1.6397	902	334	924	378
CPCM-B3LYP/6-31+G(d)	1.6488	881	328	873	372
CPCM-BLYP/6-31G(d)	1.6672	842	317	874	359
CPCM-BLYP/6-31+G(d)	1.6769	822	316	828	356
CPCM-PBE/6-31G(d)	1.6558	856	315	892	358
CPCM-PBE/6-31+G(d)	1.6650	837	315	845	357

The effect of water upon the Cr-O distances (MP2/6-31+G(d)) in $CrO_4^{2-} \bullet nH_2O$ (n = 0–6) is given in Figure 9. The net effect is a moderate shortening of the Cr-O distance by 0.01 Å (n = 6). The individual Cr-O distances vary by up to 0.05 Å (in perchlorate, the variation was smaller at 0.02 Å). The Cr...O distance is about the same as the corresponding distance in the hydrated permanganate (Figure S8), although there is less variation within a particular species. For the O...H and O...O distances, the increase of about 0.1 Å upon going from mono to hexahydrate is much more pronounced than permanganate, although the spread within a species is about half (O...H) to the same (O...O) (Figures S9 and S10). The vibrational frequencies (HF/6-31+G*) as a function of hydration number are shown in Figure 10. The frequencies are lower than in permanganate. In addition, the HF/6-31+G* level predicts that the asymmetric stretching mode is lower than the symmetric stretching mode for chromate, whereas experimentally, the modes are reversed. The effect of hydration is similar to that in permanganate, except that the splitting within degenerate modes is about twice as large, which makes it more difficult to correlate the two stretching modes because of the overlap. In an aqueous solution, the Cr-O distance is 1.660(3) Å by LAXS, and the Cr...O distance is 3.955(5) Å [48]. While the Cr-O distance is well-reproduced by the calculations, the Cr...O distance is too short in the hexahydrate. This is rectified in the dodecahydrate model, in which the Cr...O distance lies in the range of 3.78–3.88 Å, depending on the level of theory.

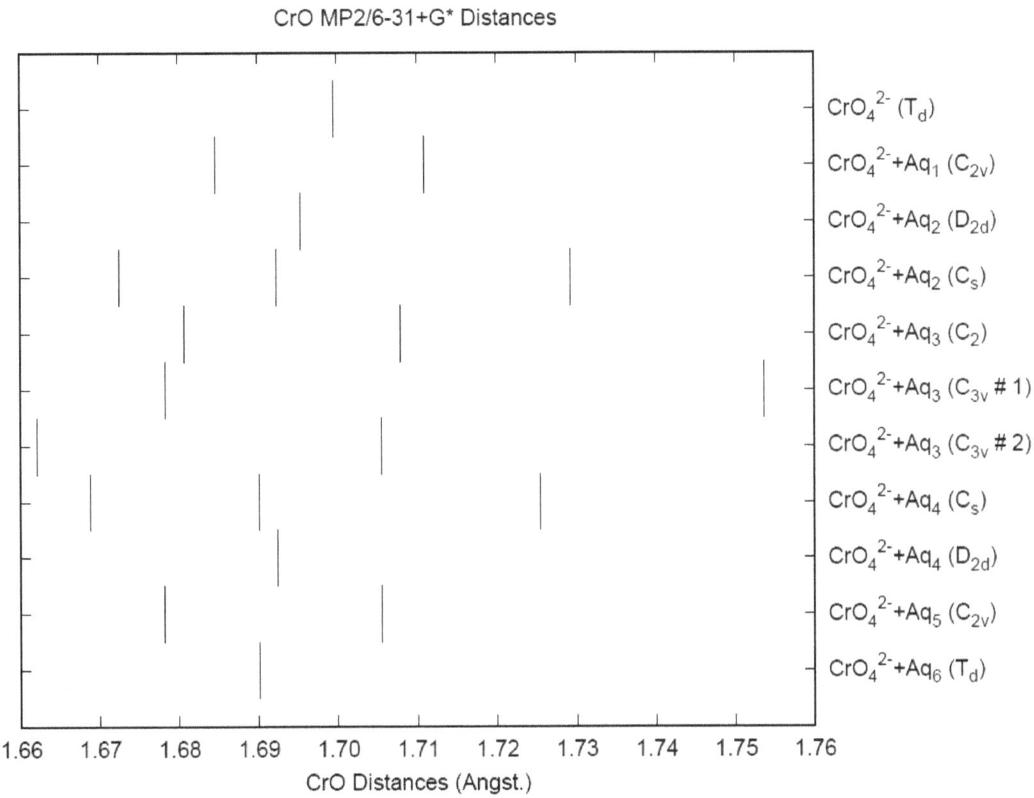

Figure 9. The Cr-O distances (MP2/6-31+G*) in hydrated chromate.

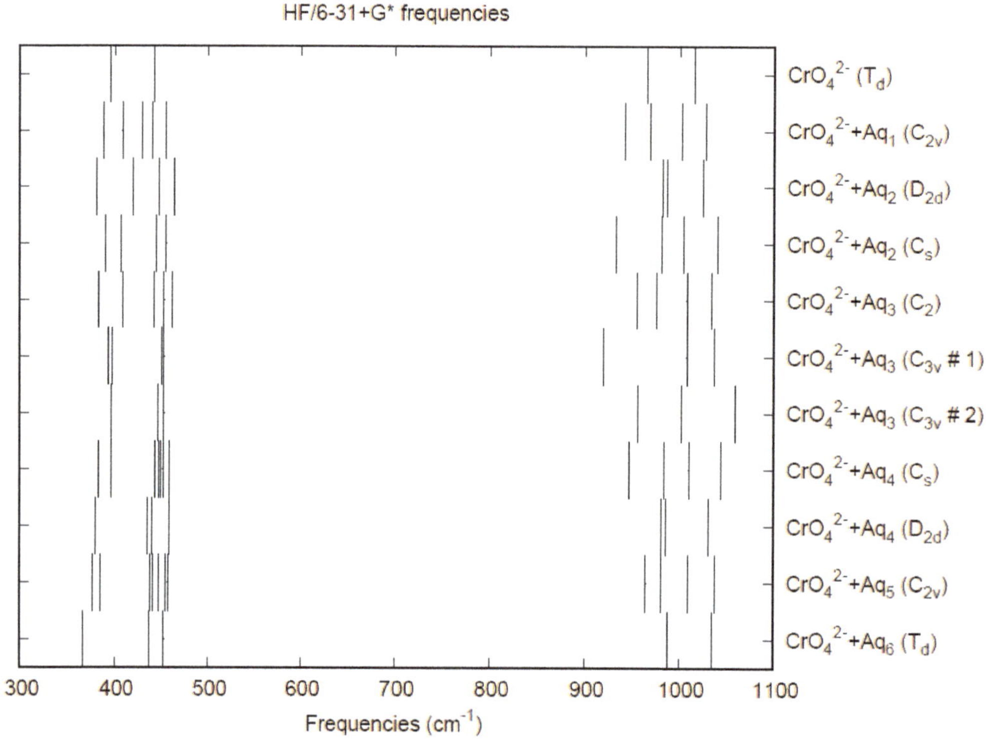

Figure 10. Vibrational frequencies (HF/6-31+G*) of hydrated chromate.

3.6. Molybdate

Numerous authors have investigated the crystal structures of alkali metal molybdate salts. The space group of anhydrous lithium molybdate was determined to be $P3_2$ by Barinova et al. [96]. The average Mo-O distance was 1.768 Å. However, Yip et al. argued [97] that the correct space group was $R\overline{3}$, as shown previously by Kolitsch [98] between 103 and 293 K (1.764 Å), whereas Zachariasen [99] suggested that $P\overline{3}$ was correct. Anhydrous sodium molybdate was shown to exist in four modifications [100] between 623 and 923 K: α (*Fd-3m*), β (unknown), γ (*Fddd*), and δ ($P6_3/mmc$) by powder diffraction. The low-temperature α-form was confirmed and refined by Fortes [101] using neutron powder diffraction (Mo-O = 1.7716 Å). Anhydrous potassium molybdate was found [102] to crystallize in the space group *C2/m*, with Mo-O distance of 1.76(1) Å. A powder diffraction study of anhydrous potassium, rubidium, and cesium molybdate [103] provided cell constants and fractional coordinates of the metal atoms, but the oxygen atoms were not located. Later, annealed anhydrous rubidium molybdate (*Pnam*) was found [104] to have an Mo-O distance of 1.75(2) Å. Anhydrous cesium molybdate was found [105] to crystallize in the space group *Pcmn*, with a Mo-O distance of 1.773 Å (corrected 1.792 Å). In addition to anhydrous salts, sodium molybdate dihydrate is also known. Mitra and Verma found [106] a space group *Pcba* with an average Mo-O distance of 1.82(3) Å (octahedrally coordinated). Other authors found tetrahedral coordination. Atovmyan et al. found [107] a space group of *Pcab* with an average Mo-O distance of 1.76(6) Å. A (re)determination by Matsumoto et al. gave [108] space group *Pbca*, Mo-O distance of 1.772(14) Å. Another redetermination by Capitelli (*Pbca*) gave Mo-O distances of 1.767(10) Å and also located the hydrogen atoms [109]. Neutron powder diffraction was used by Fortes [110] (*Pbca*) to give Mo-O distances of 1.766(10) Å. The consensus for a Mo-O distance of 1.77 Å is clear.

Early Raman work by Venkateswaran [92] gave the molybdate frequencies of 897–944 (ν_1-A_1), 841–896 (ν_3-T_2), 218–241 (ν_2-E), and 317–361 cm^{-1} (ν_4-T_2). Busey and Keller [54] found Raman bands at 897 (ν_1-A_1), 841 (ν_3-T_2), and 318 cm^{-1} (either ν_2-E or ν_4-T_2) for sodium molybdate. Weinstock et al. [33] measured sodium molybdate at 897 (ν_1-A_1), 837 (ν_3-T_2), 317 (ν_2-E, R), and 325 cm^{-1} (ν_4-T_2, IR). Dean and Wilkinson [111] gave an accurate value of 986.1 cm^{-1} for the value of ν_1-A_1 extrapolated to infinite dilution.

The calculated bond lengths and vibrational frequencies for MoO_4^{2-} are given in Table 7. The Hartree-Fock distances are slightly too short (0.01 Å) compared to the experiment, but the MP2 distances are about 0.06 Å too long. The B3LYP distances are slightly too long (0.03 Å). It seems that this example is unusual compared to the others in that the Hartree-Fock distances are closest to X-ray diffraction results. The Hartree-Fock frequencies overestimate the experiment, whereas the MP2 and B3LYP frequencies underestimate it. In addition, HF and B3LYP place the ν_1-A_1 band above the ν_3-T_2 band, whereas MP2 reverses this order. The B3LYP frequencies are closest to the experiment. The CPCM solvation model gives rise to smaller Mo-O distances and larger A_1 frequencies.

Table 7. Bond lengths (Å) and vibrational frequencies (unscaled, cm^{-1}) of molybdate, MoO_4^{2-}.

	d(Mo-O)	ν_1 (A_1)	ν_2 (E)	ν_3 (T_2)	ν_4 (T_2)
HF/6-31G(d)	1.7569	977	341	901	370
HF/6-31+G(d)	1.7601	966	342	874	365
MP2/6-31G(d)	1.8241	784	280	803	295
MP2/6-31+G(d)	1.8320	766	286	775	296
B3LYP/6-31G(d)	1.7951	863	291	825	314
B3LYP/6-31+G(d)	1.8005	849	298	801	314
BLYP/6-31G(d)	1.8182	812	276	782	298
BLYP/6-31+G(d)	1.8248	797	285	760	300
PBE/6-31G(d)	1.8070	828	280	798	301
PBE/6-31+G(d)	1.8126	814	288	776	302
CPCM-HF/6-31G(d)	1.7534	968	315	876	318
CPCM-HF/6-31+G(d)	1.7569	953	313	838	310
CPCM-MP2/6-31G(d)	1.8188	810	291	808	297
CPCM-MP2/6-31+G(d)	1.8271	787	295	764	291
CPCM-B3LYP/6-31G(d)	1.7898	881	298	829	309
CPCM-B3LYP/6-31+G(d)	1.7950	867	303	795	308
CPCM-BLYP/6-31G(d)	1.8117	835	286	794	297
CPCM-BLYP/6-31+G(d)	1.8178	819	292	762	295
CPCM-PBE/6-31G(d)	1.8007	850	289	807	299
CPCM-PBE/6-31+G(d)	1.8059	836	295	776	298

The effect of water upon the Mo-O distances (MP2/6-31+G(d)) in $MoO_4^{2-} \bullet nH_2O$ (n = 0–6) is given in Figure 11. The net effect is a moderate shortening of the Mo-O distance by 0.013 Å (n = 6). The individual Mo-O distances vary by up to 0.025 Å (in chromate, the variation was larger at 0.05 Å). The Mo...O distance is about 0.1 Å larger/smaller than the corresponding distance in the hydrated chromate/pertechnetate (Figure S11), although there is less variation within a particular species. For the O...H and O...O distances, the increase of about 0.1 Å upon going from mono to hexahydrate is much similar to chromate, although the spread within a species is about half (Figures S12 and S13). In 2M sodium molybdate solution, the Mo-O (Mo...O) distance was found to be 1.786(8) (4.06(2)) Å by LAXS [112]. The vibrational frequencies (HF/6-31+G*) as a function of hydration number are shown in Figure 12. The frequencies are lower than in pertechnetate and chromate. The effect of hydration is similar to that in chromate, except that the splitting within degenerate modes is somewhat smaller. The splitting is still large enough to obfuscate the correlation between the two deformation modes because of the overlap. In an aqueous solution, the Mo-O distance is 1.775(4) Å by LAXS, and the Mo...O distance is 4.010(3) Å [48]. While the Mo-O distance is well-reproduced by the calculations, the Mo...O distance is too short in the hexahydrate. This is rectified in the dodecahydrate model, in which the Mo…O distance lies in the range of 3.82–3.94 Å, depending on the level of theory.

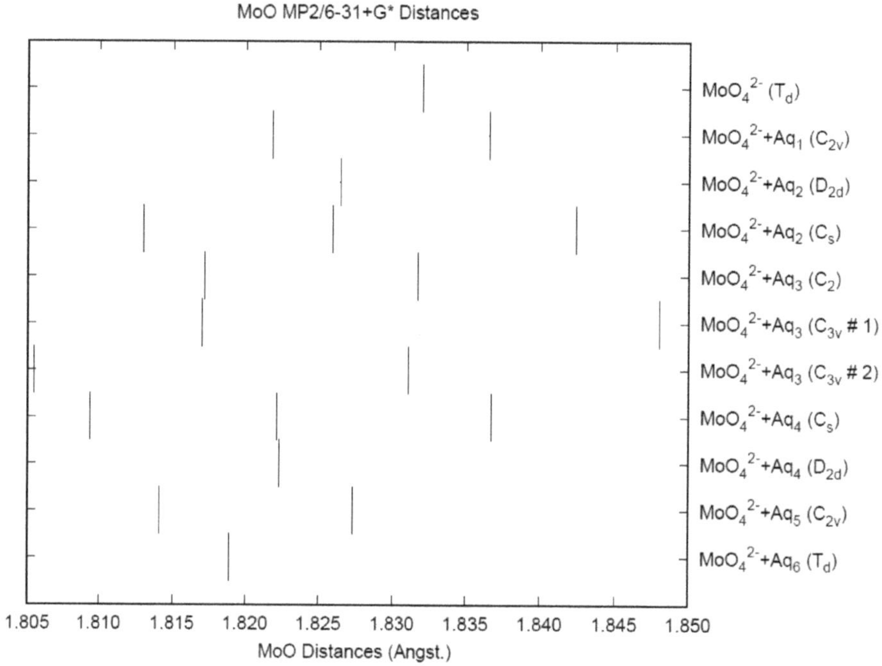

Figure 11. The Mo-O distances (MP2/6-31+G*) in hydrated molybdate.

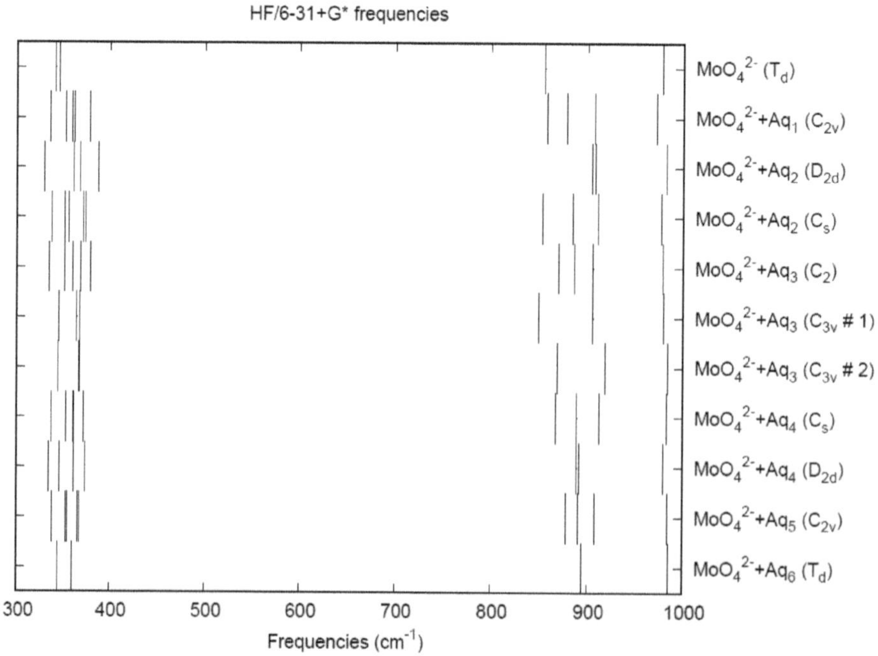

Figure 12. Vibrational frequencies (HF/6-31+G*) of hydrated molybdate.

3.7. Tungstate

The crystal structures of alkali metal tungstates have been investigated. Zachariasen and Plettinger showed [113] that anhydrous lithium tungstate has space group $R\bar{3}$, with W-O bonds of 1.79(2) Å. Okada et al. showed [114] that anhydrous sodium tungstate has space group *Fd3m*, with W-O bonds of 1.819(8) Å. This was revisited by Fortes [101] ($Fd\bar{3}m$), who found a somewhat shorter length of 1.7830(2) Å using neutron diffraction. Anhydrous potassium tungstate crystallizes [115] in the space group *C2/m*, with a W-O distance of 1.79(2) Å. Powder diffraction on anhydrous potassium, rubidium, and cesium tungstate [103] provided cell constants and fractional coordinates of the metal atoms, but as with the molybdates, the oxygen atoms were not located. By neutron powder diffraction [104], it was found that rubidium tungstate crystallizes in the space group *C2/m*, W-O = 1.775(9) Å. To the best of our knowledge, the crystal structure of cesium tungstate remains unknown.

Early Raman work by Venkateswaran [92] gave the tungstate frequencies at 934 (ν_1-A_1), 840 (ν_3-T_2), 325 (ν_2-E), and 452 cm^{-1} (ν_4-T_2). Busey and Keller [54] found IR and Raman bands at 928–931 (ν_1-A_1), 813–855 (ν_3-T_2), 335 cm^{-1} (either or both ν_2-E, ν_4-T_2) for sodium tungstate powder (anhydrous and dihydrate). Weinstock et al. [33] measured the Raman spectrum of sodium tungstate and found bands at 931 (ν_1-A_1), 838 (ν_3-T_2), and 325 cm^{-1} (either or both ν_2-E, ν_4-T_2). Dean and Wilkinson [111] gave an accurate value of 931.1 cm^{-1} for the value of ν_1-A_1 extrapolated to infinite dilution.

The calculated bond lengths and vibrational frequencies for $WO_4^{2\times}$ are given in Table 8. With the experimental distances spanning the range from 1.775–1.819 Å, the Hartree-Fock distances are at the low end of this range, and the B3LYP distances are at the high end. The MP2 distances are somewhat too long. The Hartree-Fock frequencies overestimate the experiment, whereas the B3LYP frequencies underestimate it by about the same amount. The MP2 frequencies underestimate the experiment a bit more, which matches the inverse trend expected with bond length. All methods correctly place the ν_1 band higher than ν_3 and predict the near degeneracy of the ν_2 and ν_4 bands. The CPCM solvation model gives rise to slightly smaller W-O distances and larger stretching frequencies.

Table 8. Bond lengths (Å) and vibrational frequencies (unscaled, cm^{-1}) of tungstate, WO_4^{2-}.

	d(W-O)	ν_1 (A_1)	ν_2 (E)	ν_3 (T_2)	ν_4 (T_2)
HF/6-31G(d)	1.7781	989	341	881	349
HF/6-31+G(d)	1.7806	979	341	857	345
MP2/6-31G(d)	1.8305	828	293	791	289
MP2/6-31+G(d)	1.8349	815	298	769	290
B3LYP/6-31G(d)	1.8107	883	297	807	298
B3LYP/6-31+G(d)	1.8150	870	301	786	297
BLYP/6-31G(d)	1.8315	835	282	768	283
BLYP/6-31+G(d)	1.8367	821	288	748	284
PBE/6-31G(d)	1.8212	849	287	782	284
PBE/6-31+G(d)	1.8255	837	292	763	286
CPCM-HF/6-31G(d)	1.7762	985	324	868	315
CPCM-HF/6-31+G(d)	1.7792	974	323	834	313
CPCM-MP2/6-31G(d)	1.8221	859	305	805	288
CPCM-MP2/6-31+G(d)	1.8267	843	309	771	287
CPCM-B3LYP/6-31G(d)	1.8060	903	306	816	296
CPCM-B3LYP/6-31+G(d)	1.8105	889	308	785	294
CPCM-BLYP/6-31G(d)	1.8257	854	290	779	274
CPCM-BLYP/6-31+G(d)	1.8309	838	293	749	271
CPCM-PBE/6-31G(d)	1.8156	870	295	793	281
CPCM-PBE/6-31+G(d)	1.8199	856	298	765	278

The effect of water upon the W-O distances (MP2/6-31+G(d)) in $WO_4^{2-} \bullet nH_2O$ (n = 0–6) is given in Figure S14. The net effect is a moderate shortening of the W-O distance by

0.012 Å ($n = 6$). The individual W-O distances vary by up to 0.015 Å, which is smaller than in chromate and molybdate. The W...O distance is about the same as the corresponding distance in the hydrated molybdate (Figure S15), although there is less variation within a particular species. For the O...H and O...O distances, the increase of about 0.1 Å upon going from mono to hexahydrate is much similar to molybdate, as are the actual values, although the spread within a species is about half (Figures S16 and S17). In 2M sodium tungstate solution, the W-O (W...O) distance was found to be 1.786(8) (4.06(2)) Å by LAXS [112]. The vibrational frequencies (HF/6-31+G*) as a function of hydration number are shown in Figure 13. The frequencies are slightly lower than in molybdate, except for ν_1. The effect of hydration is similar to that in molybdate, except that the splitting within degenerate modes is somewhat smaller. As with molybdate, the splitting is still large enough to obfuscate the correlation with the two deformation modes because of the overlap. In an aqueous solution, the W-O distance is 1.797(4) Å by LAXS, and the W...O distance is 4.024(4) Å [48]. While the W-O distance is well-reproduced by the calculations, the W...O distance is too short in the hexahydrate. This is rectified in the dodecahydrate model, in which the W...O distance lies in the range 3.82–3.94 Å, depending on the level of theory.

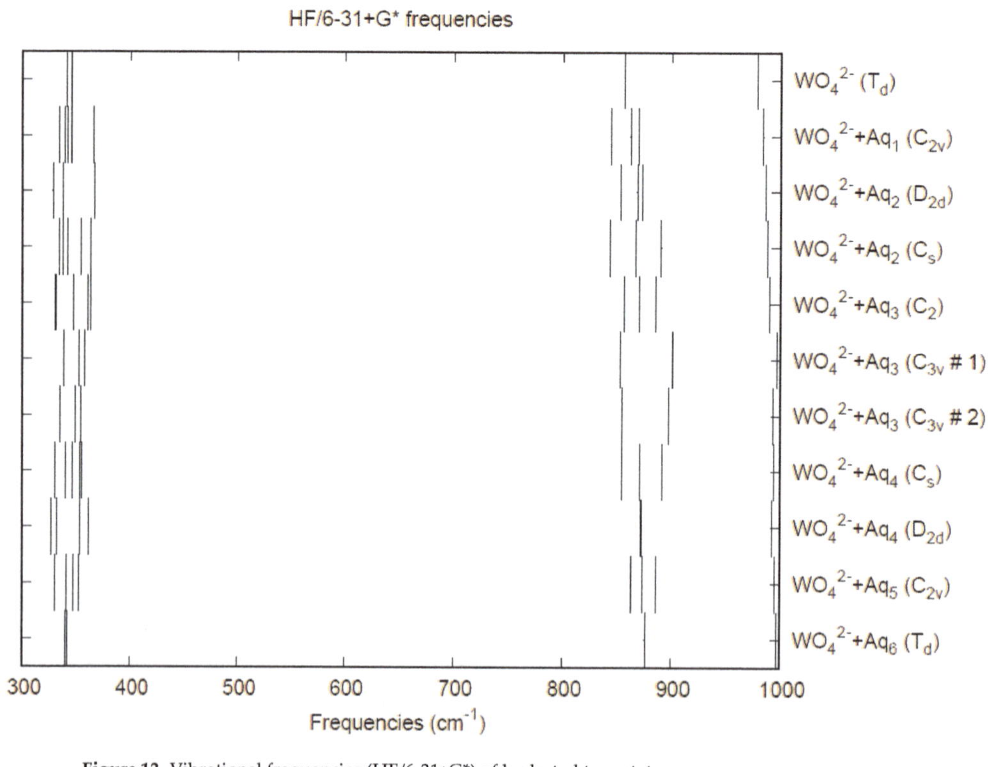

Figure 13. Vibrational frequencies (HF/6-31+G*) of hydrated tungstate.

3.8. Niobates

Unlike the previously mentioned anions, the experimental evidence for the existence of orthoniobate (NbO_4^{3-}), especially in aqueous solution, is scant. We therefore review some of the literature on the complex chemistry of the $Nb_2O_5{\cdot}M_2O{\cdot}H_2O$ (M = Li, Na, K, Rb, Cs) phase diagrams and the compounds therein.

Studies of niobium oxide [116,117] and the phase equilibria with lithium [118], sodium [119], potassium [120], rubidium [121], and cesium [122] carbonate/oxide were carried out by Reisman and coworkers. Amorphous niobium oxide, upon heating, converts to the γ-phase at 435 °C, which then converts to the α-phase at 830 °C before melting at 1491 °C. The existence of the previously identified β and δ phases as equilibrium phases was ruled out [116]. A metastable ε-phase was observed between 1300–1500 °C [117]. Lithium niobates form with a niobium to lithium ratio of 14:1 ($Li_2Nb_{28}O_{71}$), 4:1 ($Li_2Nb_8O_{21}$), 1:1 ($LiNbO_3$), and 1:3 (Li_3NbO_4) [118]. Only $LiNbO_3$ could be readily indexed from powder X-ray diffraction. Sodium niobates form at the same ratios, and all could be indexed [119]. Potassium niobates form at ratios of 22:3 ($K_6Nb_{44}O_{113}$), 3:1 (KNb_3O_8), 3:2 ($K_4Nb_6O_{17}$), 1:1 ($KNbO_3$), and 1:3 (K_3NbO_4) [120]. The metastable 7:6 and 6:7 compounds may also form [123]. Rubidium niobates form at ratios of 15:2 ($Rb_4Nb_{30}O_{77}$), 4:1 ($Rb_2Nb_8O_{21}$), 11:4 ($Rb_8Nb_{22}O_{59}$), 2:1 ($Rb_2Nb_4O_{11}$), 3:2 ($Rb_4Nb_6O_{17}$), 1:1 ($RbNbO_3$), 3:4 ($Rb_8Nb_6O_{19}$), and 1:4 ($Rb_8Nb_2O_9$) [121]. Cesium niobates form at ratios of 15:2 ($Cs_4Nb_{30}O_{77}$), 13:5 ($Cs_5Nb_{13}O_{35}$), 2:1 ($Cs_2Nb_4O_{11}$), 3:2 ($Cs_4Nb_6O_{17}$), and 1:1 ($CsNbO_3$) [122]. From these results, it appears that compounds of stoichiometry M_3NbO_4 only exist for M = Li, Na, and K.

The structures of these compounds have proven difficult to elucidate. Powder diffraction gives the intensity of the scattered waves (X-ray, neutron, electron) as a function of scattering angle 2θ, which generally uniquely characterizes the compound. If this can be indexed, then the unit cell parameters and sometimes the space group can be identified. In favorable cases, a single crystal can be grown, and the related technique of X-ray crystallography may be applied.

3.8.1. Niobium Oxides

Early work on the oxides of niobium suggested the existence of three polymorphs of Nb_2O_5, the low (T), medium (M), and high (H)-temperature forms, and the powder diffraction patterns were measured (copper-K_α) [124]. The H-form of Brauer appears to be equivalent to the α-form of Reisman [116], who was able to index to a monoclinic cell, a = 21.34 Å, b = 3.816 Å, c = 19.47 Å, β = 120.33°. A crystal structure of this form was determined by Gatehouse and coworkers (P_2, a = 21.16 Å, b = 3.822 Å, c = 19.35 Å, β = 119.83°) [125]. The structure consists of three types of building blocks: a slab consisting of edge-shared 3 × 5 NbO_6 octahedra, a slab of edge-shared 3 × 4 NbO_6 octahedra, and two tetrahedral holes, one of which consisted of NbO_4 tetrahedra. The tetrahedra contain Nb-O distances of 1.65(5) and 1.68(5) Å. The structure was refined by Kato (*P2/m*, a = 21.153 Å, b = 3.8233 Å, c = 19.356 Å, β = 119.80°) [126], who altered the space group by delocalizing the tetrahedra over both holes. The Nb-O distance in the tetrahedra was altered to 1.826 Å. Hahn's niobium(V) oxide appears to be the low-temperature T-form [127], as do the forms examined by Frevel [128] and Nolander [129] (*Pba2* or *Pbam*, a = 6.170 Å, b = 29.25 Å, c = 3.928 Å). Single-crystal X-ray diffraction was carried out by Kato and Tamura [130] (*Pbam*, a = 6.175 Å, b = 29.175 Å, c = 3.930 Å). The structure consists of sheets of 6- (octahedral) or 7-coordinated (pentagonal bipyramidal) niobium atoms, along with some nine-coordinate interstitial niobium atoms. A single-crystal X-ray diffraction study of the M-form was carried out by Mertin and coworkers [131] (*I4/mmm*, a = 20.44 Å, b = 3.832 Å), who found sheets of 4 × 4 corner linked niobium octahedra that are edge-linked to other 4x4 sheets. The crystal structure of ζ-Nb_2O_5 was solved by Laves et al. [132] (*B2/b*, a = 12.73 Å, b = 5.56 Å, c = 4.88 Å, γ = 105.08°) and consists of a dense network of NbO_6 octahedra. The crystal structure of another form, N-Nb_2O_5, was solved by Andersson [133] (*C2/m*, a = 28.51 Å, b = 3.830 Å, c = 17.48 Å, β = 120.80°). The structure is composed of 4 × 4 blocks of NbO_6 octahedra. Powder diffraction gives slightly different values [134] (*C2/m*, a = 28.53 Å, b = 3.827 Å, c = 17.58 Å, β = 125.10°). We speculate that it may be possible to make in situ small amounts of transient NbO_4^{3-} or its protonated forms by partial dissolution of the H-form, which already contains some tetrahedral niobium(V).

3.8.2. Lithium Niobates

There are several known structures for lithium niobates. Lithium orthoniobate, Li_3NbO_4, was studied by Blasse [135], who determined a cubic form from powder X-ray diffraction with a = 8.433 Å and also located the niobium atoms. It was also studied by Grenier et al. [136], who found cubic forms for both the low (*Fm3m*, a = 4.212 Å) and high (*I23*, a = 8.429 Å) temperature modifications using powder X-ray diffraction. In the low-temperature form (I), the cations were randomly distributed, whereas, in the high-temperature form (II), powder neutron diffraction gives a tetrahedral grouping of niobium octahedra, with lithium octahedra completing the 3D network. The structure might be viewed as a Nb_4O_4 core bound to 12 additional oxygens to form $Nb_4O_{16}{}^{12-}$, held together by Li^+ ions. The positions were further refined by Grenier and Bassi [137]. The unit cell was confirmed by Whiston and Smith [138] (cubic, a = 8.4300 ± 0.0008). Ukei et al. revised the space group of the high-temperature form to *I-43m*, a = 8.412 Å [139].

Lithium metaniobate, $LiNbO_3$, is the most studied by far because of its ferroelectric properties. Bailey [140] found that $LiNbO_3$ crystallizes in the space group *R3c* with a rhombohedral unit cell (a = 5.4920 Å, α = 55.88°). The corresponding hexagonal unit cell has a_H = 5.147 Å c_H = 13.856 Å. Two possible models were considered for the atomic positions. Megaw discussed the relationship of this structure to its ferroelectricity [141]. Powder neutron diffraction was carried out by Shiozaki and Mitsui [142], suggesting a disordered distribution of lithium ions. Abrahams et al. carried out a single crystal X-ray diffraction study at ambient temperature (*R3c*, a_H = 5.14829(2) Å, c_H = 13.8631(4) Å; a_R = 5.4944, α = 55.87°) [143]. They found that the unit cell consists of 6 planar sheets of oxygen atoms perpendicular to c, with the octahedral interstices filled by Nb, Li, and X, where X indicates a vacancy. Isotropic thermal parameters were sufficient. The following single-crystal neutron diffraction study [144] confirmed the results and cast doubt on the accuracy of the atomic positions found from powder neutron diffraction determined previously [142]. The powder X-ray diffraction method was then applied at 24, 250, 500, 750, 1000, and 1200 °C [145], where the atomic arrangement essentially remains unchanged. Absorption and extinction effects gave unphysical values for the isotropic B at some temperatures. The shifts in atomic positions led Abrahams to propose that at the Curie point (1210 °C), the space group changes to the paraelectric *R-3*. Stoichiometric $LiNbO_3$ melts incongruently, and the congruently melting compound corresponds to a stoichiometry of $Li_{0.946}NbO_{2.973}$ ($Li_{0.955}Nb_{1.009}O_3$). A small amount of Li_2O has been lost. This form crystallizes (*R3c*, a_H = 5.15052(6) Å, c_H = 13.86496(3) Å) nearly identically to the stoichiometric compound (*R3c*, a_H = 5.14739(8) Å, c_H = 13.85614(9) Å) [146]. In the congruent form, 4.2% of the niobium ions have migrated to the vacated lithium sites. It was found that the thermal vibrations of the congruent form were anharmonic. A single crystal study on the congruent form was carried out by Ohgaki et al. (*R3c*, a_H = 5.15020(6) Å, c_H = 13.8653(4) Å), who also incorporated anharmonic corrections for vibration [147]. There also exists an ilmenite-type polymorph of $LiNbO_3$, characterized by Kumada et al. using powder diffraction (*R3*, a = 5.212 Å, c = 14.356 Å) [148]. Neutron powder diffraction both below (*R3c*) and above (*R-3c*) the Curie point demonstrated that the high-temperature paraelectric phase of $LiNbO_3$ had disordered lithium on both sides of a LiO_3 pyramid [149]. Synchrotron X-ray studies have also been carried out [150,151].

It was argued by Whiston and Smith that the $Li_2Nb_8O_{21}$ (1:4) compound observed by Reisman and Holtzberg [118] was actually $LiNb_3O_8$ (monoclinic, a = 7.435 Å, b = 5.618 Å, c = 7.079 Å, β = 101.0°) [138]. Powder and single-crystal X-ray diffraction was used to solve the structure of $LiNb_3O_8$ by Lundberg (P21/a, a = 15.262(2) Å, b = 5.033(1) Å, c = 7.457(1) Å, β = 107.34(1)°) [152]. It is comprised of chains of edge-shared metal octahedra. The structure was also solved by Gatehouse and Leverett ($P2_1/c$, a = 7.459(4) Å, b = 5.034(4) Å, c = 15.270(7) Å, β = 107.33(3)°) [153].

The structure of Li_7NbO_6 [154] was solved by Muhle using powder X-ray diffraction (*P-1*, a = 5.37932(9) Å, b = 5.91942(11) Å, c = 5.37922(9) Å, α = 117.0033(9)°, β = 119.6023(7)°, γ = 63.2570(9)°) [155]. This structure is interesting in that it consists of isolated $NbO_6{}^{7-}$

octahedra held together by octahedrally and tetrahedrally-coordinated lithium ions. Because of the structure of Li_7NbO_6, it may be possible to generate in situ protonated forms of $NbO_6{}^{7-}$ from it before subsequent condensation.

The crystal structure of $Li_8Nb_2O_9$ was solved by Braun and Hoppe (*P-1*, a = 15.210 Å, b = 8.815 Å, c = 5.858 Å, α = 109.8°, β = 101.4°, γ = 87.0°) [156] It consists of a network of edge and vertex fused NbO_6 octahedra.

The compound $Li_2Nb_{28}O_{71}$ of Reisman and Holtzberg [118] was argued by Whiston and Smith to actually be $Li_2Nb_{24}O_{61}$ (monoclinic, a = 44.79 Å, b = 3.807 Å, c = 28.38 Å, β = 106.5°) [138]. Norin and Nolander argued that extensive twinning of crystals occurred and obtained a powder pattern consistent with the space group being either *C2*, *Cm*, or *C2/m* (a = 28.52(2) Å, b = 3.828(2) Å, c = 17.55(1) Å, β = 124.98(4)°) [134]. There were many similarities to N-N_2O_5.

3.8.3. Sodium Niobates

Sodium orthoniobate, Na_3NbO_4, was studied by Whiston and Smith (*C2* or *Cm* or *C2/m*, a = 12.23 Å, b = 12.85 Å, c = 5.782 Å, β = 121.5°) [138]. Bouillaud found two forms, a cubic form (a = 4.606 Å) [157] with random Li^+ and Nb^{5+} arrangement and a monoclinic form (a = 11.095(5) Å, b = 12.993(5) Å, c = 5.730(5) Å, β = 108.9(5)°) which was believed to be the form observed by Whiston and Smith, in spite of the differences in lattice constants. These lattice constants were confirmed by Barker and Wood, who also observed a third orthorhombic phase (a = 13.806 Å, b = 13.411 Å, c = 9.579 Å) [158]. Meyer and Hoppe determined the crystal structure of the monoclinic form (*C2/m*, a = 11.068 Å, b = 13.032 Å, c = 5.733 Å, β = 109.13°) and found it to consist of discrete $Nb_4O_{16}{}^{12-}$ ions [159].

Sodium metaniobate, $NaNbO_3$, has been investigated by numerous groups. It was found to be pseudocubic at ambient temperature by Barth (a = 3.890 Å) [160]. Wood found reversible polymorphism: 25 °C, orthorhombic (a_0 = 5.512 Å, b_0 = 5.577 Å, c_0 = 4 × 3.885 Å; also can be described as pseudotetragonal a_0 = 2 × 3.921 Å, c_0 = 4 × 3.885 Å; monoclinic axes a_0 = 2 × 3.921 Å, b_0 = 4 × 3.885 Å, c_0 = 2 × 3.921 Å, β = 90.67°); 300 °C, orthorhombic pseudotetragonal (a_0 = 2 × 3.91 Å, c_0 = 4 × 3.94 Å); 425 °C, tetragonal pseudocubic (a_0 = 2 × 3.928, c_0 = 4 × 3.928 Å); 490 °C, tetragonal pseudocubic (a_0 = 2 × 3.93, c_0 = 4 × 3.93 Å) [161]. Vousden found an orthorhombic phase at ambient temperature (a = 5.5682 Å, b = 3.8795 Å, c = 5.5052 Å) [162]. Upon heating, phase changes to tetragonal (300 °C) and cubic (600 °C) forms were noted. Reisman noted the existence of four phases (α, β, γ, δ) with phase transitions at 640, 562, and 354 °C. Single crystal diffraction of the room-temperature form (δ) gave (*P22₁2*, a = 5.5682 Å, b = 15.5180 Å, c = 5.5052 Å) [163], which was inconsistent with its ferroelectric nature [164,165]. Shirane et al. undertook optical, dielectric, and X-ray measurements from 20–670 °C [166]. In addition, Francombe examined the diffraction lines from 20–640 °C [167]. Solovev studied polycrystalline $NaNbO_3$ with X-rays between −200–700 °C [168]. The subcell lattice parameters ranged between 3.87–3.95 Å, giving six regions: pseudomonoclinic (−200–360 °C), pseudomonoclinic (360–430 °C), pseudocubic (430–470 °C), pseudocubic (470–520 °C), tetragonal (520–640 °C), and cubic (640–700 °C). Bouillaud found in a study of solid solutions that the monoclinic end member $NaNbO_3$ was pseudocubic with a = c = 3.9150(8) Å, b =3.8800(8) Å, 90.72(5)° with superstructure 2a.2b.2c [169]. A single crystal study at room temperature by Sakowski-Cowley et al. gave an orthorhombic structure (*Pbma*, a = 5.566 Å, b = 15.520 Å, c = 5.506 Å) [170]. The results of Hewat [171] using neutron diffraction matched those of Sakowski-Cowley. A study of the room temperature (P-antiferromagnetic) and a new low-temperature ferromagnetic N phase by Darlington and Megaw [172] showed that the transition occurred at about -110 °C. This structure was further refined by Seidel and Hoffmann at 20 K [173]. Further high-precision study of the high-temperature transitions up to 800 °C in a single crystal by Glazer and Megaw [174] demonstrated the existence of six high-temperature phases: P (<373 °C), R (373–480 °C), S (480–520 °C), T_1 (520–575 °C), T_2 (575–640 °C), and the cubic phase (>640 °C). The crystal structure of the tetragonal T_2 form at 600 °C was solved by Glazer and Megaw (*P4/mbm*, a = 5.5639(8), b = 5.5639(8), c = 3.9428(8)) [175] and can be described as slight tilting of the NbO_6 octahedra from the cubic perovskite structure. Essentially, the same structure was proposed by Ishida and Honjo [176]. The structure was further refined by Darlington and Knight by synchrotron powder diffraction at 615 °C [177]. The crystal structures of the orthorhombic S (500 °C) and T_1 (530 °C) phases were solved by Ahtee et al. [178]. The lattice parameters are: for T_1, *Ccmm*, a = 7.8642 Å, b = 7.8550 Å, c = 7.8696 Å; for S, *Pnmm*, a = 7.8608 Å, b = 7.8556 Å, c = 7.8606 Å. The structures can be described as successively more tilted regular NbO_6 octahedra. The T_1 structure was further refined by Darlington and Knight by synchrotron powder diffraction at 540 °C [177]. Nanoparticles 20 nm in size refined best to $Pmc2_1$, a = 7.8636(7) Å, b = 5.6306(9) Å, c = 5.5483(9) Å, according to Shanker et al. [179] This appears to be similar to the room-temperature polar phase described by Johnston et al. ($P2_1ma$, a = 5.571(1) Å, b = 7.766(1) Å, c = 5.513(1) Å) [180]. Using neutron diffraction, Darlington and Knight [181] confirmed the structures of the T_1 and T_2 phases and, based on the superlattice reflections, suggested that Phase S was a 2 × 4 × 6 multiplicity instead of the 2 × 2 × 2 reported by Ahtee et al. [178] Phase R had the same multiplicity. In addition, Phase P was suggested to be monoclinic (a_0 = 5.5101 Å, b_0 = 5.5717 Å, c_0 = 15.5181 Å, β = 89.94°). Koruza et al. used powder X-ray diffraction [182] and found: R, 420 °C, *Pmmn*, a = 7.8440(12) Å, b = 7.8459(13) Å, c = 7.8567(3) Å; 300 °C, *Pbcm*, a = 5.5327(2) Å, b = 5.5630(2) Å, c = 15.6450(5) Å; 250 °C, $Pmc2_1$, a = 7.8066(2) Å, b = 5.5279(2) Å, c = 5.5637(2) Å; and 25 °C, $Pmc2_1$, a = 7.7633(3) Å, b = 5.5143(2) Å, c = 5.5655(2) Å. Mishra et al. carried out a neutron powder diffraction study over a wide range of temperature [183] and found: 1075 K, U, cubic, *Pm-3m*, a = 3.9507(2) Å; 900 K, T_2, tetragonal, *P4/mbm*; 850 K, T_1, orthorhombic, *Cmcm*; 770 K, S, orthorhombic, *Pbnm*, a = 5.5555(2) Å, b = 5.5556(2) Å, c = 47.1489(9) Å; 680 K, R, orthorhombic, *Pbnm*; a = 5.5459(1) Å, b = 5.5505(1) Å; c = 23.5229(3) Å; 300 K, P, orthorhombic, *Pbma*. Peel and coworkers used both neutron and X-ray powder diffraction between 360 and 520 °C [184] and found that phase S (500 °C) was a 2 × 2 × 4 structure, *Pmmn*, a = 7.85684(4) Å, b = 7.86748(7) Å, c = 15.72795(13) Å, and that phase R (440 °C) was a 2x2x6 structure, *Pmmn*, a = 7.85371(6) Å, b = 23.54619(19) Å, c = 7.85677(7) Å (a 2 × 6 × 2 *Pnma* model was also close). An ilmenite form also exists (*R3*, a = 5.333(1) Å, b = 15.611(4) Å) [148].

Another sodium-rich structure, Na_5NbO_5, was found by Darriet et al. [185]. This structure is unique in that it has discrete square pyramidal NbO_5^{5-} coordinated to sodium atoms with oxygen voids.

The crystal structure of $Na_2Nb_4O_{11}$ was determined by Jahnberg (*C2/c*, a = 10.840 Å, b = 6.162 Å, c = 12.745 Å, β = 106.22°) [186], confirming the stoichiometry of Andersson [187]. It consists of sheets of edge-shared pentagonal bipyramidal NbO_7. These sheets are connected to each other by NbO_6 octahedra and NaO_7 polyhedra. The presence of ferroelectricity suggested that a noncentrosymmetric space group might be more appropriate, and the refinement of powder diffraction data was conducted by Maso and West (*Cc*, a = 10.8427(1) Å, b = 6.1664(6) Å, c = 12.7524(1) Å, β = 106.176(1)°) [188].

The structure of $NaNb_3O_8$ was examined using powder X-ray diffraction, and an orthorhombic cell was found: Andersson, *Pbam* or *Pba2*, a = 12.372 Å, b = 37.10 Å, c = 3.954 Å [187]; Bouillaud, a = 12.344(4) Å, b = 37.121(6) Å, b = 3.950(2) Å [189]; Nedjar, *Pmnm* or $Pmn2_1$, a = 8.771(14) Å, b = 10.16(2) Å, c = 3.784(3) Å [190]. The layers can be described as edge-shared NbO_6 octahedra, sharing their corners with NbO_4 tetrahedra. A high-pressure form, synthesized by Range et al., consists of both NbO_7 and NbO_8 polyhedra (*Ibam*, a = 7.3244(4) Å, b = 10.3100(5) Å, c = 7.0426(4) Å) [191].

Using single crystal X-ray diffraction, Andersson solved the structure of $NaNb_{13}O_{33}$ (C2/m, a = 22.40 Å, b = 3.834 Å, c = 15.37 Å, β = 91.47°) [192]. It consists of a 3D network of octahedral NbO_6 along with square-planar NaO_4. The powder diffraction pattern was also published [187].

3.8.4. Potassium Niobates

The potassium niobates are described next. Potassium orthoniobate, K_3NbO_4, was confirmed to exist by Guerchais, who gave the d-spacings [193]. It was also studied by

Addison, who identified that there were two different powder patterns corresponding to two phases of the purported compound, depending on the synthesis temperature [194], in agreement with one of two scenarios suggested by the observations of Reisman [120]. Stecura et al. found that a = 12.05(4) Å, b = 14.34(5) Å, c = 10.50(3) Å [195]. Meyer and Hoppe found a cubic γ structure with a = 8.605 Å, and a tetragonal β structure with a = 6.14 Å, c = 8.37 Å [196].

Potassium pyroniobate, $K_4Nb_2O_7$, was proposed by Guerchais, who provided the d-spacings [193].

Potassium metaniobate, $KNbO_3$, was found by Vousden to be orthorhombic by X-ray powder diffraction (a = 5.7203 Å, b = 3.9714 Å, c = 5.6946 Å) [162]. Wood found reversible polymorphism: 25 °C, orthorhombic (a_0 = 5.702(10) Å, b_0 = 5.739(10) Å, c_0 = 3.984(10) Å; also can be described as pseudotetragonal a_0 = 4.045 Å, c_0 = 3.984 Å; monoclinic axes a_0 = 4.045 Å, b_0 = 3.984 Å, c_0 = 4.045 Å, β = 90.35°); 260 °C, tetragonal (a_0 = 4.00(2) Å, c_0 = 4.07(2) Å); 500 °C, cubic (a_0 = 4.024(1) Å) [161]. The phase changes occur between 200–225 and 420–435 °C. These are in good agreement with the later temperature-dependent study of Shirane et al. from 25–510 °C [166]. Reisman inferred the existence of this compound from the phase diagram [120] and refined procedures for its preparation [197]. Guerchais also found a pseudocubic structure (a = 4.029 Å) [193]. A single crystal X-ray diffraction by Katz and Megaw at room temperature also gave excellent agreement (*Bmm2*, a = 5.697 Å, b = 3.971 Å, c = 5.721 Å) with the expected distorted perovskite structure [198]. The tetragonal phase was refined by Hewat at 270 °C using single crystal neutron diffraction (*P4mm*) [199]. Hewat also refined the structure using powder neutron diffraction [200] for the tetragonal (*P4mm*, a = 3.996 Å, c = 4.063 Å), orthorhombic (*Amm2*, a = 3.973 Å, b = 5.695 Å, c = 5.721 Å) and rhombohedral (−43 °C, *R3m*, a = 4.016 Å, α = 89.817°) phases. Sugimoto confirmed the structure of the room temperature form (*Amm2*, a = 3.9807(7) Å, b = 5.688(1) Å, c = 5.711(1) Å) [201]. Nanowires were synthesized by Kim et al. in a monoclinic phase, as determined by X-ray and neutron powder diffraction (*P1m1*, a = 4.04976(6) Å, b = 3.99218(6) Å, c = 4.02076 Å, β = 90.1012(27)°) [202].

A metastable orthorhombic phase, $K_{12}Nb_{14}O_{41}$, was found by Whiston and Smith (a = 8.81 Å, b = 21.55 Å, c = 3.76 Å) [138].

The compound $K_4Nb_6O_{17}$ was identified by Reisman and Holtzberg [120] and confirmed by Guerchais, who gave the d-spacings [193]. The lattice parameters of this tetragonal cell were given by Whiston and Smith as a = 12.575(2) Å, c = 3.763(1) Å [138]. However, Gasperin and Le Bihan found a lamellar structure consisting of sheets of NbO_6 octahedra (*P2*$_1$*nb*, a = 7.83 Å, b = 33.21 Å, c = 6.46 Å) [203].

The tetragonal compound $K_6Nb_{10.8}O_{30}$ was made by Becker and Hold, whose crystal structure was solved (*P4/mbm*, a = 12.537(2) Å, c =3.9730(1) Å) [204]. The tungsten-bronze-type structure contains NbO_6 octahedra vertex shared with two others to form an Nb_3O_{15} unit, and these are vertex-connected to form a sheet. Above and below these units is a site 20% occupied by tricapped trigonal prism Nb^{5+}, together with planes of potassium and oxygen ions.

The triclinic compound $K_3Nb_7O_{19}$ was synthesized by Fallon et al., and single-crystal XRD was carried out (*P-1*, a = 14.143(3), b = 9.948(2), c = 6.463(2) Å, α = 106.45(2), β = 95.82(1), γ = 97.29(1)°) [205]. It consists of double strings of seven edge-shared NbO_6 octahedra.

The tetragonal compound KNb_3O_8 was reported by Reisman et al. [120] and by Whiston and Smith (a = 37.71(3) Å, c = 3.939(5) Å) [138]. Gasperin solved the crystal structure of an orthorhombic form (*Amam*, a = 8.903(3), b = 21.16(2), c = 3.799(2) Å) [206]. It has a lamellar structure consisting of two edge-shared octahedral units connected by a vertex-shared octahedra.

The compound $K_2Nb_8O_{21}$ was found by Guerchais, who gave the d-spacings [193]. It was also indexed by Whiston and Smith [138] to be primitive tetragonal, a = 27.41, c = 3.955 Å, but they expressed doubts about stoichiometry.

The compound KNb_5O_{13} was prepared by Kwak et al. and characterized by single crystal X-ray diffraction (*Pbcm*, a = 5.672(2), b = 10.737(5), c = 16.742(6) Å) [207]. It consists of slabs of NbO_6 octahedra connected by oxygen atoms, with potassium ions occupying the tunnels produced.

The compound KNb_7O_{18} was studied by electron microscopy and electron diffraction by Hu et al., who found that it was tetragonal, with a = 27.5, c = 3.94 Å [208]. The x and y positions were located, but not the z positions.

The compound $K_6Nb_{44}O_{113}$, proposed by Reisman et al. [120], was indexed as monoclinic by Whiston and Smith (a = 19.83, b = 3.899, c = 19.03 Å, β = 115.0°) [138].

3.8.5. Rubidium Niobates

The rubidium niobates are described next. $Rb_8Nb_2O_9$ was identified by Reisman and Holtzberg [121].

Rb_3NbO_4 was found by Meyer and Hoppe [196]. At least three polymorphs exist. Two cubic forms were identified by powder diffraction: α, a = 4.45 Å and γ, a = 8.90 Å, the latter of which was isotypic with the potassium analog. In addition, there exists a tetragonal form, a = 6.30 Å, c = 8.81 Å. Annealing of the reactant mixtures at constant temperature gives the tetragonal form first, followed by the α form (2 weeks) and γ form (3 months).

$Rb_8Nb_6O_{19}$ was identified by Reisman and Holtzberg [121].

$RbNbO_3$ was identified by Reisman and Holtzberg [121]. The "pyrgom" (Greek for piled-up) structure was solved by Serafin and Hoppe as triclinic (a = 8.87 Å; b = 8.39 Å; c = 5.11 Å; α = 94.6°; β = 93.5°; γ = 113.9° [209] and *P-1*, a = 8.882(3) Å, b = 8.395(3) Å, c = 5.109(2) Å, α = 94.60(3)°, β = 93.53(3)°, γ = 113.83(3)°) [210]. Each niobium possesses a tetragonal pyramid of oxygen atoms connected by edges to others in chains. The chains are connected by rubidium atoms. A high-pressure orthorhombic perovskite form is also known (a = 3.9965 Å, b = 5.8360 Å, c = 5.8698 Å) [211].

$Rb_4Nb_6O_{17}$ was identified by Reisman and Holtzberg [121]. Iyer and Smith identified it as orthorhombic (a = 6.42 Å, b = 7.68 Å, c = 38.55 Å) [212] but suggested a slightly different composition $Rb_8Nb_{14}O_{39}$. The crystal structure of a hydrated version of this salt, $Rb_4Nb_6O_{17}\cdot 3H_2O$, was solved by Gasperin and Le Bihan (*Pmnb*, a = 7.83(1) Å, b = 39.06(5) Å, c = 6.57 (1) Å) [213].

$Rb_2Nb_4O_{11}$ was identified by Reisman and Holtzberg [121]. Iyer and Smith identified it as monoclinic (*Cm*, *C2*, or *C2/m*, a = 12.95 Å, b = 7.48 Å, c = 14.92 Å, β = 106.4°) [212] but suggested $Rb_{12}Nb_{22}O_{61}$ as the stoichiometry.

$Rb_8Nb_{22}O_{59}$ was identified by Reisman and Holtzberg [121]. Iyer and Smith suggested hexagonal symmetry (*P6₃22*, a = 7.45 Å, c = 7.66 Å) [212] but reformulated it as $RbNb_3O_8$. The structure was solved by Fallon and Gatehouse (*R-3m*, a = 7.527(6) Å, c = 43.17(2) Å) [214]. Dewan and coworkers also solved the structure (*R-3m*, a = 7.53(1) Å, c = 43.39(6) Å) [215], as did Jones and Robertson (*R-3m*, a_h = 7.52 Å, c_h = 43.17 Å; rhombohedral a_r = 15.03 Å, α_r = 29.0°) [216] and Voronkova et al. (*R-3m*, a = 7.502(1), c = 43.12(2)) [217]. This structure was referred to as 11-L (a = 7.7224(7) Å, c = 43.180(5) Å) by Minor et al. [218] Two other structures of similar composition, 16-L (a = 7.5138(5) Å, c = 65.115(9) Å) and 9-L (a = 7.5179(2) Å, c = 36.353(1) Å) were noted.

A structure identified as "$RbNb_3O_9$" was solved (hexagonal, *P6/mmm*, a = 7.39 Å, c = 3.89 Å; orthorhombic *Cmmm*, a = 13.038 Å, b = 7.502 Å, c = 3.89 Å) and described as being similar to the hexagonal tungsten bronzes [219]. The c-axis is about half of that reported by Iyer and Smith for $RbNb_3O_8$ [212]. It appears similar to the hexagonal-type bronze of Minor et al. (a = 12.991(4) Å, b = 7.550(1) Å, c = 3.8978(8) Å) [218].

$Rb_2Nb_8O_{21}$ was identified by Reisman and Holtzberg [121]. Iyer and Smith suggested tetragonal symmetry (a = 26.55 Å, c = 3.85 Å) [212]. Jones and Robertson suggested orthorhombic symmetry (*Cmcm* (more likely) or *Cmc2*, a = 7.55(1) Å, b = 13.00(2) Å, c = 7.78(1) Å) [216].

$Rb_4Nb_{30}O_{77}$ was identified by Reisman and Holtzberg [121]. Iyer and Smith suggested monoclinic symmetry (*Pm*, *P2*, or *P2/m*, a = 20.17 Å, b = 3.83 Å, c = 20.75 Å, β = 123.5°) [212]. It was suggested that this may have been a solid solution for the composition $Rb_2Nb_{26}O_{66}$ because of its similarity to the high-temperature form of Nb_2O_5. One phase present in Reisman's sample, called "$Rb_3Nb_{54}O_{146}$", was indexed as tetragonal ($P42_12$ or $P\text{-}42_1m$, a = 27.69 Å, c = 3.98 Å) [219]. It consists of octahedra surrounding a heptagonal tunnel. This appears similar to the Gatehouse tetragonal bronze (a = 27.484(3) Å, c = 3.9656(4) Å) [218].

3.8.6. Cesium Niobates

The cesium niobates are described next. Cs_3NbO_4 was synthesized by Meyer and Hoppe [196]. They found a face-centered cubic structure (a = 9.19 Å).

$CsNbO_3$ was identified by Reisman and Mineo [122]. Iyer and Smith found that it crystallized at 870 °C with orthorhombic symmetry (a = 7.24 Å, b = 15.13 Å, c = 9.77 Å) [212]. Meyer and Hoppe solved the crystal structure and found it to be monoclinic (*P2₁/c*, a = 5.148 Å, b = 15.89 Å, c = 9.143 Å, β = 93.3°) with discrete $Nb_4O_{12}^{4-}$ ions [220].

$Cs_4Nb_6O_{17}$ was identified by Reisman and Mineo [122].

$Cs_2Nb_4O_{11}$ was identified by Reisman and Mineo [122]. Iyer and Smith found that it crystallized at 1170 °C with orthorhombic symmetry (a = 13.03 Å, b = 7.60 Å, c = 14.36 Å) [212]. Gasperin also found orthorhombic symmetry (*P2nn*, a = 10.484 (2) Å, b = 28.898 (4) Å, c = 7.464 Å) [221]. A different high-temperature phase was found by Smith et al. at 450 K (*Imma*, a = 7.4729 (5) Å, b = 28.9520 (18) Å, c = 10.5080 (6) Å) [222].

$Cs_{10}Nb_{26}O_{70}$ was identified by Reisman and Mineo [122].

$Cs_8Nb_{22}O_{59}$, after reformulating as $CsNb_3O_8$, was found to be isomorphous to the rubidium compound by Dewan et al. [215]. They found the following space group and unit cell (*R-3m*, a = 7.53(1) Å, c = 43.02(6) Å), as did Voronkova and coworkers (*R-3m*, a = 7.53(2) Å, c = 42.83(4) Å) [217], and Jones and Robertson (*R-3m*, a_r = 15.02 Å, α_r = 28.9°; a_h = 7.50 Å, c_h = 43.15 Å) [216].

$Cs_4Nb_{30}O_{77}$ was identified by Reisman and Mineo [122].

3.8.7. Calculations on Orthoniobate

The calculated bond lengths and vibrational frequencies for NbO_4^{3-} are given in Table 9. The only structures containing discrete orthoniobate are the salts of potassium, rubidium, and cesium, whose structures contain a lot of disorder [196]. However, with fractional coordinates for Nb (0, 0, 0) and O ± (0, 0.211, 0.01) and ±(0.211, 0, 0.01), and cubic unit cell of dimensions 8.90 Å, K_3NbO_4 would have an Nb-O distance of 1.88 Å. The Hartree-Fock distances are slightly less than this, the B3LYP distances are slightly higher, and the MP2 distances are too long. To the best of our knowledge, the vibrational frequencies of orthoniobate have not been measured. We expect a clear separation of the symmetric and antisymmetric stretching frequencies, but it is possible that there may be an accidental degeneracy of the deformation frequencies, which are mostly predicted to be within 30 cm^{-1} of each other. The CPCM solvation model gives rise to smaller Nb-O distances and larger stretching frequencies.

The effect of water upon the Nb-O distances (MP2/6-31+G(d)) in $NbO_4^{3-}\bullet nH_2O$ (n = 0–6) is given in Figure 14. The net effect is a moderate shortening of the Nb-O distance by 0.025 Å (n = 6). The individual Nb-O distances vary by up to 0.04 Å. These changes are a little more pronounced than for the divalent ions. The Nb…O distance exhibits an odd behavior, decreasing from n = 1 to n = 3 and then leveling off or even increasing slightly. (Figure S18). For the O…H and O…O distances, the increase of about 0.1 Å upon going from mono to hexahydrate is consistent with the divalent ions (Figures S19 and S20). The vibrational frequencies (HF/6-31+G*) as a function of hydration number are shown in Figure 15. The effect of hydration is to increase the stretching frequencies by about 40 cm^{-1} going from n = 0 to n = 6, whereas the deformation frequencies only slightly increase. The splitting is large enough to obfuscate the correlation between the two deformation modes because of the overlap, and these might not be able to be observed separately in aqueous solution.

Table 9. Bond lengths (Å) and vibrational frequencies (unscaled, cm^{-1}) of orthoniobate, NbO_4^{3-}.

	d(Nb-O)	ν_1 (A_1)	ν_2 (E)	ν_3 (T_2)	ν_4 (T_2)
HF/6-31G(d)	1.8669	832	289	726	328
HF/6-31+G(d)	1.8754	812	294	699	325
MP2/6-31G(d)	1.9054	716	237	614	257
MP2/6-31+G(d)	1.9191	694	253	610	266
B3LYP/6-31G(d)	1.8796	736	248	642	278
B3LYP/6-31+G(d)	1.8966	699	256	624	278
BLYP/6-31G(d)	1.8956	739	246–257	556–605	208–237
BLYP/6-31+G(d)	1.9044	719	235–252	568–573	223–229
PBE/6-31G(d)	1.8851	752	212–235	593–602	184
PBE/6-31+G(d)	1.8929	730	241–262	547–592	217–240
CPCM-HF/6-31G(d)	1.8490	859	285	757	292
CPCM-HF/6-31+G(d)	1.8570	834	287	705	284
CPCM-MP2/6-31G(d)	1.8911	764	258	694	266
CPCM-MP2/6-31+G(d)	1.9030	735	263	641	258
CPCM-B3LYP/6-31G(d)	1.8736	786	254	708	262
CPCM-B3LYP/6-31+G(d)	1.8830	766	265	667	261
CPCM-BLYP/6-31G(d)	1.8934	752	244	678	253
CPCM-BLYP/6-31+G(d)	1.9035	729	250	641	253
CPCM-PBE/6-31G(d)	1.8830	760	245	687	253
CPCM-PBE/6-31+G(d)	1.8916	742	257	652	253

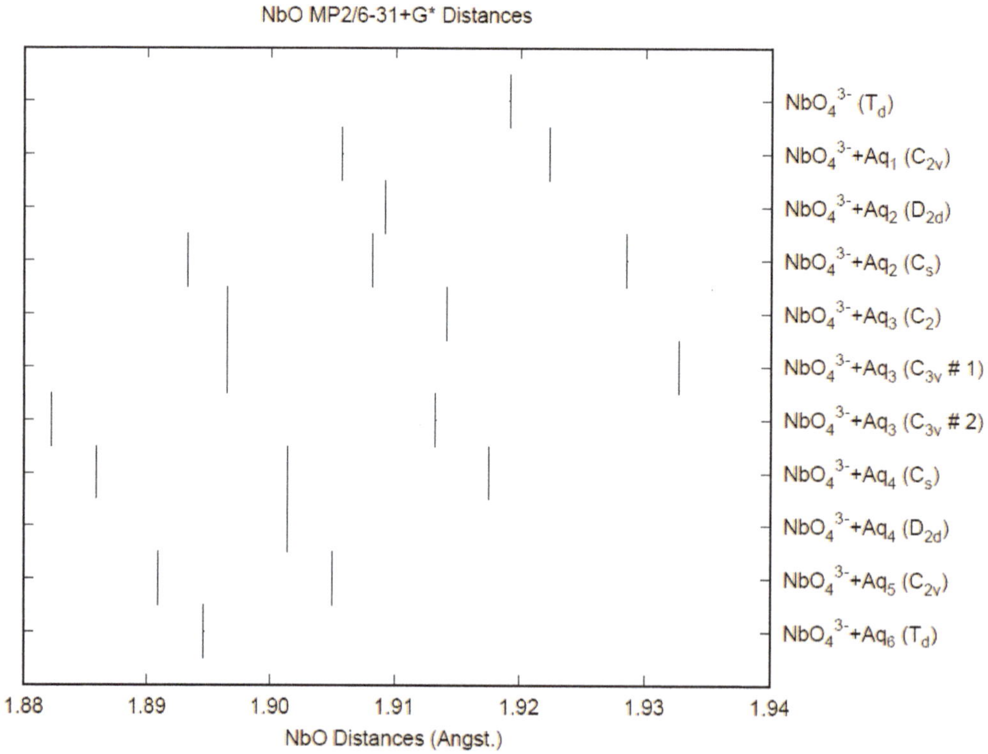

Figure 14. The Nb-O distances (MP2/6-31+G*) in hydrated orthoniobate.

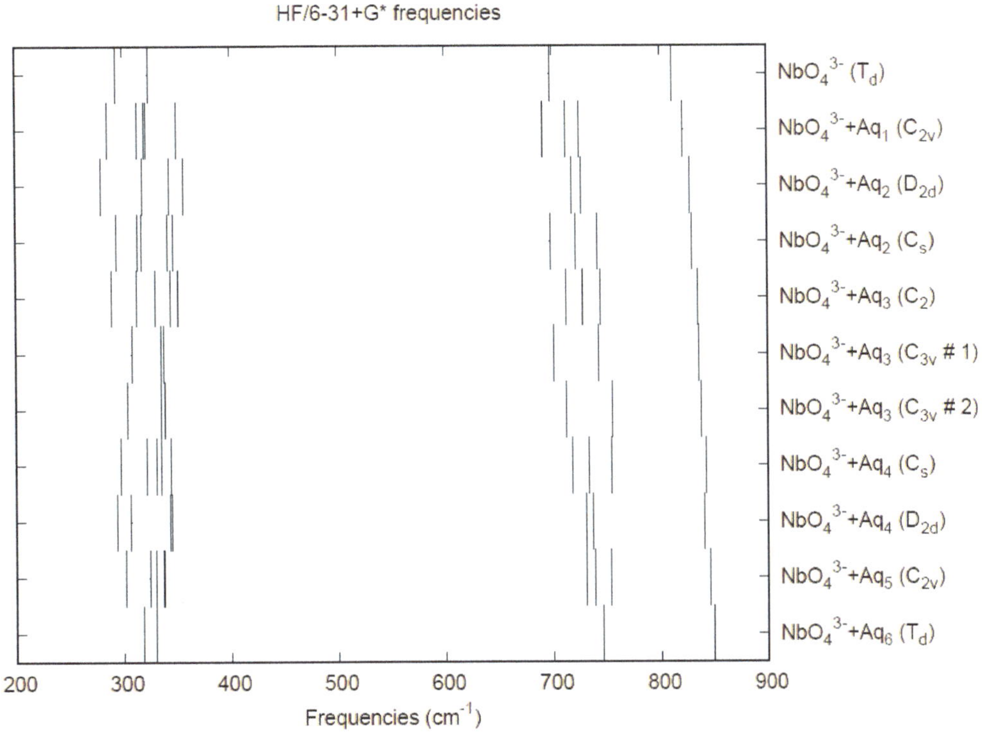

Figure 15. Vibrational frequencies (HF/6-31+G*) of hydrated orthoniobate.

3.9. *Tantalates*

Like orthoniobate, the experimental evidence for the existence of orthotantalate (TaO_4^{3-}), especially in aqueous solution, is scarce. We therefore review some of the literature on the complex chemistry of the $Ta_2O_5{\cdot}M_2O{\cdot}H_2O$ (M = Li, Na, K, Rb, Cs) phase diagrams and the compounds therein.

The phases Na_3TaO_4, $NaTaO_3$, and a compound intermediate in composition between $Na_2Ta_4O_{11}$ and $NaTa_3O_8$ were identified by King et al. [223]. Reisman studied the phase diagrams of $Li_2O{\cdot}Ta_2O_5$, $Na_2O{\cdot}Ta_2O_5$ [224], and $K_2O{\cdot}Ta_2O_5$ [225]. Minor and coworkers studied the tantalum-rich portion of the $Rb_2O{\cdot}Ta_2O_5$ phase diagram [218]. Tantalum oxide, upon heating, slowly converts from the room-temperature β-phase to the α-phase at 1360 °C before melting at 1872 °C. The melting point of the β-phase is 1785 °C [225]. Lithium tantalates form with a tantalum to lithium ratio of 3:1 ($LiTa_3O_8$), 1:1 ($LiTaO_3$), and 1:3 (Li_3TaO_4), whereas sodium tantalates form with a tantalum to lithium ratio of 3:1 ($NaTa_3O_8$), 2:1 ($Na_2Ta_4O_{11}$) and 1:1 ($NaTaO_3$) [224]. Potassium tantalates form with a tantalum-to-potassium ratio of 5:1 (KTa_5O_{13}), 2:1 ($K_2Ta_4O_{11}$), 1:1 ($KTaO_3$), and 1:3 (K_3TaO_4) [225].

3.9.1. Tantalum Oxide

For Ta_2O_5, Brauer found only a single structure that was isomorphic to the low-temperature form of Nb_2O_5 by powder diffraction [124]. However, Lagergren and Magneli found an additional high-temperature phase with a transition temperature at 1320 ± 20 °C [226], which was confirmed by Reisman et al. at 1360 ± 5 °C [224]. They could not index the lines but suggested a monoclinic or possibly triclinic system with only a small deviation from an orthorhombic system. Frevel and Rinn found a unit cell (a = 7.32 Å, b = 15.55 Å, c = 10.79 Å, β = 120.6°) and a pseudohexagonal modification (a = 3.60(2) Å, c = 3.88(1) Å) [128]. King and coworkers reported an orthorhombic (sub?)cell (a = 6.18 Å, b = 3.66 Å, c = 3.89 Å) with pseudohexagonal subcell a′ = 3.54–3.66 Å, c′ = 3.89 Å [223].

3.9.2. Lithium Tantalates

For Li_3TaO_4, Reisman was unable to index the powder diffraction patterns [224]. Blasse, however, found that it could be assigned a pseudo-tetragonal unit cell with a = 6.01 Å, c = 16.67 Å [227]. Zocchi and coworkers used X-ray and neutron diffraction to find the structures of the low-temperature β (*C2/c*, a = b = 8.500(3) Å, c = 9.344(3) Å, β = 117.05(2)°) and high-temperature α (*P2/n*, a = 6.018(1) Å, b = 5.995(1) Å, c = 12.865 Å, β = 103.53(2)°) phases [228]. Du Boulay reinvestigated the low-temperature form (*C2/c*, a = 8.508(1) Å, b = 8.516(1) Å, c = 9.338(1) Å, β = 116.869(10)°) [229].

For $LiTaO_3$, Lapitskii and Simanov indexed this salt on the basis of a rhombohedral cell [230]. Abrahams and Bernstein also found it to be rhombohedral by single-crystal X-ray diffraction (*R3c*, a_H = 5.15428(1) Å, c_H = 13.78351(2) Å) and found the atomic coordinates to be analogous to the niobate [231]. They repeated the measurements with neutron diffraction to confirm the positions of the lithium atoms between one of the two models [232]. As the temperature increases, it transforms from the ferroelectric phase (*R3c*) to a paraelectric phase (940 K, *R-3c*, a_T = 5.2203 Å, c_T = 13.7631 Å) at the Curie temperature T_c = 907 K, as followed by neutron diffraction at 298K, 760 K, 820 K, 885 K, and 940 K [233]. Hsu and coworkers found the following unit cell parameters: *R3c*, rhombohedral, a = 5.471(2) Å, α = 56.16(3)° [150].

For $LiTa_3O_8$, Reisman was unable to index the powder diffraction patterns [224]. Whiston proposed that this might actually have been $Li_2Ta_8O_{21}$ [138]. Gatehouse and Leverett showed that $LiTa_3O_8$ was isomorphous to the niobate ($P2_1/c$, a = 7.41(5), b = 5.10(4), c = 15.12(10), β = 107.2(1)°) [153]. Roth and coworkers found that there were three crystalline forms, a low-temperature form L-$LiTa_3O_8$ (monoclinic, $P2_1/c$, a = 7.41 Å, b = 5.10 Å, c = 15.12 Å, β = 107.2°) stable below 800 °C and isomorphic to $LiNa_3O_8$, an intermediate temperature form M-$LiTa_3O_8$ (monoclinic, *Cc* or *C2/c*, a = 9.420 Å, b = 11.536 Å, c = 5.055 Å, β = 91.53°) stable between 800–1100 °C, and a high-temperature form H-$LiTa_3O_8$ (orthorhombic, *Pmma*, $P2_1ma$, or *Pm2a*, a = 16.716 Å, b = 8.941 Å, c = 3.840 Å) stable between 1100°C and the melting temperature of 1600 °C [234]. Gatehouse et al. found by XRD that M-$LiTa_3O_8$ was monoclinic (*C2/c*, a = 9.413(5) Å, b = 11.522(6) Å, c = 5.050(3) Å, β = 91.05(10)°) and similar to the mineral wodginite $Mn_4B_4Ta_8O_{32}$ [235]. This was confirmed by Santoro et al. using neutron diffraction (*C2/c*, a = 9.410(2) Å, b = 11.522(2) Å, c = 5.0506(5) Å, β = 91.108(5)°) [236]. Nord and Thomas found by powder and crystal X-ray diffraction that H-$LiTa_3O_8$ was orthorhombic (*Pmma*, a = 16.705(2) Å, b = 3.836(1) Å, c = 8.928(1) Å) but could only locate the lithium atoms by powder neutron diffraction [237]. Werner et al. found by powder X-ray diffraction approximate lithium positions in a series of solid solutions of $Li_{1-x}Ta_3O_{8-x}F_x$, including x = 0 (a = 16.702(1) Å, b = 3.8485(5) Å, c = 8.9340(5) Å) [238]. Fallon et al. found by single-crystal XRD the space group *Pmma*, a = 16.702(8) Å, b = 3.840(4) Å, c = 8.929(5) Å [239]. Hodeau et al. revised the space group by electron and neutron diffraction (*Pmmn*, a = 16.718(2) Å, b = 7.696(1) Å, c = 8.931(1) Å) [240].

3.9.3. Sodium Tantalates

For Na_3TaO_4, King and coworkers found a hexagonal unit cell (a = 16.40 Å, c = 16.81 Å) [223]. Whiston found a = 12.23 Å, b = 12.85 Å, c = 5.732 Å, β = 121.5° [138].

For $NaTaO_3$, Vousden found an orthorhombic phase at ambient temperature (a = 5.5239 Å, b = 3.8831 Å, c = 5.4778 Å) [162]. King and coworkers found a larger tetragonal (or orthorhombic and pseudotetragonal) unit cell (a = 10.99 Å, b = 10.99 Å, c = 7.64 Å) [223]. Kay and Miles found the orthorhombic distorted perovskite structure from single crystal X-ray diffraction ($Pc2_1n$, a = 5.4941 Å, b = 7.7508 Å, c = 5.5130 Å) [241]. Reisman also used an orthorhombic cell (a = 5.5237 Å, b = 2(3.8961) Å, c = 5.4781 Å) [224]. Ismailzade showed that it underwent phase transitions from pseudomonoclinic (296 K, a = c = 3.8895 Å, b = 3.8859 Å, β = 90.37°) to another pseudomonoclinic (753 K, a = c = 3.918 Å, b = 3.8859 Å, β = 90.03°), to tetragonal (823 K, a = b = c = 3.923 Å, β = 90.02°), and then to cubic (903 K, a = b = c = 3.929 Å) [242]. Ahtee used powder diffraction difference reflection intensities to refine the structure to allow tilting of the TaO_6 octahedra (*Pcmn*) [243]. Later, Ahtee used neutron diffraction to reexamine the room-temperature phase (*Pcmn*, a = 5.4842(2) Å, b = 7.7952(2) Å, c = 5.5213(2) Å), and two high-temperature phases: at 803 K (*Bmmb*, a = 7.8453(5) Å, b = 7.8541(8) Å, c = 7.8633 (5) Å) and 873 K (*Bmmb*, a = 7.8560(10) Å, b = 7.8724(7) Å, c = 7.8604 (9) Å); and at 893 K (*P4/mbm*, a_T = 5.5552(3) Å; c_T = 3.9338(4) Å) [244]. Darlington re-examined these high-temperature phases using neutron diffraction every 5 K and suggested that the low-temperature structure determined previously was incorrect [177]. Shanker found unit cell parameters: *Pbnm*, a = 5.5883(4) Å, b = 5.6584(3) Å, c = 7.9309(7) Å [179]. Kennedy and coworkers used powder neutron diffraction to study the phase transitions from 298 K to 933 K [245]. The orthorhombic structure (300 K, *Pbnm*, a = 5.4768(1) Å, b = 5.5212(1) Å, c = 7.7890(2) Å) transforms to orthorhombic (773 K, *Cmcm*, a = 7.8337(2), b = 7.8485(3), c = 7.8552(3)) at ~700 K, then to tetragonal (843 K, *P4/mbm*, a = 5.5503(1) Å, c = 3.9335 (1) Å) at 835 K, and finally to cubic (893 K, *Pm-3m*, a = 3.9313(1)) above 890 K.

For $Na_2Ta_4O_{11}$, King and coworkers found a tetragonal unit cell (a = 11.00 Å, b = 11.00 Å, c = 14.6 Å) [223]. Reisman found an orthorhombic unit cell (a = 6.00 Å, b = 2(3.83) Å, c = 5.08 Å) [224]. Whiston found a tetragonal unit cell (a = 7.860(1) Å, c = 7.152(3) Å) [138]. Chaminade et al. were able to index it in the monoclinic lattice (a = 10.760 Å, b = 6.200 Å, c = 12.720 Å, β = 106.35°) and a hexagonal lattice (a = 6.208 ± 0.003 Å, c = 36.659 ± 0.010 Å) [246]. Similarities were found between this compound and the subsequently discovered mineral natrotantite, which was synthesized by Ercit and coworkers, who found the unit cell parameters (*R-3c*, a = 6.2092(1) Å, c = 36.619(1) Å) [247]. Mattes and Schaper found a very similar structure (*R-3c*, a = 6.198(3) Å, c = 36.56(2) Å) and also located the atoms [248].

For $NaTa_3O_8$, Reisman assumed an orthorhombic unit cell (a = 2(3.87) Å, b = 2(5.97) Å, c = 5.97 Å) [224]. Whiston claimed that this was actually $Na_2Ta_8O_{21}$ with a tetragonal unit cell (a = 12.5 Å, c = 3.92 Å) [138]. Chaminade indexed the latter in an orthorhombic unit cell (a = 12.430 ± 0.005 Å, b = 37.290 ± 0.015 Å, c = 3.900 ± 0.002 Å) [246].

3.9.4. Potassium Tantalates

For K_3TaO_4, the interplanar spacings were reported but not indexed by Reisman et al. [225]. Whiston also failed to index them [138]. This was identified as a corrosion product of tantalum metal upon the action of potassium by comparison of the powder patterns [249]. Stecura was finally able to index the pattern of the hygroscopic corrosion product and found a body-centered orthorhombic system with a = 14.19(5) Å, b = 17.04(5) Å, and c = 12.41(4) Å [195].

For $KTaO_3$, Vousden found a cubic phase at ambient temperature (a = 3.9885 Å) [162]. The X-ray line spacings of Reisman et al. agreed with those of Vousden [225]. Zhurova et al. also found a cubic system (*Pm-3m*, a = 3.9883(2)) [250].

For $K_2Ta_4O_{11}$, the interplanar spacings were reported but not indexed by Reisman et al. [225]. Sawaguchi observed its formation from $KTaO_3$ by slowly cooling the melt. This tetragonal bronze structure has a = 12.574 Å, c = 3.969 Å [251]. Whiston observed a tripling of one of the axes (a = 37.29 Å, c = 3.878 Å) [138].

For KTa_5O_{13}, the interplanar spacings were reported but not indexed by Reisman et al. [222]. Whiston also failed to index these [138]. Awadalla and Gatehouse were successful at determining the unit cell (*Pbcm*, a = 5.653(3), b = 10.708(5), c = 16.799(7)) and atomic positions [252].

3.9.5. Rubidium Tantalates

Rb_3TaO_4 was synthesized by Meyer and Hoppe [196]. They found a face-centered cubic structure (a = 8.90 Å).

$Rb_8Ta_6O_{19}\cdot nH_2O$, n = 4,14 was synthesized by Hartl et al. who found the unit cell parameters (n = 4, *C2/c*, a = 12.169(4) Å, b = 14.592(5) Å, c = 14.147(4) Å, β = 90.734(6)° and n = 14, $P2_1/n$, a = 10.3130(6) Å, b = 15.9072(9) Å, c = 11.5043(6) Å, β = 100.060(1)°) [253]. These compounds consist of discrete $[Ta_6O_{19}]^{8-}$ ions.

$RbTaO_3$ was found to be monoclinic by Serafin and Hoppe (*C2/m*, a = 9.589 Å, b = 8.503 Å, c = 8.135 Å, β = 94.87°) [254].

Du Boulay and coworkers identified orthorhombic $Rb_3Ta_5O_{14}$ from synchrotron X-ray data (*Pnma*, a = 7.3677(3) Å, 14.7904(19) Å, c = 25.379(3) Å) [255].

Iyer and Smith found that $RbTa_3O_8$ crystallized with orthorhombic symmetry (space group *C222*, *Cmm2*, *Cm2m*, or *Cmmm*, a = 13.07 Å, b = 7.26 Å, c = 3.85 Å, β = 105°) [212], whereas Voronkova found a different space group and unit cell (*R-3m*, a = 7.486(9) Å, c = 43.02(3) Å) [217]. Minor and coworkers found this as well (9-L, hexagonal, a = 7.508 Å, c = 36.41 Å) [218] and also identified a compound $Rb_8Ta_{22}O_{59}$ (11-L) along with some bronzes (labeled GTB—Gatehouse tetragonal bronze and HTB—hexagonal tungsten bronze). They found a hexagonal unit cell (a = 7.5065(7) Å, c = 43.194(5) Å) [218].

3.9.6. Cesium Tantalates

Cs_3TaO_4 was synthesized by Meyer and Hoppe [196]. They found a face-centered cubic structure (a = 9.29 Å).

$Cs_8Ta_6O_{19}\cdot nH_2O$, n = 0,14 was synthesized by Hartl et al. who found the unit cell parameters (n = 0, *I4/m*, a = 9.859(1) Å, c = 14.033(1) Å and n = 14, $P2_1/n$, a = 10.554(1) Å, b = 16.149(6) Å, c = 11.714(1) Å, β = 99.97(2)°) [253]. These compounds consist of discrete $[Ta_6O_{19}]^{8-}$ ions.

Iyer and Smith found that $CsTaO_3$ crystallized at 1400 °C with monoclinic symmetry (a = 12.90 Å, b = 7.51 Å, c = 14.81 Å, β = 105°) [212].

Serafin and Hoppe investigated the structure of $Cs_3Ta_5O_{14}$ (*Pbam*, a = 26.235(2) Å, b = 7.429(1) Å, c = 7.388 (1) Å) [256]. A reinvestigation by Du Boulay et al. only marginally changed the parameters (*Pbam*, a = 26.219(6) Å, b = 7.4283(10) Å, c = 7.3914 (10) Å) [257].

Voronkova found for $CsTa_3O_8$ the following unit cell (*R-3m*, a = 7.473(2), c = 42.78(2)) [217].

3.9.7. Calculations on Orthotantalate

The calculated bond lengths and vibrational frequencies for TaO_4^{3-} are given in Table 10. It is believed that the only structures containing discrete orthotantalate are the salts of rubidium and cesium [193]. To the best of our knowledge, the oxygen positions have not been precisely determined. Our predictions are quite similar for orthotantalate as for orthoniobate, with bond lengths following the trend HF < B3LYP < MP2. To the best of our knowledge, the vibrational frequencies of orthotantalate have not been measured. Like orthoniobate, we expect a clear separation of the symmetric and antisymmetric stretching frequencies but an accidental degeneracy of the deformation frequencies, which are mostly predicted to be within 30 cm^{-1} of each other. For the B3LYP frequencies, the degeneracy was not strictly maintained, most likely because the density matrix did not maintain full symmetry. The CPCM solvation model gives rise to smaller Ta-O distances and larger stretching frequencies.

The effect of water upon the Ta-O distances (MP2/6-31+G(d)) in $TaO_4^{3-}\bullet nH_2O$ (n = 0–6) is given in Figure 16. The net effect is a moderate shortening of the Ta-O distance by 0.022 Å (n = 6). The individual Ta-O distances vary by up to 0.04 Å. These changes are similar to orthoniobate. The Ta…O distance exhibits an odd behavior similar to orthoniobate, decreasing from n = 1 to n = 3 and then leveling off or even increasing slightly. (Figure S21). For the O…H and O…O distances, the increase of about 0.1 Å upon going from mono to hexahydrate is consistent with orthoniobate (Figures S22 and S23). The vibrational frequencies (HF/6-31+G*) as a function of hydration number are shown in Figure 17. The effect of hydration is to increase the stretching frequencies by about 40 cm^{-1} going from n = 0 to n = 6, whereas the deformation frequencies only slightly increase. The splitting is large enough to obfuscate the correlation between the two deformation modes because of the overlap, and these might not be able to be observed separately in aqueous solution. This behavior is similar to orthoniobate.

Table 10. Bond lengths (Å) and vibrational frequencies (unscaled, cm^{-1}) of orthotantalate, TaO_4^{3-}.

	d(Ta-O)	ν_1 (A_1)	ν_2 (E)	ν_3 (T_2)	ν_4 (T_2)
HF/6-31G(d)	1.8807	835	292	709	313
HF/6-31+G(d)	1.8894	814	296	682	309
MP2/6-31G(d)	1.9118	737	245	611	250
MP2/6-31+G(d)	1.9245	714	258	604	257
B3LYP/6-31G(d)	1.8870	784	208	575–612	235–244
B3LYP/6-31+G(d)	1.8960	766	218–220	551–589	243–252
BLYP/6-31G(d)	1.9030	753	209	597–599	220–232
BLYP/6-31+G(d)	1.9117	730	258	548–583	213–229
PBE/6-31G(d)	1.8933	762	208	604–607	221–233
PBE/6-31+G(d)	1.9009	741	259–261	559–593	212–228
CPCM-HF/6-31G(d)	1.8634	858	288	744	2872
CPCM-HF/6-31+G(d)	1.8721	833	288	698	264
CPCM-MP2/6-31G(d)	1.8907	789	266	697	252
CPCM-MP2/6-31+G(d)	1.9012	762	269	650	247
CPCM-B3LYP/6-31G(d)	1.8831	798	261	696	248
CPCM-B3LYP/6-31+G(d)	1.8925	775	269	659	246
CPCM-BLYP/6-31G(d)	1.9009	765	248	667	236
CPCM-BLYP/6-31+G(d)	1.9109	740	255	632	234
CPCM-PBE/6-31G(d)	1.8912	773	252	675	238
CPCM-PBE/6-31+G(d)	1.8998	752	261	642	237

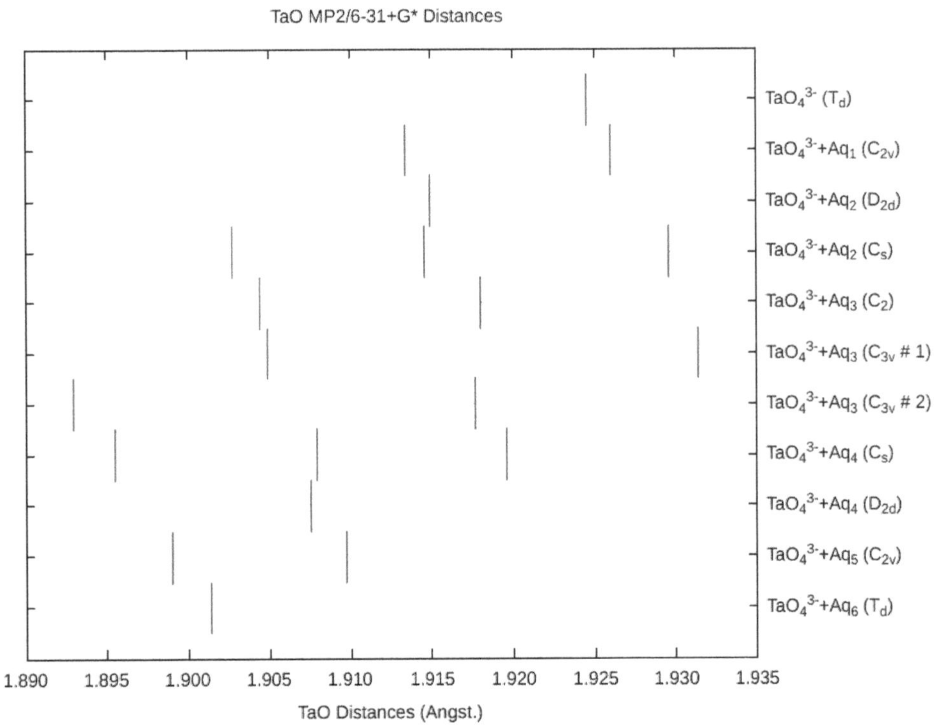

Figure 16. The Ta-O distances (MP2/6-31+G*) in hydrated orthotantalate.

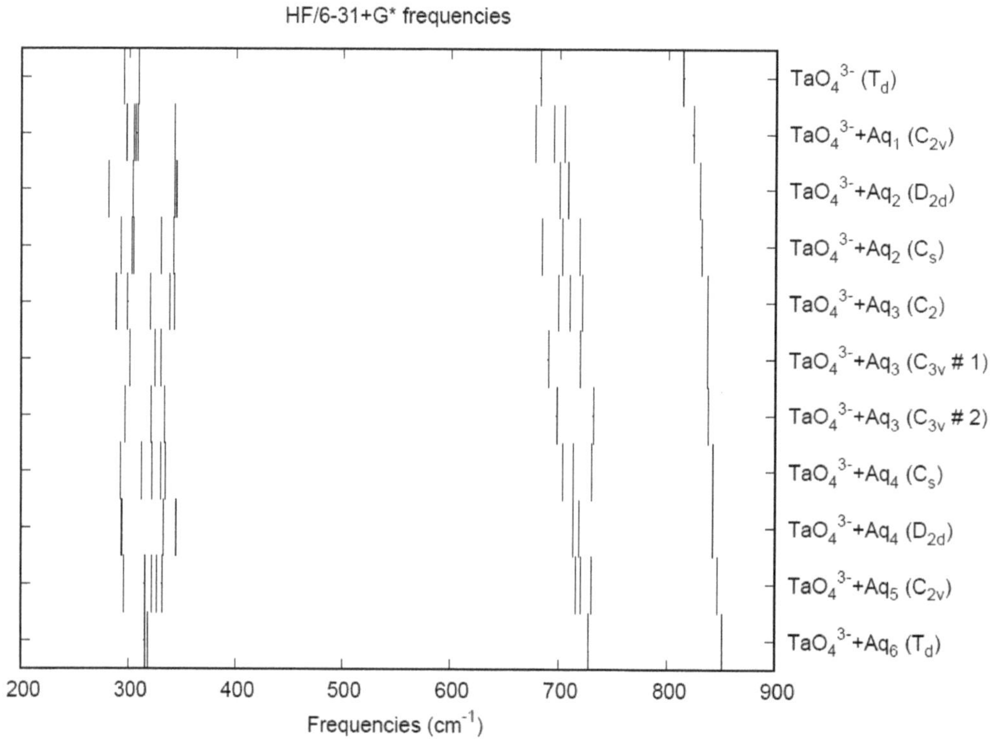

Figure 17. Vibrational frequencies (HF/6-31+G*) of hydrated orthotantalate.

4. Discussion

For most of the tetrahedral species discussed above, the B3LYP/6-31+G* level of theory gives excellent values of the metal-oxygen bond distance and the vibrational frequencies. This result is encouraging, given the small basis set size and the lack of a solvation model. The overall trends for the anions upon hydration are, for the most part, consistent with our previous work on similar anions [8]. In some cases, the MP2-FC level severely overestimates the vibrational frequencies. Our standard assumption of the structures of the hydrated anions, based on our prior calculations, can sometimes give small imaginary frequencies, usually at the MP2/6-31G* and B3LYP/6-31G* levels, which may be due to the lack of diffuse functions on these basis sets. These assumptions were shown to be incorrect for the neutral species RuO_4 and OsO_4. For the highly charged orthoniobate and orthotantalate anions, convergence difficulties were observed for some structures for the B3LYP levels.

Based on these calculations, we predict that it will be difficult or impossible to prepare FeO_4, especially in aqueous solution, as attempts to hydrate this molecule often led to hydrated forms of $FeO_2(\mu_2\text{-}O_2)$, $FeO(\mu_3\text{-}O_3)$, O + FeO_3, or even O + $FeO_2(OH)_2$. The extremely basic nature of NbO_4^{3-} and TaO_4^{3-} suggests that these could only exist in extremely basic aqueous solutions. Ions derived from niobium(V) and tantalum(V) tend to be octahedrally coordinated and/or strongly condensed into more complex ions such as NbO_5^{5-}, NbO_6^{7-}, $Nb_4O_{12}^{4-}$, $Nb_4O_{16}^{12-}$, and $Ta_6O_{19}^{8-}$. We believe that the best chance to observe the NbO_4^{3-} and TaO_4^{3-} ions in aqueous solution would be by flowing concentrated rubidium or cesium hydroxide over the corresponding orthoniobate or tantalate salt and quickly observing downstream before these coordination expansion/condensation reactions can happen.

5. Conclusions

A series of ab initio and density functional calculations at several levels of theory on several hydrated permanganate, pertechnetate, perrhenate, chromate, molybdate, tungstate, orthoniobate, orthotantalate ions; and of iron(VIII), ruthenium(VIII), and osmium(VIII) oxide have been carried out and structural and vibrational characteristics determined. Based on a careful comparison of the experimental X-ray diffraction and solution vibrational spectra of known species to our calculations, the B3LYP/6-31+G* level of theory can reproduce experiment fairly well and allows us to make predictions for the as yet unknown FeO_4, NbO_4^{3-}, and TaO_4^{3-} species.

Supplementary Materials: The following supporting information can be downloaded at: https://www.mdpi.com/article/10.3390/liquids4030031/s1, Figures S1–S23: Distances in the hydrated species; Tables S1 and S2: Total electronic energies; Table S3: Geometric and vibrational parameters.

Author Contributions: Conceptualization, C.C.P.; methodology, C.C.P.; validation, C.C.P.; formal analysis, C.C.P. and B.L.G.; investigation, C.C.P., B.L.G. and J.P.F.; data curation, C.C.P.; writing—original draft preparation, C.C.P.; writing—review and editing, C.C.P.; visualization, C.C.P.; supervision, C.C.P.; project administration, C.C.P.; funding acquisition, C.C.P. All authors have read and agreed to the published version of the manuscript.

Funding: This research was funded by Saint Mary's University Faculty of Graduate Studies and Research Internal Research Grants Summer 2011 (BLG) and Summer 2013 (JPF) and by the Natural Sciences and Engineering Research Council of Canada.

Data Availability Statement: The total energies of all species are contained in the Supplementary Materials.

Acknowledgments: The authors thank ACEnet, Compute Canada, and the Digital Research Alliance of Canada for computational support.

Conflicts of Interest: The authors declare no conflicts of interest. The funders had no role in the design of the study, in the collection, analyses, or interpretation of data, in the writing of the manuscript, or in the decision to publish the results.

References

1. Richens, D.T. *The Chemistry of Aqua Ions*; Wiley: Chichester, UK, 1997.
2. Applegarth, L.M.S.G.A.; Corbeil, C.R.; Mercer, D.J.W.; Pye, C.C.; Tremaine, P.R. Raman and ab Initio Investigation of Aqueous Cu(I) Chloride Complexes from 25 to 80 °C. *J. Phys. Chem. B* **2014**, *118*, 204–214, and references 41, 42, 44 therein. [CrossRef] [PubMed]
3. Pye, C.C.; Gunasekara, C.M. An Ab Initio Investigation of the Hydration of Thallium(III) and Mercury(II). *J. Sol. Chem.* **2020**, *49*, 1419–1429. [CrossRef]
4. Pye, C.C.; Whynot, D.C.M.; Corbeil, C.R.; Mercer, D.J.W. Desymmetrization in geometry optimization: Application to an ab initio study of copper(I) hydration. *Pure Appl. Chem.* **2020**, *92*, 1643–1654. [CrossRef]
5. Pye, C.C.; Gunasekara, C.M. An Ab Initio Investigation of the Hydration of Lead(II). *Liquids* **2022**, *2*, 39–49. [CrossRef]
6. Pye, C.C.; Gunasekara, C.M. An Ab Initio Investigation of the Hydration of Tin(II). *Liquids* **2022**, *2*, 465–473. [CrossRef]
7. Pye, C.C.; Gunasekara, C.M. An Ab Initio Investigation of the Hydration of Antimony(III). *Liquids* **2024**, *4*, 322–331. [CrossRef]
8. Pye, C.C.; Walker, V.E.J. Ab Initio Investigation of the Hydration of the Tetrahedral Perchlorate, Perbromate, Selenate, Arsenate, and Vanadate Anions. *J. Phys. Chem. A* **2011**, *115*, 13007–13015, and references 9, 10 therein. [CrossRef] [PubMed]
9. Pye, C.C.; Michels, M.R. An ab Initio Investigation of Hydrogen Phosphate Ion Hydration. *Can. J. Anal. Sci. Spectrosc.* **2004**, *49*, 175–184.
10. Pye, C.C.; Michels, M.R. An ab Initio Investigation of Dihydrogen Phosphate Ion Hydration. *Can. J. Anal. Sci. Spectrosc.* **2005**, *50*, 70–86.
11. Pye, C.C. An ab initio study of the effect of hydration on the vibrational spectrum of hydrogen sulfate. *Comp. Theor. Chem.* **2020**, *1176*, 112749. [CrossRef]
12. Pye, C.C.; Akbari, P.; Goodall, B.L.; Shah Alam, M.Y. An ab initio study of the effect of hydration on the vibrational spectrum of hydrogen selenate ion. *J. Mol. Liq.* **2022**, *359*, 119383. [CrossRef]
13. Pye, C.C.; Chiome, T.; Goodall, B.L. An ab initio study of the effect of hydration on the vibrational spectrum of hydrogen arsenate ion. *Comp. Theor. Chem.* **2022**, *1215*, 113838. [CrossRef]
14. Pye, C.C.; Berryman, V.E.J.; Goodall, B.L. An Ab Initio Study of the Effect of Hydration on the Vibrational Spectrum of Hydrogen Vanadate Ion. *J. Solution Chem.* **2024**, *53*, 626–641. [CrossRef]

15. Applegarth, L.M.S.G.A.; Pye, C.C.; Cox, J.S.; Tremaine, P.R. Raman Spectroscopic and ab Initio Investigation of Aqueous Boric Acid, Borate, and Polyborate Speciation from 25 to 80 °C. *Ind. Eng. Chem. Res.* **2017**, *56*, 13983–13996. [CrossRef]
16. Pye, C.C. A Crystallographic Review of Alkali Borate Salts and Ab Initio Study of Borate Ions/Molecules. In *Concepts, Methods and Applications of Quantum Systems in Chemistry and Physics, Progress in Theoretical Chemistry and Physics*; Wang, Y.A., Thachuk, M., Krems, R., Maruani, J., Eds.; Springer: Cham, Switzerland, 2018; Volume 31, pp. 107–142.
17. Pye, C.C. An Ab Initio Study of Boric Acid, Borate, and their Interconversion. In *Concepts, Methods and Applications of Quantum Systems in Chemistry and Physics, Progress in Theoretical Chemistry and Physics*; Wang, Y.A., Thachuk, M., Krems, R., Maruani, J., Eds.; Springer: Cham, Switzerland, 2018; Volume 31, pp. 143–177.
18. Sasidharanpillai, S.; Cox, J.S.; Pye, C.C.; Tremaine, P.R. A Raman spectroscopic and ab initio investigation of aqueous boron speciation under alkaline hydrothermal conditions: Evidence for the structure and thermodynamic stability of the diborate ion. *Dalton Trans.* **2023**, *52*, 18391–18406. [CrossRef] [PubMed]
19. Goodall, B.L. An Ab Initio Investigation of the Hydration of the d^0 Transition Metal Tetraoxo Complexes: CrO_4^{2-}, MnO_4^-, FeO_4. Honours Thesis, Saint Mary's University, Halifax, UK, 2012.
20. Frisch, M.J.; Trucks, G.W.; Schlegel, H.B.; Scuseria, G.E.; Robb, M.A.; Cheeseman, J.R.; Montgomery, J.A., Jr.; Vreven, T.; Kudin, K.N.; Burant, J.C.; et al. *Gaussian 03*, Revision D.02; Gaussian, Inc.: Wallingford, CT, USA, 2004.
21. Frisch, M.J.; Trucks, G.W.; Schlegel, H.B.; Scuseria, G.E.; Robb, M.A.; Cheeseman, J.R.; Scalmani, G.; Barone, V.; Petersson, G.A.; Nakatsuji, H.; et al. *Gaussian 16*, Revision C.01; Gaussian, Inc.: Wallingford, CT, USA, 2019.
22. Mooney, R.C.L. The Crystal Structure of Potassium Permanganate. *Phys. Rev.* **1931**, *37*, 1306–1310. [CrossRef]
23. Ramaseshan, S.; Venkatesan, K.; Mani, N.V. The use of anomalous scattering for the determination of crystal structures—$KMnO_4$. *Proc. Indian Acad. Sci. A* **1957**, *46*, 95–111. [CrossRef]
24. Palenik, G.J. Crystal Structure of Potassium Permanganate. *Inorg. Chem.* **1967**, *6*, 503–507. [CrossRef]
25. Hoppe, R.; Fischer, D.; Schneider, J. Zur Kenntnis von Oxyden A[MO_4]: Über $LiMnO_4$, $KMnO_4$, $RbMnO_4$, $CsMnO_4$ sowie $RbIO_4$ und $CsIO_4$. (-Was heißt eigenlich „Die Kristallstruktur von..."?-) [On the Knowledge of Oxides A[MO_4]: On $LiMnO_4$, $KMnO_4$, $RbMnO_4$, $CsMnO_4$ as well as $RbIO_4$, $CsIO_4$. (– What does "The Crystal Structure of..." mean? –)]. *Z. Anorg. Allg. Chem.* **1999**, *625*, 1135–1142. (In German)
26. Marabello, D.; Bianchi, R.; Gervasio, G.; Cargnoni, F. An experimental (120 K) and theoretical electron-density study of $KMnO_4$ and $KClO_4$. *Acta Crystallogr. A* **2004**, *60*, 494–501. [CrossRef]
27. Fischer, D.; Hoppe, R.; Schaefer, W.; Knight, K.S. Koordinationszahl 4 oder 6 für Lithium?: Die Kristallstruktur von Wasserfreiem Lithiumpermanganat, Li[MnO_4] [Coordination number 4 or 6 for Lithium?: The Crystal Structure of Lithium Permanganate Li[MnO_4]]. *Z. Anorg. Allg. Chem.* **1993**, *619*, 1419–1425. (In German) [CrossRef]
28. Bauchert, J.M.; Henning, H.; Schleid, T. Synthesis and crystal structure of anhydrous Na[MnO_4]. *Z. Naturforsch. B* **2016**, *71*, 993–995. [CrossRef]
29. Ketelaar, J.A.A. Die Kristallstruktur von K-, Rb-, Cs-, und Tl-Silicofluorid und von $LiMnO_4 \cdot 3H_2O$ [The crystal structure of K-, Rb-, Cs-, and Tl-hexafluorosilicate, and of $LiMnO_4 \cdot 3H_2O$]. *Z. Krist.* **1935**, *92*, 155–156. (In German)
30. Fischer, D.; Hoppe, R. Uber den Aufbau von $LiMnO_4 \cdot 3H_2O$ = [$Li(OH_2)_{6/2}$][MnO_4] [On the structure of $LiMnO_4 \cdot 3H_2O$ = [$Li(OH_2)_{6/2}$][MnO_4]]. *Z. Anorg. Allg. Chem.* **1990**, *590*, 18–22. (In German) [CrossRef]
31. Kiefer, W.; Bernstein, H.J. Resonance Raman effect in MnO_4^- with 5145 Å laser excitation. *Chem. Phys. Lett.* **1971**, *8*, 381–383. [CrossRef]
32. Kiefer, W.; Bernstein, H.J. The resonance Raman effect of the permanganate and chromate ions. *Mol. Phys.* **1972**, *23*, 835–851. [CrossRef]
33. Weinstock, N.; Schulze, H.; Müller, A. Assignment of $\nu_2(E)$ and $\nu_4(F_2)$ of tetrahedral species by the calculation of the relative Raman intensities: The vibrational spectra of VO_4^{3-}, CrO_4^{2-}, MoO_4^{2-}, WO_4^{2-}, MnO_4^-, TcO_4^-, ReO_4^-, RuO_4, and OsO_4. *J. Chem. Phys.* **1973**, *59*, 5063–5067. [CrossRef]
34. Dong, J.; Wang, Y.; Zhou, M. The vibrational spectrum of the MnO_2^- and MnO_4^- anions in solid argon. *Chem. Phys. Lett.* **2002**, *364*, 511–516. [CrossRef]
35. Wolfsberg, M.; Helmholz, L. The Spectra and Electronic Structure of the Tetrahedral Ions MnO_4^-, CrO_4^{2-}, and ClO_4^-. *J. Chem. Phys.* **1952**, *20*, 837–843. [CrossRef]
36. Ballhausen, C.J. A vibrational analysis of the MnO_4^- bands. *Theor. Chim. Acta* **1963**, *1*, 285–293. [CrossRef]
37. Holt, S.L.; Ballhausen, C.J. Low temperature absorption spectra of $KMnO_4$ in $KClO_4$. *Theor. Chim. Acta* **1967**, *7*, 313–320. [CrossRef]
38. Hillier, I.H.; Saunders, V.R. Ab initio Molecular Orbital Calculations of the Ground and Excited States of the Permanganate and Chromate Ions. *Proc. Roy. Soc. London A* **1970**, *320*, 161–173.
39. Wood, M.H. An ab initio study of the ground and low-lying excited states of the permanganate ion. *Theor. Chim. Acta* **1975**, *36*, 309–315. [CrossRef]
40. Dickson, R.M.; Ziegler, T. A Density Functional Study of the Electronic Spectrum of Permanganate. *Int. J. Quant. Chem.* **1996**, *58*, 681–687. [CrossRef]
41. Stuckl, A.C.; Daul, C.A.; Gudel, H.U. Excited-state energies and distortions of *d* transition metal tetraoxo complexes: A density functional study. *J. Chem. Phys.* **1997**, *107*, 4606–4617. [CrossRef]

42. Buhl, M. Structure, Dynamics, and Magnetic Shieldings of Permanganate Ion in Aqueous Solution. A Density Functional Study. *J. Phys. Chem. A* **2002**, *106*, 10505–10509. [CrossRef]
43. Carbonniere, P.; Ciofini, I.; Adamo, C.; Pouchan, C. Vibrational behavior of tetrahedral d^0 oxo-compounds: A theoretical study. *Chem. Phys. Lett.* **2006**, *429*, 52–57. [CrossRef]
44. Jose, L.; Seth, M.; Ziegler, T. Molecular and Vibrational Structure of Tetroxo d^0 Metal Complexes in their Excited States. A Study Based on Time-Dependent Density Functional Calculations and Franck-Condon Theory. *J. Phys. Chem. A* **2012**, *116*, 1864–1876. [CrossRef] [PubMed]
45. Su, J.; Xu, W.-H.; Xu, C.-F.; Schwarz, W.H.E.; Li, J. Theoretical Studies on the Photoelectron and Absorption Spectra of MnO_4^- and TcO_4^-. *Inorg. Chem.* **2013**, *52*, 9867–9874. [CrossRef]
46. Almeida, N.M.S.; McKinlay, R.G.; Paterson, M.J. Excited electronic states of MnO_4^-: Challenges for wavefunction and density functional response theories. *Chem. Phys.* **2015**, *446*, 86–91. [CrossRef]
47. Olsen, J.M.H.; Hedegard, E.D. Modeling the absorption spectrum of the permanganate ion in vacuum and in aqueous solution. *Phys. Chem. Chem. Phys.* **2017**, *19*, 15870–15875. [CrossRef] [PubMed]
48. Smiechowski, M.; Persson, I. Hydration of Oxometallate Ions in Aqueous Solution. *Inorg. Chem.* **2020**, *59*, 8231–8239. [CrossRef] [PubMed]
49. Krebs, B.; Hasse, K.-D. Refinements of the Crystal Structures of $KTcO_4$, $KReO_4$ and OsO_4. The Bond Lengths in Tetrahedral Oxo-Anions and Oxides of d^0 Transition Metals. *Acta Crystallogr. B* **1976**, *32*, 1334–1337. [CrossRef]
50. German, K.F.; Grigoriev, M.S.; Garashchenko, B.L.; Kopytin, A.V.; Tyupina, E.A. Redetermination of the crystal structure of $NaTcO_4$ at 100 and 296 K based on single-crystal X-ray data. *Acta Crystallogr. E* **2017**, *73*, 1037–1040. [CrossRef]
51. Weaver, J.; Soderquist, C.Z.; Washton, N.M.; Lipton, A.D.; Gassman, P.L.; Lukens, W.W.; Kruger, A.A.; Wall, N.A.; McCloy, J.S. Chemical Trends in Solid Alkali Pertechnetates. *Inorg. Chem.* **2017**, *56*, 2533–2544. [CrossRef]
52. Meyer, G.; Hoppe, R. Die Kristallstruktur von $CsTcO_4$ [The crystal structure of $CsTcO_4$]. *Z. Anorg. Allg. Chem.* **1976**, *420*, 40–50. (In German) [CrossRef]
53. Maruk, A.Y.; Grigorev, M.S.; German, K.E. Lithium Pertechnetate Trihydrate $LiTcO_4{\cdot}3H_2O$: Synthesis and Crystal Structure. *Russ. J. Coord. Chem.* **2011**, *37*, 444–446.
54. Busey, R.H.; Keller, O.L., Jr. Structure of the Aqueous Pertechnetate Ion by Raman and Infrared Spectroscopy. Raman and Infrared Spectra of Crystalline $KTcO_4$, $KReO_4$, Na_2MoO_4, Na_2WO_4, $Na_2MoO_4{\cdot}2H_2O$, and $Na_2WO_4{\cdot}2H_2O$. *J. Chem. Phys.* **1964**, *41*, 215–225. [CrossRef]
55. Morrow, J.C. The crystal structure of $KReO_4$. *Acta Crystallogr.* **1960**, *13*, 443–445. [CrossRef]
56. Lock, C.J.L.; Turner, G. A Reinvestigation of the Crystal Structure of Potassium Perrhenate. *Acta Crystallogr. B* **1975**, *31*, 1764–1765.
57. Brown, R.J.C.; Powell, B.M.; Stuart, S.N. Thermal Effects in the Structure of Potassium Perrhenate. *Acta Crystallogr. C* **1993**, *49*, 214–216. [CrossRef]
58. Atzesdorfer, A.; Range, K.-J. Sodium Metaperrhenate, $NaReO_4$: High Pressure Synthesis of Single Crystals and Structure Refinement. *Z. Naturforsch. B* **1995**, *50*, 1417–1418. [CrossRef]
59. Beintema, J. Die Kristallstruktur der Alkaliperrhenate und -perjodate [The crystal structure of alkali perrhenate and periodate]. *Z. Kristallogr.* **1937**, *97*, 300–322. (In German) [CrossRef]
60. Rogner, P.; Range, K.-J. Strukturverfeinerung von Rubidiumperrhenat bei 297 und 159 K [Structure Refinement of Rubidium Perrhenate at 297 and 159 K]. *Z. Naturforsch. B* **1993**, *48*, 233–234. (In German) [CrossRef]
61. Jaeger, F.M.; Beintema, J. The Crystalstructure of Cesium-, Thallium- and Rubidium-Perrhenates. *Proc. K. Ned. Akad. Wet.* **1933**, *36*, 523–528.
62. Rogner, P.; Range, K.-J. The Crystal Structure of β-$CsReO_4$, the Room-Temperature Modification of Cesium Perrhenate. *Z. Naturforsch. B* **1993**, *48*, 685–687. [CrossRef]
63. Range, K.-J.; Rogner, P.; Heyns, A.M.; Prinsloo, L.C. An X-Ray, Raman and IR Study of α-$CsReO_4$, the High-Temperature Modification of Cesium Perrhenate. *Z. Naturforsch. B* **1992**, *47*, 1513–1520. [CrossRef]
64. Betz, T.; Hoppe, R. Über Perrhenate.1. Zur Kenntnis von $LiReO_4$ [About Perrhenates. 1. On $LiReO_4$]. *Z. Anorg. Allg. Chem.* **1983**, *500*, 23–30. (In German) [CrossRef]
65. Abakumov, A.M.; Rozova, M.G.; Shpanchenko, R.V.; Mironov, A.V.; Antipov, E.V.; Bramnik, K.G. Synthesis and crystal structure of the lithium perrhenate monohydrate $LiReO_4{\cdot}H_2O$. *Solid State Sci.* **2001**, *3*, 581–586. [CrossRef]
66. Khrustalev, V.N.; Varfolomeev, M.B.; Shamrai, N.B.; Lindeman, S.V.; Struchkov, Y.T. The Crystal Structure of Lithium Perrhenate Sesquihydrate. *Russ. J. Inorg. Chem.* **1995**, *40*, 197–201, Translation of *Zh. Neorg. Khim.* pp. 208–213.
67. Pley, M.; Wickleder, M.S. Two crystalline modifications of RuO_4. *J. Solid State Chem.* **2005**, *178*, 3206–3209. [CrossRef]
68. Trehoux, J.; Thomas, D.; Nowogrocki, G.; Tridot, G. Étude cristallographique du tétroxyde de ruthenium, du nitrosopentachlorure d'ammonium et du dichlorure de ruthenium chloropentammine [Crystallographic study of ruthenium tetroxide, ammonium nitrosopentachlororuthenide(III) and chloropentammineruthenium(III) dichloride]. *Compt. Rend. C* **1969**, *268*, 246–249. (In French)
69. Ueki, T.; Zalkin, A.; Templeton, D.H. The Crystal Structure of Osmium Tetroxide. *Acta Crystallogr.* **1965**, *19*, 157–160. [CrossRef]
70. Seip, H.M.; Stolevik, R. Studies on the Failure of the First Born Approximation In Electron Diffraction. II. Osmium Tetroxide. *Acta Chem. Scand.* **1966**, *20*, 385–394. [CrossRef]

71. Dodd, R.E. Infra-red Spectra of Ruthenium and Osmium Tetroxides. *Trans. Faraday Soc.* **1959**, *55*, 1480–1483. [CrossRef]
72. Ortner, M.H. Infrared Spectrum and Thermodynamic Properties of Ruthenium Tetroxide. *J. Chem. Phys.* **1961**, *34*, 556–558. [CrossRef]
73. Griffith, W.P. Raman spectra of ruthenium tetroxide and related species. *J. Chem. Soc. A* **1968**, 1663–1664. [CrossRef]
74. Woodward, L.A.; Roberts, H.L. The Raman and Infra-red Absorption Spectra of Osmium Tetroxide: Relation to the Structure of the Perrhenate and Tungstate Ions in Aqueous Solution. *Trans. Faraday Soc.* **1956**, *52*, 615–619. [CrossRef]
75. Huston, J.L.; Claassen, H.H. Raman Spectra and Force Constants for OsO_4 and XeO_4. *J. Chem. Phys* **1970**, *52*, 5646–5648. [CrossRef]
76. McDowell, R.S.; Goldblatt, M. The Vibrational Spectrum and Force Field of Osmium Tetroxide. *Inorg. Chem.* **1971**, *10*, 625–630. [CrossRef]
77. Huang, W.; Xing, D.-H.; Lu, J.-B.; Long, B.; Schwarz, W.H.E.; Li, J. How Much Can Density Functional Approximations (DFA) Fail? The Extreme Case of the FeO_4 Species. *J. Chem. Theory Comput.* **2016**, *12*, 1525–1533. [CrossRef] [PubMed]
78. Miller, J.J. The Crystal Structure of Anhydrous Sodium Chromate, Na_2CrO_4. *Z. Kristallogr.* **1936**, *94*, 131–136. [CrossRef]
79. Zachariasen, W.H.; Ziegler, G.E. The Crystal Structure of Potassium Chromate, K_2CrO_4. *Z. Kristallogr.* **1931**, *80*, 164–173. [CrossRef]
80. Smith, H.W., Jr.; Colby, M.Y. The Crystal Structure of Rubidium Chromate, Rb_2CrO_4. *Z. Kristallogr.* **1941**, *103*, 90–95. [CrossRef]
81. Miller, J.J. The Crystal Structure of Caesium Chromate, Cs_2CrO_4. *Z. Kristallogr.* **1938**, *99*, 32–37. [CrossRef]
82. McGinnety, J.A. Redetermination of the Structures of Potassium Sulphate and Potassium Chromate: The Effect of Electrostatic Crystal Forces upon Observed Bond Lengths. *Acta Crystallogr. B* **1972**, *28*, 2845–2852. [CrossRef]
83. Toriumi, K.; Saito, Y. Electron-Density Distribution in Crystals of α-K_2CrO_4. *Acta Crystallogr. B* **1978**, *34*, 3149–3156. [CrossRef]
84. Brown, I.D.; Faggiani, R. Refinement of the Crystal Structure of Lithium Chromate. *Acta Crystallogr. B* **1975**, *31*, 2364–2365. [CrossRef]
85. Nimmo, J.K. Sodium Chromate (II) at 296 K (Neutron). *Acta Crystallogr. B* **1981**, *37*, 431–433. [CrossRef]
86. Aleksovska, S.; Nyburg, S.C.; Pejov, L.; Petrusevski, V.M. β-K_2SO_4-Type Isomorphs: Prediction of Structures and Refinement of Rb_2CrO_4. *Acta Crystallogr. B* **1998**, *54*, 115–120. [CrossRef]
87. Morris, A.J.; Kennard, C.H.L.; Moore, F.H.; Smith, G.; Montgomery, H. Cesium Chromate, $CrCs_2O_4$ (Neutron). *Cryst. Struct. Comm.* **1981**, *10*, 529–532.
88. Ruben, H.; Olovsson, I.; Zalkin, A.; Templeton, D.H. Sodium Chromate Tetrahydrate. *Acta Crystallogr. B* **1973**, *29*, 2963–2964. [CrossRef]
89. Kahlenberg, V. Crystal structures of two hydrated sodium chromates: $Na_2CrO_4 \times 10H_2O$ and $Na_2CrO_4 \times 1.5H_2O$. *Z. Kristallogr.* **2012**, *227*, 621–628. [CrossRef]
90. Cadbury, W.E., Jr.; Meldrum, W.B.; Lucasse, W.W. Transition Phenomena Involving Decahydrated Mixed Crystals of Sodium Sulfate and Sodium Chromate. *J. Am. Chem. Soc.* **1941**, *63*, 2262–2267. [CrossRef]
91. Richards, T.W.; Kelley, G.L. The Transition Temperatures of Sodium Chromate as Convenient Fixed Points in Thermometry. *J. Am. Chem. Soc.* **1911**, *33*, 847–863. [CrossRef]
92. Venkateswaran, C.S. The Raman Spectra of Some Inorganic Compounds. *Proc. Indian Acad. Sci.* **1938**, *7*, 144–155. [CrossRef]
93. Stammreich, H.; Bassi, D.; Sala, O. Raman spectrum and force constants of the chromate ion. *Spectrochim. Acta* **1958**, *12*, 403–405. [CrossRef]
94. Michel, G.; Machiroux, R. Raman Spectroscopic Investigations of the $CrO_4{}^{2-}/Cr_2O_7{}^{2-}$ Equilibrium in Aqueous Solution. *J. Raman. Spectrosc.* **1983**, *14*, 22–27. [CrossRef]
95. Hoffmann, M.M.; Darab, J.G.; Fulton, J.L. An Infrared and X-ray Absorption Study of the Equilibria and Structures of Chromate, Bichromate, and Dichromate in Ambient Aqueous Solutions. *J. Phys. Chem. A* **2001**, *105*, 1772–1782. [CrossRef]
96. Barinova, A.V.; Rastsvetaeva, R.K.; Nekrasov, Y.V.; Pushcharovskii, D.Y. Crystal structure of Li_2MoO_4. *Dokl. Chem.* **2001**, *376*, 16–19, translation of *Dokl. Akad. Nauk* **2001**, *376*, 343–346. [CrossRef]
97. Yip, T.W.S.; Cussen, E.J.; Wilson, C. Spontaneous formation of crystalline lithium molybdate from solid reagents at room temperature. *Dalton Trans.* **2010**, *39*, 411–417. [CrossRef] [PubMed]
98. Kolitsch, U. The crystal structures of phenacite-type $Li_2(MoO_4)$, and scheelite-type $LiY(MoO_4)_2$ and $LiNd(MoO_4)_2$. *Z. Kristallogr.* **2001**, *216*, 449–454. [CrossRef]
99. Zachariasen, W. Notiz Über die Kristallstruktur von Phenacit, Willemit und Verwandten Verbindungen [Note about the crystal structure of phenacite, willemite, and related compounds]. *Nor. Geol. Tidsskr.* **1926**, *9*, 65–73. (In German)
100. Bramnik, K.G.; Ehrenberg, H. Study of the Na_2O-MoO_3 System. $Na_5Mo_{11}O_{36}$—A New Oxide with Anatase-related Structure, and the Crystal Structures of Na_2MoO_4. *Z. Anorg. Allg. Chem.* **2004**, *630*, 1336–1341. [CrossRef]
101. Fortes, A.D. Crystal structures of spinel-type Na_2MoO_4 and Na_2WO_4 revisited using neutron powder diffraction. *Acta Crystallogr. E* **2015**, *71*, 592–596. [CrossRef] [PubMed]
102. Gatehouse, B.M.; Leverett, P. Crystal structure of potassium molybdate, K_2MoO_4. *J. Chem. Soc. A* **1969**, 849–854. [CrossRef]
103. Kools, F.X.N.M.; Koster, A.S.; Rieck, G.D. The Structures of Potassium, Rubidum and Cesium Molybdate and Tungstate. *Acta Crystallogr. B* **1970**, *26*, 1974–1977. [CrossRef]
104. Shigematsu, H.; Nomura, K.; Nishiyama, K.; Tojo, T.; Kawaji, H.; Atake, T.; Kawamura, Y.; Miyoshi, T.; Matsushita, Y.; Tanaka, M.; et al. Structures and Phase Transitions in Rb_2MoO_4 and Rb_2WO_4. *Ferroelectrics* **2011**, *414*, 195–200. [CrossRef]

105. Gonschorek, W.; Hahn, T. Die Kristallstruktur des Caesiummolybdats, Cs_2MoO_4 [The Crystal Structure of Cesium Molybdate, Cs_2MoO_4]. *Z. Kristallogr.* **1973**, *138*, 167–176. (In German) [CrossRef]
106. Mitra, R.P.; Verma, H.K.L. Crystal Structures of Dihydrates of Sodium Tungstate & Sodium Molybdate. *Indian J. Chem.* **1969**, *7*, 598–602.
107. Atovmyan, L.O.; Dyachenko, O.A. X-ray Structural Investigation of $Na_2MoO_4{\cdot}2H_2O$ Crystals. *J. Struct. Chem.* **1969**, *10*, 416–418, translation of *Zh. Strukt. Khim.* **1969**, *10*, 504–507. [CrossRef]
108. Matsumoto, K.; Kobayashi, A.; Sasaki, Y. The Crystal Structure of Sodium Molybdate Dihydrate, $Na_2MoO_4{\cdot}2H_2O$. *Bull. Chem. Soc. Jpn.* **1975**, *48*, 1009–1013. [CrossRef]
109. Capitelli, F.; Selim, M.; Mukherjea, K.K. Synthesis and Crystal Structure Determination of Sodium Molybdate Dihydrate. *Asian J. Chem.* **2006**, *18*, 2856–2860.
110. Fortes, A.D. Crystal structures of deuterated sodium molybdate dihydrate and sodium tungstate dihydrate from time-of-flight neutron powder diffraction. *Acta Crystallogr. E* **2015**, *71*, 799–806. [CrossRef] [PubMed]
111. Dean, K.J.; Wilkinson, G.R. Precision Raman Investigation of the ν_1 Mode of Vibration of SO_4^{2-}, WO_4^{2-} and MoO_4^{2-} in Aqueous Solutions of Different Concentrations. *J. Raman. Spectrosc.* **1983**, *14*, 130–134. [CrossRef]
112. Johansson, G.; Caminiti, R. The Hydration of Tungstate and Molybdate Ions in Aqueous Solution. *Z. Naturforsch. A* **1986**, *41*, 1325–1329. [CrossRef]
113. Zachariasen, W.H.; Plettinger, H.A. The Crystal Structure of Lithium Tungstate. *Acta Crystallogr.* **1961**, *14*, 229–230. [CrossRef]
114. Okada, K.; Morikawa, H.; Marumo, F.; Iwai, S. Sodium Tungstate. *Acta Crystallogr. B* **1974**, *30*, 1872–1873. [CrossRef]
115. Koster, A.S.; Kools, F.X.N.M.; Rieck, G.D. The Crystal Structure of Potassium Tungstate, K_2WO_4. *Acta Crystallogr. B* **1969**, *25*, 1704–1708. [CrossRef]
116. Holtzberg, F.; Reisman, A.; Berry, M.; Berkenblit, M. Chemistry of the Group VB Pentoxides. VI. The Polymorphism of Nb_2O_5. *J. Am. Chem. Soc.* **1957**, *79*, 2039–2043, Correction, p. 6579. [CrossRef]
117. Reisman, A.; Holtzberg, F. Further Comments on the Polymorphism of Nb_2O_5. The High Temperature Metastable Phase. *J. Am. Chem. Soc.* **1959**, *81*, 3182–3184. [CrossRef]
118. Reisman, A.; Holtzberg, F. Heterogeneous Equilibria in the Systems Li_2O-, Ag_2O-Nb_2O_5 and Oxide-Models. *J. Am. Chem. Soc.* **1958**, *80*, 6503–6507. [CrossRef]
119. Reisman, A.; Holtzberg, F.; Banks, E. Reactions of the Group VB Pentoxides with Alkali Oxides and Carbonates. VII. Heterogeneous Equilibria in the system Na_2O or Na_2CO_3-Nb_2O_5. *J. Am. Chem. Soc.* **1958**, *80*, 37–42. [CrossRef]
120. Reisman, A.; Holtzberg, F. Phase Equilibria in the System K_2CO_3-Nb_2O_5 by the Method of Differential Thermal Analysis. *J. Am. Chem. Soc.* **1955**, *77*, 2115–2119. [CrossRef]
121. Reisman, A.; Holtzberg, F. Equilibria in the System Rb_2O-Nb_2O_5 and Sequential Trends in Oxide-Oxide Interactions: The Prediction of Compound Retention. *J. Phys. Chem.* **1960**, *64*, 748–753. [CrossRef]
122. Reisman, A.; Mineo, J. Compound Repetition in Oxide-Oxide Interactions. The System Cs_2O-Nb_2O_5. *J. Phys. Chem.* **1961**, *65*, 996–998. [CrossRef]
123. Reisman, A.; Holtzberg, F.; Berkenblit, M. Metastability in Niobate Systems. *J. Am. Chem. Soc.* **1959**, *81*, 1292–1295. [CrossRef]
124. Brauer, G. Die Oxyde des Niobs [The Oxide of Niobium]. *Z. Anorg. Allg. Chem.* **1941**, *248*, 1–31. (In German) [CrossRef]
125. Gatehouse, B.M.; Wadsley, A.D. The Crystal Structure of the High Temperature form of Niobium Pentoxide. *Acta Crystallogr.* **1964**, *17*, 1545–1553. [CrossRef]
126. Kato, K. Structure Refinement of H-Nb_2O_5. *Acta Crystallogr. B* **1976**, *32*, 764–767. [CrossRef]
127. Hahn, R.B. Phosphates of Niobium and Tantalum. *J. Am. Chem. Soc.* **1951**, *73*, 5091–5093. [CrossRef]
128. Frevel, L.K.; Rinn, H.W. Powder Diffraction Standards for Niobium Pentoxide and Tantalum Pentoxide. *Anal. Chem.* **1955**, *27*, 1329–1330. [CrossRef]
129. Nolander, B.; Norin, R. Note on the Crystal Structure of T-Nb_2O_5 and of an Isostructural High-Temperature Zr-Nb-oxide. *Acta Chem. Scand.* **1972**, *26*, 3814–3816. [CrossRef]
130. Kato, K.; Tamura, S. Die Kristallstruktur von T-Nb_2O_5. [The Crystal Structure of T-Nb2O5]. *Acta Crystallogr. B* **1975**, *31*, 673–677. (In German) [CrossRef]
131. Mertin, W.; Andersson, S.; Gruehn, R. Über die Kristallstructur von M-Nb_2O_5 [On the Crystal Structure of M-Nb_2O_5]. *J. Solid State Chem.* **1970**, *1*, 419–424. (In German) [CrossRef]
132. Laves, F.; Petter, W.; Wulf, H. Die Kristalstruktur von ζ-Nb_2O_5 [The Crystal Structure of ζ-Nb_2O_5]. *Naturwissenschaften* **1964**, *51*, 633–634. [CrossRef]
133. Andersson, S. The Crystal Structure of N-Nb_2O_5, Prepared in the Presence of Small Amounts of LiF. *Z. Anorg. Allg. Chem.* **1967**, *351*, 106–112. [CrossRef]
134. Norin, R.; Nolander, B. Note on the Crystal Structures of a Niobium-rich Phase in the Li_2O-Nb_2O_5 System and of the Isostructural Compound N-Nb_2O_5. *Acta Chem. Scand.* **1971**, *25*, 741–743. [CrossRef]
135. Blasse, G. New Types of Cation-Order in the Rocksalt Lattice: The Structure of Li_3SbO_4 and Li_3NbO_4. *Z. Anorg. Allg. Chem.* **1963**, *326*, 44–46. [CrossRef]
136. Grenier, J.-C.; Martin, C.; Durif, A. Étude cristallographique des orthoniobates et orthotantalates de lithium [Crystallographic study of the orthoniobates and orthotantalates of lithium]. *Bull Soc. Fr. Min. Crist.* **1964**, *87*, 316–320. (In French)

137. Grenier, J.-C.; Bassi, G. Affinement de la structure de $NbLi_3O_4$ [Refinement of the structure of $NbLi_3O_4$]. *Bull Soc. Fr. Min. Crist.* **1965**, *88*, 345–346. (In French)
138. Whiston, C.D.; Smith, A.J. Double Oxides Containing Niobium or Tantalum. I. Systems Including Alkali Metals. *Acta Crystallogr.* **1965**, *19*, 169–173. [CrossRef]
139. Ukei, K.; Suzuki, H.; Shishido, T.; Fukuda, T. Li_3NbO_4. *Acta Crystallogr. C* **1994**, *50*, 655–656. [CrossRef]
140. Bailey, P. An X-ray Diffraction Investigation of the Structures of Anti-Ferroelectric Calcium Titanate and Ferroelectric Lithium Columbate. Ph.D. Thesis, University of Bristol, Bristol, UK, 1952.
141. Megaw, H.D. Ferroelectricity and Crystal Structure. II. *Acta Crystallogr.* **1954**, *7*, 187–194. [CrossRef]
142. Shiozaki, Y.; Mitsui, T. Powder Neutron Diffraction Study of $LiNbO_3$. *J. Phys. Chem. Solids* **1963**, *24*, 1057–1061. [CrossRef]
143. Abrahams, S.C.; Reddy, J.M.; Bernstein, J.L. Ferroelectric Lithium Niobate. 3. Single Crystal X-ray Diffraction Study at 24 °C. *J. Phys. Chem. Solids* **1966**, *27*, 997–1012. [CrossRef]
144. Abrahams, S.C.; Hamilton, W.C.; Reddy, J.M. Ferroelectric Lithium Niobate. 4. Single Crystal Neutron Diffraction Study at 24 °C. *J. Phys. Chem. Solids* **1966**, *27*, 1013–1018. [CrossRef]
145. Abrahams, S.C.; Levinstein, H.J.; Reddy, J.M. Ferroelectric Lithium Niobate. 5. Polycrystal X-ray Diffraction Study between 24 °C and 1200 °C. *J. Phys. Chem. Solids* **1966**, *27*, 1019–1026. [CrossRef]
146. Abrahams, S.C.; Marsh, P. Defect Structure Dependence on Composition in Lithium Niobate. *Acta Crystallogr. B* **1986**, *42*, 61–68. [CrossRef]
147. Oghaki, M.; Tanaka, K.; Marumo, F. Structure Refinement of lithium (I) niobium (V) trioxide, $LiNbO_3$, with anharmonic thermal vibration model. *Miner. J.* **1992**, *16*, 150–160.
148. Kumada, N.; Kinomura, N.; Muto, F. イルメナイト型 $LiNbO_3$ 及び $NaNbO_3$ の結晶構造 [Crystal Structures of Ilmenite Type $LiNbO_3$ and $NaNbO_3$]. *J. Ceramic Soc. Jpn.* **1990**, *98*, 384–388. (In Japanese) [CrossRef]
149. Boysen, H.; Altorfer, F. A Neutron Powder Investigation of the High-Temperature Structure and Phase Transition in $LiNbO_3$. *Acta Crystallogr. B* **1994**, *50*, 405–414. [CrossRef]
150. Hsu, R.; Maslen, E.N.; Du Boulay, D.; Ishizawa, N. Synchrotron X-ray Studies of $LiNbO_3$ and $LiTaO_3$. *Acta Crystallogr. B* **1997**, *53*, 420–428. [CrossRef]
151. Etschmann, B.; Ishikawa, N.; Streltsov, V.; Oishi, S. A Synchrotron X-ray diffraction analysis of near-stoichiometric $LiNbO_3$. *Z. Kristallogr.* **2001**, *216*, 455–461. [CrossRef]
152. Lundberg, M. The Crystal Structure of $LiNb_3O_8$. *Acta Chem. Scand.* **1971**, *25*, 3337–3346. [CrossRef]
153. Gatehouse, B.M.; Leverett, P. Lithium Triniobate(V), $LiNb_3O_8$. *Cryst. Struct. Commun.* **1972**, *1*, 83–85.
154. Scholder, R.; Glaser, H. Über Lithium- und Natriumuranate(V) und über strukturelle Beziehungem zwischen den Verbindungstypen Li_7AO_6 und Li_8AO_6 [About lithium and sodium uranates (V) and about structural relationships between the compound types Li_7AO_6 and Li_8AO_6]. *Z. Anorg. Allg. Chem.* **1964**, *327*, 15–27. (In German) [CrossRef]
155. Muhle, C.; Dinnebier, R.E.; van Wullen, L.; Schwering, G.; Jansen, M. New Insights into the Structural and Dynamical Features of Lithium Hexaoxometalates Li_7MO_6 (M = Nb, Ta, Sb, Bi). *Inorg. Chem.* **2004**, *43*, 874–881. [CrossRef]
156. Braun, R.M.; Hoppe, R. Zur Kenntnis von $Li_{16}Nb_4O_{18}$ [On the knowledge of $Li_{16}Nb_4O_{18}$]. *Z. Anorg. Allg. Chem.* **1982**, *493*, 7–16. (In German) [CrossRef]
157. Bouillaud, Y. Polymorphisme de l'orthoniobate de sodium [Polymorphism of sodium orthoniobate]. *Bull. Soc. Chim. Fr.* **1967**, *34*, 3879–3880. (In French)
158. Barker, M.G.; Wood, D.J. Preparation and Characterisation of the Compounds Sodium Tetraoxoniobate(V) Na_3NbO_4 and Sodium Tetraoxotantalate(V) Na_3TaO_4. *J. Chem. Soc. Dalton Trans.* **1972**, 9–13. [CrossRef]
159. Meyer, G.; Hoppe, R. Na_3NbO_4, das erste Orthioniobat mit Inselstruktur: $Na_{12}[Nb_4O_{16}]$ [Na_3NbO_4, the first orthoniobate with an island structure: $Na_{12}[Nb_4O_{16}]$]. *Naturwissenschaften* **1974**, *61*, 501. (In German) [CrossRef]
160. Barth, T. Die Kristallstruktur von Perowskit und Verwandten Verbindungen [The crystal structure of perovskite and related compounds]. *Norsk. Geolog. Tidsskr.* **1925**, *8*, 201–216.
161. Wood, E.A. Polymorphism in Potassium Niobate, Sodium Niobate, and other ABO_3 Compounds. *Acta Crystallogr.* **1951**, *4*, 353–362. [CrossRef]
162. Vousden, P. A Study of the Unit-cell Dimensions and Symmetry of certain Ferroelectric Compounds of Niobium and Tantalum at Room Temperature. *Acta Crystallogr.* **1951**, *4*, 373–376. [CrossRef]
163. Vousden, P. The Structure of Ferroelectric Sodium Niobate at Room Temperature. *Acta Crystallogr.* **1951**, *4*, 545–551. [CrossRef]
164. Pepinsky, R. Remarks on Vousden's structure of ferroelectric sodium niobate. *Acta Crystallogr.* **1952**, *5*, 288. [CrossRef]
165. Vousden, R. The non-polarity of sodium niobate. *Acta Crystallogr.* **1952**, *5*, 690. [CrossRef]
166. Shirane, G.; Newnham, R.; Pepinsky, R. Dielectric Properties and Phase Transitions of $NaNbO_3$ and $(Na,K)NbO_3$. *Phys. Rev.* **1954**, *96*, 581–588. [CrossRef]
167. Francombe, M.H. High-Temperature Structure Transitions in Sodium Niobate. *Acta Crystallogr.* **1956**, *9*, 256–259. [CrossRef]
168. Solovev, S.P.; Venevtsev, Y.N.; Zhdanov, G.S. An X-ray Study of Phase Transitions in $NaNbO_3$. *Sov. Phys. Crystallogr.* **1961**, *6*, 171–175, translated from *Kristallographiya*, 218–224.
169. Bouillaud, Y. Métaniobate de sodium non stœchiométrique [Nonstoichiometric sodium metaniobate]. *Bull. Soc. Fr. Miner. Crist.* **1969**, *92*, 347–351. (In French)

170. Sakowski-Cowley, A.C.; Lukaszewicz, K.; Megaw, H.D. The Structure of Sodium Niobate at Room Temperature, and the Problem of Reliability in Pseudosymmetric Structures. *Acta Crystallogr. B* **1969**, *25*, 851–865. [CrossRef]
171. Hewat, A.W. Neutron powder profile refinement of ferroelectric and antiferroelectric crystal structures: Sodium niobate at 22 °C. *Ferroelectrics* **1974**, *7*, 83–85. [CrossRef]
172. Darlington, C.N.; Megaw, H.D. The Low-Temperature Phase Transition of Sodium Niobate and the Structure of the Low-Temperature Phase. *N. Acta Crystallogr. B* **1973**, *29*, 2171–2185. [CrossRef]
173. Seidel, P.; Hoffmann, W. Verfeinerung der Kristallstruktur von $NaNbO_3$ N. Bestimmung der absoluten Konfiguration und des Zwillingsgesetzes [Refinement of the crystal structure of $NaNbO_3$ N. Determination of the absolute configuration and twin's law]. *Z. Kristallogr.* **1976**, *143*, 444–459. (In German) [CrossRef]
174. Glazer, A.M.; Megaw, H.D. Studies of the Lattice Parameters and Domains in the Phase Transitions of $NaNbO_3$. *Acta Crystallogr. A* **1973**, *29*, 489–495. [CrossRef]
175. Glazer, A.M.; Megaw, H.D. The structure of sodium niobate (T_2) at 600 °C, and the cubic-tetragonal transition in relation to soft-phonon modes. *Phil. Mag.* **1972**, *25*, 1119–1135. [CrossRef]
176. Ishida, K.; Honjo, G. Soft modes and Superlattice Structures in $NaNbO_3$. *J. Phys. Soc. Jpn.* **1973**, *34*, 1279–1288. [CrossRef]
177. Darlington, C.N.W.; Knight, K.S. High-temperature phases of $NaNbO_3$ and $NaTaO_3$. *Acta Crystallogr. B* **1999**, *55*, 24–30. [CrossRef]
178. Ahtee, M.; Glazer, A.M.; Megaw, H.D. The structures of sodium niobate between 480 ° and 575 °C, and their relevance to soft-phonon modes. *Philos. Mag.* **1972**, *25*, 995–1014. [CrossRef]
179. Shanker, V.; Samal, S.L.; Pradhan, G.K.; Narayana, C.; Ganguli, A.K. Nanocrystalline $NaNbO_3$ and $NaTaO_3$: Rietveld studies, Raman spectroscopy and dielectric properties. *Solid State Sci.* **2009**, *11*, 562–569. [CrossRef]
180. Johnston, K.E.; Tang, C.C.; Parker, J.E.; Knight, K.S.; Lightfoot, P.; Ashbrook, S.E. The Polar Phase of $NaNbO_3$: A Combined Study by Powder Diffraction, Solid-State NMR, and First-Principles Calculations. *J. Am. Chem. Soc.* **2010**, *132*, 8732–8746. [CrossRef]
181. Darlington, C.N.W.; Knight, K.S. On the lattice parameters of sodium niobate at room temperature and above. *Physica B* **1999**, *266*, 368–372. [CrossRef]
182. Koruza, J.; Tellier, J.; Malic, B.; Bobnar, V.; Kosec, M. Phase transitions of sodium niobate powder and ceramics, prepared by solid state synthesis. *J. Appl. Phys.* **2010**, *108*, 113509-1–113509-9. [CrossRef]
183. Mishra, S.K.; Mittal, R.; Pomjakushin, V.Y.; Chaplot, S.L. Phase stability and structural temperature dependence in sodium niobate: A high-resolution powder neutron diffraction study. *Phys. Rev. B* **2011**, *83*, 134105-1–1134105-9. [CrossRef]
184. Peel, M.D.; Thompson, S.P.; Daoud-Aladine, A.; Ashbrook, S.E.; Lightfoot, S. New Twists on the Perovskite Theme: Crystal structures of the Elusive Phases R and S of $NaNbO_3$. *Inorg. Chem.* **2012**, *51*, 6876–6889. [CrossRef]
185. Darriet, J.; Maazaz, A.; Bouloux, J.C.; Delman, C. Les Phases Na_5NbO_5 et Na_5TaO_5: Structure Cristalline de Na_5NbO_5 [The Phases Na_5NbO_5 and Na_5TaO_5: Crystal Structure of Na_5NbO_5]. *Z. Anorg. Allg. Chem.* **1982**, *485*, 115–121. (In French) [CrossRef]
186. Jahnberg, L. Crystal Structures of $Na_2Nb_4O_{11}$ and $CaTa_4O_{11}$. *J. Solid State Chem.* **1970**, *1*, 454–462. [CrossRef]
187. Andersson, S. Phase Analysis Studies on the $NaNbO_3$-Nb_2O_5, NaF-Nb_2O_5, and $NaNbO_3$-Nb_2O_5-H_2O Systems. *Acta Chem. Scand.* **1967**, *21*, 1777–1782. [CrossRef]
188. Masó, N.; West, A.R. A new family of ferroelectric materials: $Me_2Nb_4O_{11}$ (Me = Na and Ag). *J. Mater. Chem.* **2010**, *20*, 2082–2084. [CrossRef]
189. Bouillaud, Y. Mise en evidence et étude cristallographique d'une solution solide entre les composés $NaNb_3O_8$ et $Nb_2W_3O_{14}$ [Demonstration and crystallographic study of a solid solution between the compounds $NaNb_3O_8$ and $Nb_2W_3O_{14}$]. *Bull. Soc. Fr. Mineral. Crist.* **1968**, *91*, 289–291. (In French)
190. Nedjar, R.; Borel, M.M.; Leclaire, A.; Raveau, B. The Sodium Niobate $NaNb_3O_8$: A Novel Lamellar Oxide Synthesized by Soft Chemistry. *J. Solid State Chem.* **1987**, *71*, 182–188. [CrossRef]
191. Range, K.-J.; Wildenauer, M.; Heyns, A.M. Extremely short non-Bonding Oxygen-Oxygen Distances: The Crystal Structures of $NbBO_4$, $NaNb_3O_8$, and $NaTa_3O_8$. *Angew. Chem. Int. Ed.* **1988**, *27*, 969–971. [CrossRef]
192. Andersson, S. The Crystal Structure of $NaNb_{13}O_{33}$ and the Geometrical Principles of the Homologous Series $NaM_{3n+1}O_{8n+1}$. *Acta Chem. Scand.* **1965**, *19*, 557–563. [CrossRef]
193. Guerchais, J.-E. Réactions entre l'oxyde de niobium et le carbonate de potassium par voie sèche; hydrolyse des produits obtenus [Reactions between niobium oxide and potassium carbonate by dry method: Hydrolysis of the products obtained]. *Bull. Soc. Chim. Fr.* **1962**, *82*, 103–107.
194. Addison, C.C.; Barker, M.G.; Lintonbon, R.M. Reactions of the Oxides of Niobium with Liquid Potassium. *J. Chem. Soc. A* **1970**, 1465–1468. [CrossRef]
195. Stecura, S. Recovery and Determination of Crystallographic Modifications of K_3TaO_4 and K_3NbO_4. *J. Less-Common Met.* **1971**, *25*, 1–10. [CrossRef]
196. Meyer, G.; Hoppe, R. Zur Kenntnis der Strukturfamilie A_3MO_4 (A = K, Rb, Cs): Darstellung, kristallographische und magnetische Eigenschaften [Knowledge of the structure family A_3MO_4 (A = K, Rb, Cs): Representation, crystallographic and magnetic properties]. *Rev. Chim. Miner.* **1975**, *12*, 454–465. (In German)
197. Reisman, A.; Holtzberg, F.; Triebwasser, S.; Berkenblit, M. Preparation of Pure Potassium Metaniobate. *J. Am. Chem. Soc.* **1956**, *78*, 719–720. [CrossRef]

198. Katz, L.; Megaw, H.D. The Structure of Potassium Niobate at Room Temperature: The Solution of a Pseudosymmetric Structure by Fourier Methods. *Acta Crystallogr.* **1967**, *22*, 639–648. [CrossRef]
199. Hewat, A.W. Soft modes and the structure, spontaneous polarization and Curie constants of perovskite ferroelectrics: Tetragonal potassium niobate. *J. Phys. C Solid State Phys.* **1973**, *6*, 1074–1084. [CrossRef]
200. Hewat, A.W. Cubic-tetragonal-orthorhombic-rhombohedral ferroelectric transitions in perovskite potassium niobate: Neutron powder profile refinement of the structures. *J. Phys. C Solid State Phys.* **1973**, *6*, 2559–2572. [CrossRef]
201. Sugimoto, W.; Mimuro, K.; Sugahara, Y.; Kuroda, K. Synthesis and Structural Study of the KNb_4O_6-Type Compound. *J. Ceramic Soc. Jpn.* **1999**, *107*, 318–321. [CrossRef]
202. Kim, S.; Lee, J.-H.; Lee, J.; Kim, S.-W.; Kim, M.H.; Park, S.; Chung, H.; Kim, Y.-I.; Kim, W. Synthesis of Monoclinic Potassium Niobate Nanowires That Are Stable at Room Temperature. *J. Am. Chem. Soc.* **2012**, *135*, 6–9. [CrossRef] [PubMed]
203. Gasperin, M.; Le Bihan, M.T. Mecanisme d'hydratation des niobates alcalins lamellaires de formule $A_3Nb_4O_{17}$ (A = K, Rb, Cs) [Mechanism of hydration of lamellar alkaline niobates of formula $A_3Nb_4O_{17}$ (A = K, Rb, Cs)]. *J. Solid State Chem.* **1982**, *43*, 346–353. (In French) [CrossRef]
204. Becker, P.; Held, P. Crystal structure of potassium niobate, $K_6Nb_{10.80}O_{30}$, a partially filled tetragonal tungsten bronze-type structure. *Z. Kristallogr. NCS* **2000**, *215*, 319–320. [CrossRef]
205. Fallon, G.D.; Gatehouse, B.M.; Guddat, L. Crystal Structures of Some Niobium and Tantalum Oxides. IX. $K_3Nb_7O_{19}$: A New Potassium Niobium Oxide Tunnel Structure. *J. Solid State Chem.* **1986**, *61*, 181–187. [CrossRef]
206. Gasperin, M. Structure du Triniobate(V) de Potassium KNb_3O_8, un Niobate Lamellaire [Structure of Potassium Triniobate(V) KNb_3O_8, a Lamellar Niobate]. *Acta Crystallogr. B* **1982**, *38*, 2024–2026. (In French) [CrossRef]
207. Kwak, J.; Yun, H.; Chae, H.K. A new potassium niobate, KNb_5O_{13}. *Acta Crystallogr. E* **2005**, *61*, i132–i134. [CrossRef]
208. Hu, J.J.; Li, F.H.; Fan, H.F. Crystal structure determination of $K_2O{\cdot}7Nb_2O_5$ by combining high-resolution electron microscopy and electron diffraction. *Ultramicroscopy* **1992**, *41*, 387–397. [CrossRef]
209. Serafin, M.; Hoppe, R. Die Koordinationszahl 5 bei $RbNbO_3$: Eine Pyrgomstruktur [The Coordination Number 5 for $RbNbO_3$: A tower structure]. *Naturwissenschaften* **1979**, *66*, 50–51. (In German) [CrossRef]
210. Serafin, M.; Hoppe, R. Zur Kenntnis von $RbNbO_3$—Ein Metaniobat mit Pyrgomstruktur [About $RbNbO_3$—A metaniobate with a tower structure]. *J. Less Common Met.* **1980**, *76*, 299–316. (In German) [CrossRef]
211. Kafalas, J.A. Influence of Madelung Energy and Covalency on the Structure of $A^{+}B^{5+}O_3$ Compounds. In Proceedings of the 5th Materials Research Symposium Sponsored by the Institute for Materials Research, National Bureau of Standards, Gaithersburg, Maryland, 18–21 October 1972; pp. 287–293.
212. Iyer, P.N.; Smith, A.J. Double Oxides Containing Niobium, Tantalum, or Protactinium. IV. Further Systems Involving Alkali Metals. *Acta Crystallogr. B* **1971**, *27*, 731–734. [CrossRef]
213. Gasperin, M.; Le Bihan, M.T. Un niobate de rubidium d'un type structural nouveau: $Rb_4Nb_6O_{17}{\cdot}3H_2O$ [A rubidium niobate of a new structural type: $Rb_4Nb_6O_{17}{\cdot}3H_2O$]. *J. Solid State Chem.* **1980**, *33*, 83–89. (In French) [CrossRef]
214. Fallon, G.D.; Gatehouse, B.M. Crystal Structures of Some Niobium and Tantalum Oxides. II. The $4Rb_2O$:$11Nb_2O_5$ Phase—A Tunnel Structure. *J. Solid. State Chem.* **1977**, *22*, 405–409. [CrossRef]
215. Dewan, J.C.; Edwards, A.J.; Jones, G.R. Crystal Structures of Octacesium and Octarubidium Docosaniobates. *J. Chem. Soc. Dalton Trans.* **1978**, 968–972. [CrossRef]
216. Jones, G.R.; Robertson, D.S. The Growth and Properties of Crystalline Rubidium and Cesium Niobates. *J. Crystal Growth* **1978**, *43*, 115–119. [CrossRef]
217. Voronkova, V.I.; Yanovskii, V.K.; Leontyeva, I.N.; Kharitonova, E.P.; Sorokina, N.I.; Simonov, V.I. Growth and Properties of RM_3O_8 (R = Rb, Cs; M = Nb, Ta) Crystals. *Cryst. Rep.* **1998**, *43*, 345–347, translated from *Kristallographiya*, 379–381.
218. Minor, D.B.; Roth, R.S.; Parker, H.S.; Brower, W.S. Phase Equilibria and Crystal Chemistry of Rubidium Niobates and Rubidium Tantalates. *J. Res. Nat. Bur. Stds.* **1977**, *82*, 151–165. [CrossRef] [PubMed]
219. Gatehouse, B.M.; Lloyd, D.J.; Miskin, B.K. The Single Crystal X-ray Structure Determination of Some Alkali Metal Molybdates and Niobates. In Proceedings of the 5th Materials Research Symposium Sponsored by the Institute for Materials Research, National Bureau of Standards, Gaithersburg, Maryland, 18–21 October 1972; pp. 15–27.
220. Meyer, G.; Hoppe, R. Über Oxoniobate(V): Die Kristallstructur von $CsNbO_3$ [On Oxoniobate(V): The Crystal Structure of $CsNbO_3$]. *Z. Anorg. Allg.Chem.* **1977**, *436*, 75–86. (In German) [CrossRef]
221. Gasperin, M. Structure du Niobate de Césium $Cs_2Nb_4O_{11}$ [Structure of Cesium Niobate $Cs_2Nb_4O_{11}$]. *Acta Crystallogr. B* **1982**, *37*, 641–643. (In French) [CrossRef]
222. Smith, R.W.; Luo, G.; Mei, W.-M. High-temperature crystal structure and electronic properties of cesium niobate $Cs_2Nb_4O_{11}$. *J. Phys. Chem. Solids* **2010**, *71*, 1357–1361. [CrossRef]
223. King, B.W.; Schultz, J.; Durbin, E.A.; Duckworth, W.H. *Some Properties of Tantala Systems*; Report #BMI-1106, Metallurgy and Ceramics (TID-4500, 11th Ed.); Battelle Memorial Institute, U.S. Government: Columbus, OH, USA, 1956; pp. 1–39.
224. Reisman, A. Compound Repetition in Oxide Systems. Solid Phases in the System Li_2O-Ta_2O_5 and Na_2O-Ta_2O_5. *J. Phys. Chem.* **1962**, *66*, 15–21. [CrossRef]
225. Reisman, A.; Holtzberg, F.; Berkenblit, M.; Berry, M. Reactions of the Group VB Pentoxides with Alkali Oxides and Carbonates. III. Thermal and X-ray Phase Diagrams of the System K_2O or K_2CO_3 with Ta_2O_5. *J. Am Chem. Soc.* **1956**, *78*, 4514–4520. [CrossRef]
226. Lagergren, S.; Magneli, A. On the Tantalum—Oxygen System. *Acta Chem. Scand.* **1952**, *6*, 444. [CrossRef]

227. Blasse, G. On the Structure of some Compounds $Li_3Me^{5+}O_4$ and some other Mixed Metal Oxides Containing Lithium. *Z. Anorg. Allg. Chem.* **1964**, *331*, 44–50. [CrossRef]
228. Zocchi, M.; Gatti, M.; Santoro, A.; Roth, R.S. Neutron and X-ray Diffraction Study on Polymorphism in Lithium Orthotantalate, Li_3TaO_4. *J. Solid State Chem.* **1983**, *48*, 420–430, Correction, Ibid. **1984**, *53*, 277–278. [CrossRef]
229. Du Boulay, D.; Sakaguchi, A.; Suda, K.; Ishizawa, N. Reinvestigation of β-Li_3TaO_4. *Acta Crystallogr.* *E* **2003**, *59*, i80–i82. [CrossRef]
230. Lapitskii, A.V.; Simanov, Y.P. МЕТАНИОБАТ И МЕТАТАНТАЛАТ ЛИТЛЯ [Lithium Metaniobate and Metatantalate]. *Zh. Fiz. Khim. (Russ. J. Phys. Chem.)* **1955**, *29*, 1201. (In Russian)
231. Abrahams, S.C.; Bernstein, J.L. Ferroelectric Lithium Tantalate-2. Single Crystal X-ray Diffraction Study at 24 °C. *J. Phys. Chem. Solids* **1967**, *28*, 1685–1692. [CrossRef]
232. Abrahams, S.C.; Hamilton, W.C.; Sequeira, A. Ferroelectric Lithium Tantalate-2. Single Crystal Neutron Diffraction Study at 24 °C. *J. Phys. Chem. Solids* **1967**, *28*, 1693–1698. [CrossRef]
233. Abrahams, S.C.; Buehler, E.; Hamilton, W.C.; Laplaca, S.J. Ferroelectric Lithium Tantalate-III. Temperature Dependence of the Structure in the Ferroelectric Phase and the Paraelectric Structure at 940 °K. *J. Phys. Chem. Solids* **1973**, *34*, 521–532. [CrossRef]
234. Roth, R.S.; Parker, H.S.; Brewer, W.S.; Minor, D. *Alkali Oxide-Tantalum Oxide and Alkali Oxide-Niobium Oxide Ionic Conductors*; NASA Report CR-134599; National Bureau of Standards: Washington, DC, USA, 1974; p. 54.
235. Gatehouse, B.M.; Negas, T.; Roth, R.S. The Crystal Structure of M-$LiTa_3O_8$ and Its Relationship to the Mineral Wodginite. *J. Solid State Chem.* **1976**, *18*, 1–7. [CrossRef]
236. Santoro, A.; Roth, R.S.; Minor, D. Neutron Powder Diffraction Study of the Intermediate-Temperature Form of Lithium Tantalate. *Acta Crystallogr. B* **1977**, *33*, 3945–3947. [CrossRef]
237. Nord, A.G.; Thomas, J.O. Structural Studies of the Solid Electrolyte High-$LiTa_3O_8$. *Acta Chem. Scand. A* **1978**, *32*, 539–544. [CrossRef]
238. Werner, P.-E.; Marinder, B.-O.; Magneli, A. Structural Studies on $Li_{1-x}Ta_3O_{8-x}F_x$ Solid Solutions by Full-profile Refinements of Guinier-Hagg X-ray Film Data. *Mat. Res. Bull.* **1978**, *13*, 1371–1378. [CrossRef]
239. Fallon, G.D.; Gatehouse, B.M.; Roth, R.S.; Roth, S.A. Crystal Structures of Some Niobium and Tantalum Oxides. Part VI. The Structure of H-$LiTa_3O_8$. *J. Solid State Chem.* **1979**, *27*, 255–259. [CrossRef]
240. Hodeau, J.L.; Marezio, M.; Santoro, A.; Roth, R.S. Neutron Diffraction Structure Determination of the High-Temperature Form of Lithium Tritantalate, H-$LiTa_3O_8$. *J. Solid State Chem.* **1984**, *51*, 275–292. [CrossRef]
241. Kay, H.F.; Miles, J.L. The Structure of Cadmium Titanate and Sodium Tantalate. *Acta Crystallogr.* **1957**, *10*, 213–218. [CrossRef]
242. Ismailzade, I.G. An X-ray Diffraction Study of Phase Transitions in Sodium Tantalate. *Sov. Phys. Cryst.* **1963**, *7*, 584–587, Translation of *Kristallografiya* **1962**, *7*, 718–723.
243. Ahtee, M.; Unonius, L. The Structure of $NaTaO_3$ by X-ray Powder Diffraction. *Acta Crystallogr. A* **1977**, *33*, 150–154. [CrossRef]
244. Ahtee, M.; Darlington, C.N.W. Structures of $NaTaO_3$ by Neutron Powder Diffraction. *Acta Crystallogr. B* **1980**, *36*, 1007–1014. [CrossRef]
245. Kennedy, B.J.; Prodjosantoso, A.K.; Howard, C.J. Powder neutron diffraction study of the high temperature phase transitions in $NaTaO_3$. *J. Phys. Condens. Matter* **1999**, *11*, 6319–6327. [CrossRef]
246. Chaminade, J.-P.; Pouchard, M.; Hagenmuller, P. Tantalates et oxyfluorotantalates de sodium [Tantalates and oxyfluorotantalates of sodium]. *Rev. Chim. Miner.* **1972**, *9*, 381–402. (In French)
247. Ercit, T.S.; Hawthorne, F.C.; Cerny, P. The crystal structure of synthetic natrotantite. *Bull. Mineral.* **1985**, *108*, 541–549. [CrossRef]
248. Mattes, R.; Schaper, J. Die Struktur von $Na_2Ta_4O_{11}$ [The Structure of $Na_2Ta_4O_{11}$]. *Rev. Chim. Miner.* **1985**, *22*, 817–820. (In German)
249. Hickam, C.W., Jr. Corrosion Product of the Tantalum-Interstitial Oxygen-Potassium System at 1800 °F (982 °C). *J. Less-Common Met.* **1968**, *14*, 315–322. [CrossRef]
250. Zhurova, E.A.; Ivanov, Y.; Zavodnik, V.; Tsirelson, V. Electron density and atomic displacements in $KTaO_3$. *Acta Crystallogr. B* **2000**, *56*, 594–600. [CrossRef]
251. Sawaguchi, E.; Kikuchi, A. Crystal Structure and the Dielectric Properties of $K_2O{\cdot}2Ta_2O_5$. *J. Phys. Soc. Jpn.* **1964**, *19*, 579. [CrossRef]
252. Awadalla, A.A.; Gatehouse, B.M. Crystal structure of Some Niobium and Tantalum Oxides. IV. The Structure of KTa_5O_{13} and its relationship to the α-PbO_2 structure. *J. Solid State Chem.* **1978**, *24*, 183–187. [CrossRef]
253. Hartl, H.; Pickhard, F.; Emmerling, F.; Rohr, C. Rubidium- und Caesium-Verbindungen mit dem Isopolyanion $[Ta_6O_{19}]^{8-}$ — Synthesen, Kristallstructuren, thermische und schwingungensspektroskopische Untersuchungen der Oxotantalate $A_8[Ta_6O_{19}]{\cdot}nH_2O$ (A = Rb, Cs; n = 0, 4, 14) [Rubidium and Caesium Compounds with the Isopolyanion $[Ta_6O_{19}]^{8-}$ —Synthesis, Crystal Structures, Thermogravimetric and Vibrational Spectroscopic Analys of the Oxotantalates $A_8[Ta_6O_{19}]{\cdot}nH_2O$ (A = Rb, Cs; n = 0, 4, 14)]. *Z. Anorg. Allg. Chem.* **2001**, *627*, 2630–2638. (In German)
254. Serafin, M.; Hoppe, R. Coordination Number 5 and 6 in $RbTaO_3$: $Rb_4{}_\infty{}^2[Ta_4O_{12}]$. *Angew. Chem. Int. Ed.* **1978**, *17*, 354–355. [CrossRef]
255. Du Boulay, D.; Yamashita, R.; Ishizawa, N. Synchrotron X-ray study of orthorhombic $Rb_3Ta_5O_{14}$ with a modified pyrochlore structure. *Acta Crystallogr. C* **2002**, *58*, i40–i44. [CrossRef]

256. Serafin, M.; Hoppe, R. Ein Oxotantalat mit Gerüststruktur: $Cs_3Ta_5O_{14}$. [A scaffolded oxotantalate: $Cs_3Ta_5O_{14}$.]. *Z. Anorg. Allg. Chem.* **1982**, *493*, 77–92. (In German) [CrossRef]
257. Du Boulay, D.; Oono, A.; Ishizawa, N. A reinvestigation of the structure of $Cs_3Ta_5O_{14}$. *Acta Crystallogr. E* **2003**, *59*, i86–i88. [CrossRef]

MDPI AG
Grosspeteranlage 5
4052 Basel
Switzerland
Tel.: +41 61 683 77 34

Liquids Editorial Office
E-mail: liquids@mdpi.com
www.mdpi.com/journal/liquids

www.ingramcontent.com/pod-product-compliance
Lightning Source LLC
Chambersburg PA
CBHW061454210926
51850CB00008B/1709
* 9 7 8 3 7 2 5 8 7 2 0 4 6 *